PHYSICAL CHEMISTRY FOR CHEMISTS AND CHEMICAL ENGINEERS

Multidisciplinary Research Perspectives

Innovations in Physical Chemistry: Monograph Series

PHYSICAL CHEMISTRY FOR CHEMISTS AND CHEMICAL ENGINEERS

Multidisciplinary Research Perspectives

Edited by

Alexander V. Vakhrushev, DSc

Reza Haghi, PhD

J. V. de Julián-Ortiz, PhD

AAP | APPLE ACADEMIC PRESS

Apple Academic Press Inc.
3333 Mistwell Crescent
Oakville, ON L6L 0A2 Canada

Apple Academic Press Inc.
9 Spinnaker Way
Waretown, NJ 08758 USA

International Standard Book Number-13: 978-1-77188-655-0 (Hardcover)
International Standard Book Number-13: 978-1-351-14658-4 (eBook)

Library and Archives Canada Cataloguing in Publication

Physical chemistry for chemists and chemical engineers : multidisciplinary research perspectives / edited by Alexander V. Vakhrushev, DSc, Reza Haghi, PhD, J.V. de Julián-Ortiz, PhD.

(Innovations in physical chemistry : monograph series)
Includes bibliographical references and index.
Issued in print and electronic formats.
ISBN 978-1-77188-655-0 (hardcover).--ISBN 978-1-315-14658-4 (PDF)

1. Chemical engineering--Research. 2. Chemistry, Physical and theoretical--Research.
I. Haghi, Reza K., editor II Vakhrushev, Alexander V., editor III. Julián-Ortiz, J. V. de, editor
IV. Series: Innovations in physical chemistry. Monograph series

TP155.P59 2018 660 C2018-902851-3 C2018-902852-1

Library of Congress Cataloging-in-Publication Data

Names: Vakhrushev, Alexander V., editor. | Haghi, Reza K., editor. | Juliāan-Ortiz, J. V. de (Jesāus Vicente), editor.

Title: Physical chemistry for chemists and chemical engineers : multidisciplinary research perspectives / Alexander V. Vakhrushev, DSc, Reza Haghi, PhD, J. V. de Juliāan-Ortiz, PhD, editors.

Description: First edition. | Toronto, Ontario, Canada ; Waretown, NJ, USA : Apple Academic Press, 2018. | Includes bibliographical references and index.

Identifiers: LCCN 2018022489 (print) | LCCN 2018023948 (ebook) | ISBN 9781315146584 (ebook) | ISBN 9781771886550 (hardcover : alk. paper)

Subjects: LCSH: Chemical engineering. | Chemistry, Physical and theoretical.

Classification: LCC TP155 (ebook) | LCC TP155 .P47 2018 (print) | DDC 660--dc23

LC record available at https://lccn.loc.gov/2018022489

Apple Academic Press also publishes its books in a variety of electronic formats. Some content that appears in print may not be available in electronic format. For information about Apple Academic Press products, visit our website at **www.appleacademicpress.com** and the CRC Press website at **www.crcpress.com**

ABOUT THE EDITORS

Alexander V. Vakhrushev, DSc

Professor, M.T. Kalashnikov Izhevsk State Technical University, Izhevsk, Russia; Head, Department of Nanotechnology and Microsystems of Kalashnikov Izhevsk State Technical University, Russia

Alexander V. Vakhrushev, DSc, is a professor at the M.T. Kalashnikov Izhevsk State Technical University in Izhevsk, Russia, where he teaches theory, calculating, and design of nano- and microsystems. He is also the Chief Researcher of the Department of Information-Measuring Systems of the Institute of Mechanics of the Ural Branch of the Russian Academy of Sciences and Head of the Department of Nanotechnology and Microsystems of Kalashnikov Izhevsk State Technical University. He is a Corresponding Member of the Russian Engineering Academy. He has over 400 publications to his name, including monographs, articles, reports, reviews, and patents. He has received several awards, including an Academician A. F. Sidorov Prize from the Ural Division of the Russian Academy of Sciences for significant contribution to the creation of the theoretical fundamentals of physical processes taking place in multi-level nanosystems and Honorable Scientist of the Udmurt Republic. He is currently a member of editorial board of several journals, including *Computational Continuum Mechanics, Chemical Physics and Mesoscopia*, and *Nanobuild*. His research interests include multiscale mathematical modeling of physical-chemical processes into the nano-hetero systems at nano-, micro- and macro-levels; static and dynamic interaction of nanoelements; and basic laws relating the structure and macro characteristics of nano-hetero structures. E-mail: vakhrushev-a@yandex.ru

Reza Haghi, PhD

Research assistant, Institute of Petroleum Engineering, Heriot-Watt University, Edinburgh, United Kingdom

Reza Haghi, PhD, is a research assistant at the Institute of Petroleum Engineering at Heriot-Watt University, Edinburgh, Scotland, United Kingdom. Dr. Haghi has published several papers in international peer-reviewed scientific journals and has published several papers in conference proceedings, technical reports, and lecture notes. Dr. Haghi is expert in the development and application of spectroscopy techniques for monitoring hydrate and corrosion risks and developed techniques for early detection of gas hydrate risks. He conducted integrated experimental modeling in his studies and extended his research to monitoring system to pH and risk of corrosion. During his PhD work at Heriot-Watt University, he has developed various novel flow assurance techniques based on spectroscopy as well as designed and operated test equipment. He received his MSc in advanced control systems from the University of Salford, Manchester, England, United Kingdom. E-mail: RKHaghi@gmail.com

J. V. de Julián-Ortiz, PhD
Assistant lecturer, Physical Chemistry, University of Valencia, Burjassot, Spain

J. V. de Julián-Ortiz, PhD, was a postdoctoral scholarship holder at the Institute of Computational Chemistry of the University of Girona, Spain (2002–2003) at the molecular engineering group directed by Prof. Ramon Carbó-Dorca. He worked as a researcher of the Network of Research of Centers of Tropical Diseases-Faculty of Pharmacy, University of Valencia, Spain (2003–2006). He worked in the design and molecular dynamics of structure-directing agents for the synthesis of zeotype materials (2007–2009) at the Institute of Chemical Technology—Spanish Scientific Council (CSIC)—Polytechnic University of Valencia at the group of Prof. Avelino Corma. From 2010 to 2015 he was part-time professor at the University of Valencia and project technician at several private foundations, co-financed by a grant from the Spanish Ministery of Innovation. He has also been the founder of MOLware SL and is scientific director in ProtoQSAR SL. This last contract was partially financed by the "Torres Quevedo" Program, a competitive basis grant that belongs to the Program for the Promotion of Talent and Employability in R&D of the Spanish Ministry of Economy and Finance and the European Social Fund. E-mail: jesus.julian@uv.es

ABOUT THE INNOVATIONS IN PHYSICAL CHEMISTRY: MONOGRAPH SERIES

This book series aims to offer a comprehensive collection of books on physical principles and mathematical techniques for majors, non-majors, and chemical engineers. Because there are many exciting new areas of research involving computational chemistry, nanomaterials, smart materials, high-performance materials, and applications of the recently discovered graphene, there can be no doubt that physical chemistry is a vitally important field. Physical chemistry is considered a daunting branch of chemistry—it is grounded in physics and mathematics and draws on quantum mechanics, thermodynamics, and statistical thermodynamics.

Editors-in-Chief

A. K. Haghi, PhD
Editor-in-Chief, International Journal of Chemoinformatics and Chemical Engineering and Polymers Research Journal; Member, Canadian Research and Development Center of Sciences and Cultures (CRDCSC), Montreal, Quebec, Canada Email: AKHaghi@Yahoo.com

Lionello Pogliani, PhD
University of Valencia-Burjassot, Spain
Email: lionello.pogliani@uv.es

Ana Cristina Faria Ribeiro, PhD
Researcher, Department of Chemistry, University of Coimbra, Portugal
Email: anacfrib@ci.uc.pt

Books in the Series

- **Applied Physical Chemistry with Multidisciplinary Approaches**
  Editors: A. K. Haghi, PhD, Devrim Balköse, PhD, and Sabu Thomas, PhD
- **Chemical Technology and Informatics in Chemistry with Applications**
  Editors: Alexander V. Vakhrushev, DSc, Omari V. Mukbaniani, DSc, and Heru Susanto, PhD

- **Engineering Technologies for Renewable and Recyclable Materials: Physical-Chemical Properties and Functional Aspects**
  Editors: Jithin Joy, Maciej Jaroszewski, PhD, Praveen K. M., and Sabu Thomas, PhD, and Reza Haghi, PhD
- **Engineering Technology and Industrial Chemistry with Applications**
  Editors: Reza Haghi, PhD, and Francisco Torrens, PhD
- **High-Performance Materials and Engineered Chemistry**
  Editors: Francisco Torrens, PhD, Devrim Balköse, PhD, and Sabu Thomas, PhD
- **Methodologies and Applications for Analytical and Physical Chemistry**
  Editors: A. K. Haghi, PhD, Sabu Thomas, PhD, Sukanchan Palit, and Priyanka Main
- **Modern Physical Chemistry: Engineering Models, Materials, and Methods with Applications**
  Editors: Reza Haghi, PhD, Emili Besalú, PhD, Maciej Jaroszewski, PhD, Sabu Thomas, PhD, and Praveen K. M.
- **Physical Chemistry for Chemists and Chemical Engineers: Multidisciplinary Research Perspectives**
  Editors: Alexander V. Vakhrushev, DSc, Reza Haghi, PhD, and J. V. de Julián-Ortiz, PhD
- **Physical Chemistry for Engineering and Applied Sciences: Theoretical and Methodological Implication**
  Editors: A. K. Haghi, PhD, Cristóbal Noé Aguilar, PhD, Sabu Thomas, PhD, and Praveen K. M.
- **Theoretical Models and Experimental Approaches in Physical Chemistry: Research Methodology and Practical Methods**
  Editors: A. K. Haghi, PhD, Sabu Thomas, PhD, Praveen K. M., and Avinash R. Pai

CONTENTS

LIST OF CONTRIBUTORS

Devrim Balköse
Department of Chemical Engineering, Izmir Institute of Technology, Gülbahçe Köyü—Urla, 35430 Izmir, Turkey

Emili Besalú
Institut de Química Computacional i Catàlisi (IQCC) and Departament de Química, Universitat de Girona, 17003 Girona, Catalonia, Spain. E-mail: emili.besalu@udg.edu

Margarida Castel-Branco
Laboratory of Pharmacology, Faculty of Pharmacy, University of Coimbra, Azinhaga de Santa Comba, 3000548 Coimbra, Portugal; Institute for Biomedical Imaging and Life Sciences, University of Coimbra, Coimbra, Portugal

Gloria Castellano
Departamento de Ciencias Experimentales y Matemáticas, Facultad de Veterinaria y Ciencias Experimentales, Universidad Católica de Valencia San Vicente Mártir, Guillem de Castro-94, 46001 València, Spain

E. Chkhaidze
Department of Chemical and Biological Engineering, Faculty of Chemical Technology and Metallurgy, Georgian Technical University, 77, Kostava str., Tbilisi 0175, Georgia. E-mail: ekachkhaidze@yahoo.com

Saurab Dhar
Department of Physics, National Institute of Technology, Agartala 799046, India

A. Y. Fedotov
Institute of Mechanics, Ural Branch, Russian Academy of Sciences, Izhevsk, Russia
Kalashnikov Izhevsk State Technical University, Izhevsk, Russia

João Ferreira
Laboratory of Pharmacology, Faculty of Pharmacy, University of Coimbra, Azinhaga de Santa Comba, 3000-548 Coimbra, Portugal

Isabel Vitória Figueiredo
Laboratory of Pharmacology, Faculty of Pharmacy, University of Coimbra, Azinhaga de Santa Comba, 3000548 Coimbra, Portugal; Institute for Biomedical Imaging and Life Sciences, University of Coimbra, Coimbra, Portugal

A. V. Givotkov
Join Stocks Company "Nord", Perm, Russia. E-mail: felistigris@mail.ru

V. B. Golubchikov
Join Stocks Company "Nord", Perm, Russia

Mehmet Gönen
Department of Chemical Engineering, Engineering Faculty, Süleyman Demirel University, Batı Yerleşkesi, 32260 Isparta, Turkey. E-mail: mehmetgonen@sdu.edu.tr

W. S. Jesus
CQC, Department of Chemistry, University of Coimbra, 3004-535 Coimbra, Portugal; Instituto de Física, Universidade Federal da Bahia, Salvador, BA 40170115, Brazil

J. Vicente Julian-Ortiz
Unidad de Investigación de Diseño de Fármacos y Conectividad Molecular, Departamento de Química Física, Facultad de Farmacia, Universitat de València, Burjassot, València, Spain, and MOLware SL, Valencia, Spain

Ivo Juránek
Institute of Experimental Pharmacology and Toxicology, SK-84104 Bratislava, Slovakia. E-mail: jejuor@uv.es

D. Kharadze
Ivane Beritashvili Center of Experimental Biomedicine, 14, Gotua st. Tbilisi 0160, Georgia

N. Lourenço
Centro de Informática e Sistemas da Universidade de Coimbra (CISUC), 3030290 Coimbra, Portugal; Department of Informatics Engineering, University of Coimbra, 3030290 Coimbra, Portugal

Tanmoy Majumder
Department of Physics, National Institute of Technology, Agartala 799046, India

J. M. C. Marques
CQC, Department of Chemistry, University of Coimbra, 3004535 Coimbra, Portugal

Suvra Prakash Mondal
Department of Physics, National Institute of Technology, Agartala 799046, India
E-mail: suvraphy@gmail.com; suvra.phy@nita.ac.in

Milan Nagy
Department of Pharmacognosy and Botany, Faculty of Pharmacy, Comenius University in Bratislava, SK-83232 Bratislava, Slovakia

Sukanchan Palit
Department of Chemical Engineering, University of Petroleum and Energy Studies, Post Office Bidholi via Premnagar, Dehradun 248007, India; 43, Judges Bagan, Post Office Haridevpur, Kolkata 700082, India. E-mail: sukanchan68@gmail.com, sukanchan92@gmail.com

F. B. Pereira
Instituto Superior de Engenharia de Coimbra, Quinta da Nora, 3030199 Coimbra, Portugal; Centro de Informática e Sistemas da Universidade de Coimbra (CISUC), 3030290 Coimbra, Portugal

Lionello Pogliani
Unidad de Investigación de Diseño de Fármacos y Conectividad Molecular, Departamento de Química Física, Facultad de Farmacia, Universitat de València, Burjassot, València, Spain, and MOLware SL, Valencia, Spain E-mail: liopo@uv.es

F. V. Prudente
Instituto de Física, Universidade Federal da Bahia, Salvador, BA 40170115, Brazil

Mahsa Khadem Sadigh
Research Institute for Applied Physics and Astronomy, University of Tabriz, Tabriz, Iran

Telmo Santos
Laboratory of Pharmacology, Faculty of Pharmacy, University of Coimbra, Azinhaga de Santa Comba, 3000548 Coimbra, Portugal

A. V. Severyukhin
Institute of Mechanics, Ural Division, Russian Academy of Sciences, Izhevsk, Russia

O. Yu. Severyukhina
Institute of Mechanics, Ural Division, Russian Academy of Sciences, Izhevsk, Russia; Kalashnikov Izhevsk State Technical University, Izhevsk, Russia

Ladislav Šoltés
Institute of Experimental Pharmacology and Toxicology, SK-84104 Bratislava, Slovakia

Dominika Topoľská
Institute of Experimental Pharmacology and Toxicology, SK-84104 Bratislava, Slovakia

Francisco Torrens
Institut Universitari de Ciència Molecular, Universitat de València, Edifici d'Instituts de Paterna, P. O. Box 22085, 46071 València, Spain

Semra Ülkü
Department of Chemical Engineering, Izmir Institute of Technology, Gülbahçe Köyü—Urla, 35430 Izmir, Turkey

Regina Ravilevna Usmanova
Ufa State Technical University of Aviation, Ufa 450000, Bashkortostan, Russia
E-mail: Usmanovarr@mail.ru

A. V. Vakhrushev
Institute of Mechanics, Ural Division, Russian Academy of Sciences, Izhevsk, Russia; Kalashnikov Izhevsk State Technical University, Izhevsk, Russia. E-mail: vakhrushev-a@yandex.ru

Katarína Valachová
Institute of Experimental Pharmacology and Toxicology, SK-84104 Bratislava, Slovakia

R. G. Valeev
Physical-Technical Institute Ural Branch of Russian Academy of Science, Izhevsk, Russia
Gennady Efremovich Zaikov
N. M. Emanuel Institute of Biochemical Physics, Russian Academy of Sciences, Moscow 119991, Russia. E-mail: chembio@chph.ras.ru

M. S. Zakerhamidi
Research Institute for Applied Physics and Astronomy, University of Tabriz, Tabriz, Iran.
E-mail: zakerhamidi@yahoo.com, Zakerhamidi@tabrizu.ac.ir

LIST OF ABBREVIATIONS

AAO	anodic aluminum oxide
AFC	aerosol-forming compositions
AU-ROC	area under the receiver operating characteristic
BNF	basalt needle felt
BSSE	basis set superposition error
CNT	carbon nanotube
CQV	coefficient in quartile variation
DFT	density functional theory
DoE	design of experiments
DPPH	2,2-diphenyl-1-picrylhydrazyl
EAM	embedded atom method
EAs	evolutionary algorithms
EDIP	environment-dependent interatomic potential
ELS	electroluminescent light sources
EPR	electron paramagnetic resonance
FCA	Freund's complete adjuvant
FCCU	fluid catalytic cracking unit
FTIR	Fourier transform infrared
GA	genetic algorithms
HCO	heavy cycle oil
IFN-γ	interferon-gamma
IgE	immunoglobulin E
IL-4	interleukin-4
LAMMPS	large-scale atomic/molecular massively parallel simulator
LCO	light cycle oil
LED	light-emitting diode
LEV	local exhaust ventilation
MD	molecular dynamics
MEAM	modified embedded atom method
MSU	monosodium urate
MWCNT	multi-walled carbon nanotube
NCs	nanocomposites
NGs	nanogenerators

NMR	nuclear magnetic resonance
NMs	nanomaterials
NPs	nanoparticles
PDMS	polydimethylsiloxane
PEA	poly(ester amide)
PES	potential energy surface
PET	polyethylene terephthalate
PMMA	poly(methyl methacrylate)
POM	polyoxometalate
PPM	porous powder material
PPV	poly(p-phenylene vinylene)
PTE	periodic table of the elements
PVA	polyvinyl alcohol
PVDF	polyvinylidene fluoride
RGA	replacement genetic algorithm
ROS	reactive oxygen species
SAR	structure-activity relationships
SEM	scanning electron microscope
SGA	simple genetic algorithm
SOC	spin–orbit coupling
SSGA	steady state genetic algorithm
SSIR	superposing significant interaction rules
TNF-α	tumor necrosis factor α
TPA	12-O-tetradecanoyl-phorbol-acetate
USC	ultrastrong coupling
VLS	vapor–liquid–solid
WGs	waveguides

LIST OF SYMBOLS

C_V	volumetric heat capacity
C_{oi}	concentration of corpuscles in departing gas
d_p	diameter of corpuscles
E	criterion of ecological efficiency
k_B	Boltzmann's constant
m_i	mass of ith atom
μ_g	dynamic viscosity of gas
N	number of atoms in the system
θ	technological parameter
ρ_p	corpuscle density
r	a geometry of the apparatus
$\vec{r}_{i0}$	initial radius vector of ith atom
y_m	a damage to a circumambient

PREFACE

Interdisciplinary studies involve two or more academic subjects and aim to blend together broad perspectives, knowledge, skills, and epistemology in an educational setting. By focusing on topics or questions too broad for a single discipline to cover, these studies strive to draw connections between seemingly different fields.

This volume is based on different aspects of chemical technology that are associated with research and the development of theories for chemical engineers linked to research. It would also prove pivotal for the professional development of all those associated with applied chemistry who make extensive and diverse use of such research. This book bridges the gap between classical analysis and modern applications.

The book clarifies the terminology used and explains the systems methodology approach to process design and operation for chemists with limited chemical engineering knowledge. The book also provides practical insights into many areas of chemical engineering.

With an emphasis on techniques most commonly used in laboratories, the book enables students to understand practical aspects of the methods and derive the maximum possible information from the experimental results obtained.

This book not only summarizes the classical theories but it also exhibits their applications in response to the current key issues.

This timely volume provides an overview of new methods and presents experimental research in physical chemistry using modern approaches. The readers will be able to apply the concepts as described in the book to their own experiments.

This book introduces the current state-of-the-art technology in key materials with an emphasis on the rapidly growing technologies. It takes a unique approach by presenting specific materials, then progresses into a discussion of the ways in which these novel materials and processes are integrated into modern functioning manufacturing industry.

—Alexander V. Vakhrushev
Reza Haghi
J. V. de Julián-Ortiz

PART I
Bioscience and Technology

CHAPTER 1

UNSATURATED BIODEGRADABLE POLY (ESTER AMIDE) COMPOSED OF FUMARIC ACID, L-LEUCINE, AND 1,6-HEXANEDIOL

E. CHKHAIDZE[1,*] and D. KHARADZE[2]

[1]*Faculty of Chemical Technology and Metallurgy, Department of Chemical and Biological Engineering, Georgian Technical University, 77, Kostava Str., 0175 Tbilisi, Georgia*

[2]*Ivane Beritashvili Center of Experimental Biomedicine, 14, Gotua St., 0160 Tbilisi, Georgia*

[*]*Corresponding author. E-mail: ekachkhaidze@yahoo.com*

CONTENTS

ABSTRACT

Unsaturated poly (ester amide) (PEA) containing unsaturated double bonds in the backbones has been synthesized by the interaction of di-p-nitrophenyl fumarate with di-p-toluenesulfonic acid salt of bis-(L-leucine)-1,6-hexylene diester in solution under conditions of active poly-condensation. Optimal reaction conditions of polymer synthesis have been studied and unsaturated L-leucine-based PEA soluble in organic solvents has been obtained for the first time.

1.1 INTRODUCTION

Poly (ester amide) (PEA) is a relatively new family of biodegraded polymers on the basis of natural amino acids, aliphatic diols, and dicarboxylic acids. The complex of positive properties inherent for aliphatic polyesters and polyamides is combined in these polymers, namely: biodegradation skills (polyesters), hydrophilic property, high biocompatibility with tissues, and desirable mechanical properties in case of average (30,000–50,000) molecular masses (polyamides)[1,2] PEAs are prospective materials from the viewpoint of their use in surgery, pharmacology, and tissue engineering. Further improvement of PEA properties and respectively, extension of the sphere of their application is possible through their functionalization and insertion of chemically active groups[3] or hydrophobic groups[4] into polymeric chains. Functionalization of polymers gives an opportunity to bind medications and bioactive substances with them using chemical bonds and to carry out their multiple transformations aimed to further upgrading of their properties, and so forth.

One of the prospective ways of functionalization of polymers is the insertion of unsaturated bonds both into the main chain and lateral chains of macromolecules.[5,6] Through adjoining to unsaturated bonds is possible to carry out insertion of desirable functional groups into macromolecules, as well as multiple grafting reactions, structurization (cross-linking) of polymers, copolymerization, hybridization with other unsaturated polymers, for example, with unsaturated polysaccharides (acryloyl dextran, etc.) for receipt of multifunctional biodegraded hydrogels, and so forth.

Recently, we have synthesized unsaturated PEA (UPEA) on the basis of fumaric acid, L-phenylalanine, and 1,6-hexanediol.[7] Obtained polymer after separation from reaction solution was dissolved only in m-cresol and trifluoroisopropanol and was not dissolved in solvents of saturated PEAs (dimethylformamide, chloroform)[1] that can be ascribed to increased chain rigidity, which is caused by existence of double bonds in it and strong intermolecular hydrophobic interaction between benzyl groups of phenylalanine. At the same time, we have established earlier[1] that PEAs received on the basis of amino acid L-leucine were better dissolved in organic solvents on the basis of L-phenylalanine compared with obtained analogs. That is why with the purpose of increase in solubility, we have decided to replace L-phenylalanine with L-leucine and to receive UPEA on the basis of corresponding derivatives.

1.2 EXPERIMENTAL PART

Synthesis of L-leucine-containing UPEA included three stages: (1) synthesis of basic unsaturated monomer, di-p-nitrophenyl fumarate (I); (2) synthesis of di-p-toluenesulfonate of bis-(L-leucine)-1,6-hexylene diester (II); and (3) polycondensation of (I) and (II) monomers.

The unsaturated active diester, di-p-nitrophenyl fumarate, and bis-electrophilic monomer (**I**) were synthesized from fumaryl chloride and p-nitrophenol as starting materials.

In the present study, we have applied the interfacial synthesis method: into 1 l three-neck flask equipped with mechanical stirrer, 16.68 g (0.12 mol) of p-nitrophenol and 12.72 g (0.12 mol) of Na_2CO_3 were dissolved in 300 ml of water at the room temperature. The prepared solution was vigorously stirred and then 6.5 ml (0.06 mol) of fumaryl chloride in 100 ml of chloroform was added. Stirring was continued for additional 15 min until a solid precipitate was formed, which then was filtered off and washed thoroughly with water and dried in vacuum at 60°C. The crude product (I) was obtained in 86% yield with melting point (m.p.; 232–234°C. Single recrystallization from acetone was sufficient to obtain monomer with "polycondensation grade" purity (m.p. 235–237°C).

We have synthesized bis-nucleophilic monomer, di-p-toluenesulfonate of bis-(L-leucine)-1,6-hexylene diester (II), and purified them using previously described method (see ref 7 and 1, respectively).

1.3 DISCUSSION OF RESULTS

At the first stage of the study, the synthesis of UPEA was implemented according to Scheme 1.1.

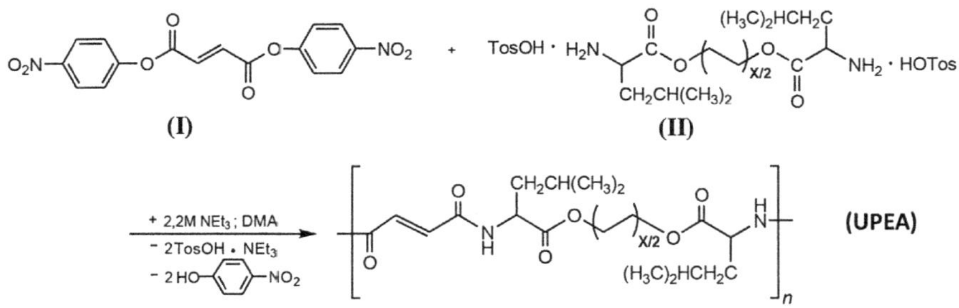

(UPEA)

SCHEME 1.1 Polycondensation reaction of di-p-nitrophenyl fumarate (I) and di-p-toluenesulfonate of bis-(L-leucine)-1,6-hexylene diester (II) monomers.

Conditions for the synthesis of a saturated PEA under specified optimum conditions of polycondensation: solvent—dimethylacetamide, acid acceptor—triethylamine, monomers concentration—1.2 mol/l, and reaction temperature—80°C. After completion of reaction, in all cases, regardless of the fact whether the soluble polymer is formed or not, reaction mass was moved to water and precipitated polymer was thoroughly washed with water, dried, and finally washed by ethyl acetate at room temperature up to negative test on p-nitrophenol (ethyl acetate should not be discolored to yellow when adding the alkali).

At first, the polycondensation proceeded homogeneously, but after 10–12 min since the beginning of reaction, the reaction solution was heavily thickened and the gel was formed. The obtained polymer was not dissolved in organic solvents, including 1,1,2,2-tetrachloroethane/phenol (3:1), m-cresol, and hexafluoroisopropanol, in which the similar phenylalanine-containing polymers are soluble (as well as not cross-linked UPEAs, see below). Addition of 5% of LiCl to reaction area (which as a rule substantially increases the solubility of amide bond-containing polymers in amid solvents) has no desired effect (see Table 1.1). All this points at formation of cross-linked polymers that could be caused by intermolecular interaction of terminal amino groups of macromolecule with strongly electrophilic double bonds of residues of fumaric acid in polymer chain (that, in our opinion, takes place due to more flexibility of macromolecules of leucine polymer, compared with phenylalanine polymer) (Scheme 1.2).

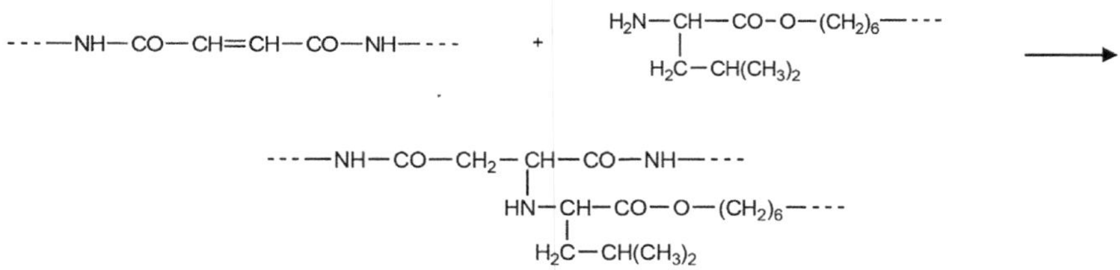

SCHEME 1.2 Intermolecular reaction between less hindered terminal amino groups of the growing macromolecules.

That is why the more detailed study of reaction became necessary for the synthesis of UPEA soluble in organic solvents on the basis of L-leucine. Note in advance that in case of receipt of homogeneous reaction solvent (i.e., soluble polymer), we added p-nitrophenyl acetate in the amount of 20% of initial diester after completion of reaction (in 24 h) in order to get rid of abovementioned undesirable intermolecular reactions, and respectively for preservation of polymer solubility.

At first, we studied temperature effect on the process of UPEA synthesis (Table 1.1). We established that reduction of reaction temperature down to 25°C only lengthens (from 10–12 to 30 min) the gelation process of the reaction solution.

TABLE 1.1 Effect of Reaction Temperature on UPEA Synthesis in Dimethylacetamide Area at Monomer Concentration 1.2 mol/l.

N	T (°C)	Annexal	Result
1	80°C	not or 5% LiCl	Jellying in 10–12 min
2	25°C	not or 5% LiCl	Jellying in 30 min
3	0–5°C	not or 5% LiCl	Jellying in 35 min

Further reduction of reaction temperature down to 0.5°C did not have the desired effect, only gelation time was increased insufficiently, up to 35 min. That is why we decided to study the effect of concentration of reaction solution on progress of polycondensation in the same dimethyl acetamide area. We selected 25°C as the reaction temperature in these experiments. Concentration of reaction solution was altered in the range of 1.2–0.15 mol/l. Obtained results are given in Table 1.2.

TABLE 1.2 Effect of Concentration of Reaction Solution on UPEA Synthesis in the Dimethyl Acetamide Area at 25°C.

N	C (mol/l)	Time	η_{red} (dl/g)[c]
1	1.2	Jellying in 30 min	
2	0.6	Jellying in 1 h	
3	0.6–0.3[a]	40 min + 23[b]	1.90
4	0.3[a]	24	1.00
5	0.15[a]	24	0.35

[a]Solution was treated by p-nitrophenyl acetate at the end of reaction.

[b]Reaction solution became viscous in 40 min and it was diluted to 0.3 mol/l and stirred within 23 h.

[c]Reduced viscosity in tetrachloroethane/phenol mixture (3:1) is C=0.5 g/dl at t=25°C.

As is seen from data of Table 1.2, diluting of solution from 1.2–0.6 mol/l has no desirable effect, only gelation time increases from 0.5–1 h. But if we start the reaction with 0.6 mol/l concentration and after 40 min when solution viscosity has begun to increase rapidly, then it is diluted by dimethyl acetamide to 0.3 mol/l and reaction solution preserves homogeneity even after 23 h. In cases, when the initial concentration of reaction solutions was 0.3 and 0.15 mol/l, gelation process did not occur at all and soluble polymers were received. As we mentioned above, all homogeneous reaction solutions were processed after completion of reaction (in 24 h) by p-nitrophenyl acetate within 48 h and afterward we implemented polymers' separation and purification using the method described above. UPEA treated in such a way has preserved its solubility in hexafluoroisopropanol for 1.5 years, while untreated polymers lose their solubility within 1–2 weeks. Obtained leucine-containing UPEA were distinguished by high viscosity (especially polymers synthesized at 0.3 mol/l concentration—N3 and N4; (Table 1.2) and good film-forming properties. Presumably, the part of macromolecules of obtained polymers, despite the solubility, should not have a linear structure and they may have loop structure that is caused by terminal amino groups and intramolecular interactions of double bonds.[8]

In the end, we would like to mention that leucine-containing UPEA is not solved in chloroform, tetrahydrofuran, and ethyl alcohol (in which leucine-containing saturated PEA is highly soluble) that have to be predetermined by the rigid structure of UPEA macromolecules caused by high concentration of double bonds. Improvement of solubility of polymers of this group is the subject of our further study.

KEYWORDS

- **fumaric acid**
- **di-p-nitrophenyl fumarate**
- **L-leucine**
- **unsaturated poly (ester amide)**
- **active polycondensation**
- **biodegradable polymers**

REFERENCES

1. Katsarava, R.; Beridze, V.; Arabuli, N; Kharadze, D.; Chu, C. C.; Won, C. Y. Amino Acid-Based Bioanalogous Polymers. Synthesis and Study of Regular Poly(ester amide)s Based Bis(α-amino acid) α,ω-alkylene Diesters, and Aliphatic Dicarboxylic Acids. *J. Polym. Sci.: Part A: Polym. Chem.* **1999,** *37*, 391–407.
2. Villuendas, I.; Iribabarren, J. I.; Munoz-Guerra, S. Poly (Ester Amide) S Derived from L-Tartaric Acid and Amino Alcohols. 1. Regic Polymers. *Macromolecules* **1999,** *32*, 8015–8023.
3. Jokhadze, G.; Chu, C. C.; Tugushi, D.; Katsarava, R. Synthesis of Biodegradable Copoly- (esteramide) Containing L-Lysine Benzyl Ester Moieties in the Backbones. *Georgian Eng. News* **2006,** *2*, 220–223.
4. Neparidze, N.; Machaidze, M.; Zavradashvili, N.; Mazanashvili, N.; Tabidze, V.; Tugushi, D.; Katsarava, R. *Biodegradable Hydrophobic Copoly(ester amide) Lateral Substituents "Polymers and medicine";* Saint-Petersburg, Russia, 2006; vol. 2, pp 27–33.
5. Lou, X.; Detrembleur, C.; Lecomte, P.; Jerome, R., Novel Unsaturated εcaprolactone Polymerizable by Ring Opening and Ring Opening Metathesis Mechanisms. *e-Polymers* **2002,** 034.

6. Domb, A. J.; Martinowitz, E.; Ron, E.; Giannos, S.; Langer, R. Polyanhydrides. IV Unsaturated and Crosslinked Polyanhydrides. *J. Polym. Sci. Polym. Chem.* **1991,** *29*, 571–579.
7. Guo, K.; Chu, C. C.; Chkhaidze E.; Katsarava, R. Synthesis and Characterization of Novel Biodegradable Unsaturated Poly(Ester Amide)s. *J. Polym. Sci. Part A: Polym. Chem.* **2005,** *43*, 1463–1477.
8. Chkhaidze, E.; Tugushi, D.; Kharadze, D.; Gomurashvili, Z.; Chu, C. C.; Katsarava, R. *J. Macrom. Sci.* **2011,** *A48*(7), 455–555.

CHAPTER 2

THE USE OF ANIMAL MODELS OF INFLAMMATION AND PAIN IN BIOMEDICAL RESEARCH

TELMO SANTOS[1], JOÃO FERREIRA[1], MARGARIDA CASTEL-BRANCO[1,2], and ISABEL VITÓRIA FIGUEIREDO[1,2,*]

[1]*Laboratory of Pharmacology, Faculty of Pharmacy, University of Coimbra, Azinhaga de Santa Comba, 3000548 Coimbra, Portugal*

[2]*Institute for Biomedical Imaging and Life Sciences, University of Coimbra, Coimbra, Portugal*

[*]*Corresponding author. E-mail: isabel.vitoria@netcabo.pt*

CONTENTS

ABSTRACT

This chapter aims to provide the readers with an overview of the main techniques used in experimental research to reproduce inflammatory episodes in in vivo models, as well and to measure the consequent inflammatory response. Considering the interconnected inflammatory and associated pain mechanisms, we will also describe the central methods to induce and measure pain. The use of experimental model of inflammation and pain is core to test new potential anti-inflammatory and analgesic drugs and quantify their efficacy.

The presented techniques will be classified according to the type of inflammation and pain that they allow to study. Therefore, we will introduce methods to study acute and chronic inflammation, as well as central and peripheral pain. Considering the scope of this book, a special focus will be paid to the chemical methods to both induce and characterize inflammation.

The chapter is organized in a first introductory section (Section 2.1) followed by a description of the main experimental techniques to induce and evaluate inflammation in vivo (Section 2.2), as well as to induce and evaluate pain (Section 2.3).

2.1 INTRODUCTION

An uncontrolled inflammatory response contributes to the development of acute and chronic inflammatory and pain diseases, being considered not only as a symptom but also as an initial phase of other complications in organisms. With this in mind, sometimes when the immune system itself is unable to stop the reaction by itself, the use of compounds with anti-inflammatory and analgesic properties is required. To develop new and more effective compounds, researchers need to understand the molecular mechanisms of inflammation and pain so that they can develop drugs that target them, on one side, and to have in vivo models to test those drugs, on the other.

This chapter is organized in a first introductory section (Section 2.1) followed by a description of the main experimental techniques to induce and evaluate inflammation in vivo (Section 2.2), as well as to induce and evaluate pain (Section 2.3).

2.1.1 INFLAMMATION

Every day and every hour living beings are subjected to assaults capable of injuring their integrity. For their protection, human beings, like the other animals, have created a set of mechanisms that allow their defense. In our case, those mechanisms count as the primary physical barriers of the immune system. The paradigm of the nonspecific immune system is the inflammatory response. Inflammation is the response of the living tissues to aggression, using cellular and molecular mediators and their orientation to the assaulted site. Depending on the type and location of the aggression, inflammation can have different forms such as allergic reactions, rheumatic, fibrinous, necrotic, toxic, and thermal, among many others.[16,33]

Inflammation may be acute or chronic. The first one is exuberant and rapid, extending only for a few days at most. Chronic inflammation, on the other hand, occurs later and prolongs for extended periods of time and can reach years.[20]

The inflammatory response, in general, is a normal phenomenon necessary to the good and healthy functioning of the organism. However, in some cases of acute or chronic inflammation, this becomes undesirable since its effects damage healthy tissues, causing complications for the organism. In these cases the so-called bad inflammation must be combated and counteracted by the use of anti-inflammatory compounds.[16,33]

2.1.2 PAIN

Pain maybe identified as being a complex and varied set of perceptions that are characterized by their unpleasant and tense nature.[56] It functions both as a warning sign and triggering defense and preservation reactions, thus having an important defense function.[30] Pain, at its upper limit, should not interfere with the normal activities of the individual, but cannot be so unnoticed that it is not felt before serious damage occurs to tissues.[41]

Pain can be categorized into two types: neuropathic and nociceptive. The neuropathic pain is not related to direct threats to the integrity of tissues, but it originates from previous lesions that have led to dysfunction of the nerve pathways, for example, in the synapses or inhibitory mechanisms.[56] The second type, nociceptive pain, responds to mechanical, thermal, electrical, and chemical nociceptive stimuli usually associated with aggressions.[10]

Regarding the chemical stimuli, the mechanisms through which they induce the emergence of a difference in membrane potential in nociceptors generally involve their interaction with ionic channels that are sensitive to voltage, with receptors associated with tyrosine kinase or with G protein, which is the most common.[5] The mediators that resort to this mechanism have specific receptors that are coupled to G proteins that differ in the α subunit, activating effector pathways.[12] Among these endogenous substances, the one that is the most potent to induce pain is bradykinin.

On the other hand, the stimulation of the primary afferent nociceptors results in the release of neuropeptides, namely substance P and neurokinin A, which contribute to a greater stimulation of the inflammatory process.

2.1.3 THE USE OF ANIMAL MODELS IN BIOMEDICAL RESEARCH

As discussed above, although inflammation is a naturally occurring phenomenon and is central to the individual's internal defense, it sometimes becomes disastrous. For these situations, pharmacology and medicine have created compounds that reduce inflammatory effects. Over the last few decades, in vitro and in vivo laboratory studies of inflammation have proven to be of great importance for the development of novel anti-inflammatory drugs. These studies have allowed the development of therapies that have greatly improved the quality of life of those who are affected by chronic inflammation and make it possible to alleviate some of the adverse effects of acute inflammations. However, this research is far from finished; the medical community and its patients crave new molecules that have more anti-inflammatory efficiency with fewer adverse effects. For this reason, it is consistent to continue in vivo research with the aim of concluding on the anti-inflammatory capacity of the new molecules with anti-inflammatory potential. To be successful, it is necessary to choose the right animal models and to dominate the techniques and compounds that can be used to induce and subsequently quantify the inflammation.

There are several animals used in scientific laboratories for in vivo experimentation. These include, for example, rats and mice, rabbits, chickens, cats, dogs, pigs, horses and nonhuman primates. The choice of the animal should be prudent and the species, strain, sex, weight and age that best suit the objectives of the work should be taken into account.

2.2 EXPERIMENTAL TECHNIQUES FOR INFLAMMATION STUDY IN VIVO

2.2.1 INFLAMMATION INDUCTION

To enable in vivo research of new anti-inflammatory drugs and their characterization, there is a need to stimulate the inflammatory response, which will then be attenuated (or not) by the molecules under study. The methods for use must be selected considering the purpose of the work and the type of inflammation desired. Among the first, there is the ability to induce localized inflammation that is not resolved in a timely manner by the body without the aid of anti-inflammatory compounds, to be simple to execute, to allow the quantification of results and to be reproducible. Most of the methods described to induce inflammation are based on the administration of biochemical compounds which, through various mechanisms, trigger the organism's response.

Due to the large variety of animals and methods to induce inflammation that could be used in these studies, those that are applicable in rats and mice and that are most commonly used were chosen for this work.

2.2.1.1 ACUTE INFLAMMATION

2.2.1.1.1 Ear Edema

Edema induction and analysis techniques are used for topical anti-inflammatory studies, that is, studies where the anti-inflammatory agent is administered to the injured site. Edema allows the study of the anti-inflammatory process in cutaneous and some subcutaneous tissues. In the following table some of the main techniques are presented organized according the prevalence of use.

Inflammation-inducing agent	12-*O*-tetradecanoyl-phorbol-13-acetate (TPA) or phorbol myristate acetate (MPA) 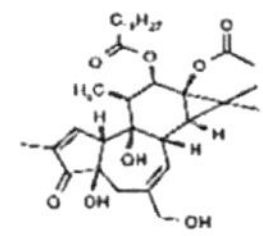
Other information	TPA ($C_{36}H_{56}O_8$) is the most commonly used phorbol ester. These compounds are tumor and inflammation promoters and bind strongly and activate C-protein (PKC) (Cell Signaling Technology, December 22, 2009). This family of proteins and their different isoforms play several roles in the cells in which they are present.
Induction of inflammation mechanism	As an inflammatory agent in in vivo experiments, TPA is used to cause acute inflammation in epidermal tissues which are stratified epithelia consisting largely of a set of cells called keratinocytes. These cells have some characteristics that allow them to respond to environmental stimuli, such as toxic and irritating agents, ultraviolet radiation, cancerous molecules, among others. Keratinocytes are highly rich in various cytokines (Interleukin-1 (IL-1), IL-6, IL-7, tumor necrosis factor-α (TNF-α), etc.), prostaglandins and leukotrienes which, as mentioned above, are involved in the inflammatory process.[59] Several forms of PKC were also found in these cells in both humans and rats.[35]. Relating these three facts, the inflammatory mechanism of TPA was experimentally proposed and demonstrated. Thus, an isoform of PKC (PKC-α) is activated by the inflammatory stimulus and increases the levels of TNF-α that induces the activity of phospholipase A2 and releasing arachidonic acid existent in the membrane phospholipids. Simultaneously, TPA increases the amount of cyclooxygenase-2 (COX-2) and macrophage inflammatory protein 2 (MIP-2)—analogues of IL-8 in mice—and leads to neutrophil accumulation in the epidermal tissue.[59]
Laboratorial methodology	Injection of 2–2.5 μg TPA in 20 μl of acetone into one of the animal's ears. Acetone is administered in the other ear, serving as control. Measurements are performed at 0, 4, 6 and 24 h after administration. The inflammation's peak is observed 4–6 h after administration of TPA.[9,59]
Inflammation inducing agent	Xylene 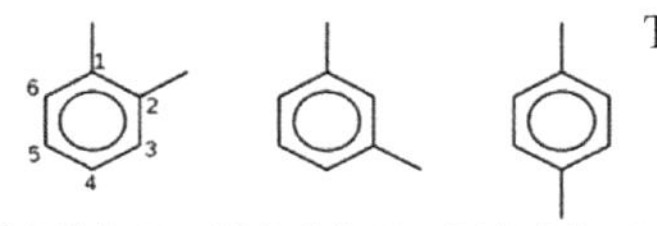 1,2-dimethylbenzene (xylene)*ortho-* 1,3-dimethylbenzene (xylene)*meta-* 1,4-dimethylbenzene (xylene)*para-* Toluene 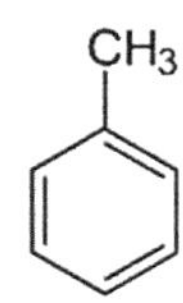
Other information	Xylene (1,2-, 1,3-, 1,4-dimethylbenzene) with the formula ($CH_3C_6H_4CH_3$) exists in three forms (*ortho, meta, para*) and is the term used for the mixture of the three aromatic hydrocarbons. Xylene and toluene ($C_6H_5CH_3$), which is also an aromatic hydrocarbon, are compounds widely used in industries, mainly as paint and rubbers solvents and have many common features in the inflammation mechanisms and in the laboratorial studies for which they are used.

Induction of inflammation mechanism	The inflammatory properties of xylene and toluene are due to the action of neurochemical mediators. Neurogenic inflammation is a type of inflammation in which the stimulation of nonmyelinated C-fibers occurs. In this case, some neuropeptides such as substance P and calcitonin gene-related peptide (CGRP) are released from their axon terminals to control the action of adjacent tissue, such as blood vessels. Substance P is present throughout the nervous system and in normal situations acts as a neurotransmitter. Several studies show that they are released in peripheral tissues in the presence of xylene and toluene, leading to increased vascular permeability, vasodilation and consequent extravasation, and accumulation of plasma in the extravascular space with the formation of edema. In this way, the inflammatory process begins.[7,37] At the experimental level, these compounds are mainly used to cause edema in the ears of rats, although they are also sometimes used elsewhere, such as the back or belly.
Laboratorial methodology	Topical application on the skin of the animal's ear of 25 μl of xylene solution or toluene in acetone (10, 50, 100%). Acetone is administered in the other ear, serving as control. Measurements are performed at 0, 1, 2, 4 and 24 h after induction. Peak inflammation is observed 1 h after application.[7,37]
Inflammation inducing agent	Arachidonic acid 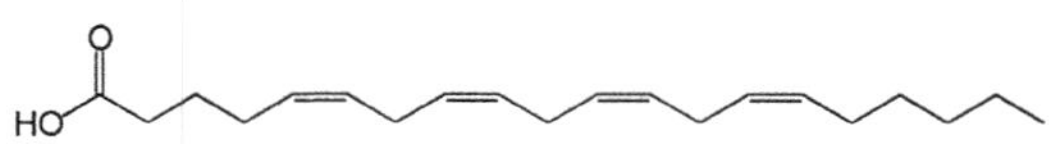
Other information	Arachidonic acid is an essential fatty acid, omega-6, formed by a chain of 20 carbons with four double bonds at positions 5, 8, 11, and 14. The presence of the double bonds causes this molecule to have several sites that can be oxidized, allowing the formation of different lipid molecules with distinct biological activities.
Induction of inflammation mechanism	Arachidonic acid is found in the phospholipid membranes where, in case of aggression, it undergoes the action of COX 1 and 2 enzymes and lipoxygenase that are activated by membranes of the damaged tissues, degrading in prostaglandins and leukotrienes. These hormones are fundamental in inflammation, since they are involved in the processes of vasodilation and pain induction, among others. When arachidonic acid is administered to the animal, an increase in its concentration and induction of inflammation is observed. Specifically, some studies show that on injecting this compound into tissues, there is an initial increase of prostaglandin E2 (PGE2) and leukotriene C4 (LTC4), followed by the remaining characteristic intermediates of inflammation.[6,17]
Laboratorial methodology	Topical application of 2–3 mg of arachidonic acid in 20 μl of acetone on the animal's ear. Acetone is administered in the other ear, serving as control. The inflammation is analyzed approximately 1 h after the induction.[6,17]

Inflammation inducing agent	Oxazolone 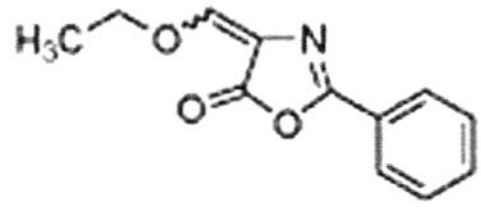
Other information	Oxazolone (4-ethoxymethylene-2-phenyl-2-oxazolin-5-one) is an allergen compound having the empirical formula$C_{12}H_{11}NO_3$).[26]
Induction of inflammation mechanism	Oxazolone causes hypersensitivity on contact, that is, when contact occurs between the compound and the skin, there is an inflammatory allergic reaction.26 Oxazolone leads to increased levels of interferon-γ, associated with T helper (T_h) cells.[13]
Laboratorial methodology	The induction of acute inflammation with oxazolone comprises two steps: 1. Initial sensitization—application of 300 μl of oxazolone solution in ethanol (2%) on the abdomen of the animal 13,26,51 or 10 μl of oxazolone also in ethanol (1.6%) on the ear.[42] 2. Inflammation test—3–7 days later, a solution of oxazolone in ethanol (0.8%) or olive oil (4:1) is again applied to the ear. The second ear will be used as control. The inflammation assessment should be done 0, 24 and 72 h after the last induction.[13,26,42,52]
Inflammation inducing agent	Croton seed oil Adapted from https://commons.wikimedia.org/wiki/File: Codiaeum_variegatum_-_seeds_-_Kroton_001.jpg
Other information	Croton is a shrub found in the American and African continents and is widely used in gardening. For scientific use, the oil of its seeds (CO) is extracted and used in in vivo experiments mainly related to inflammation and tumors (Goulart, 18-12-2009).
Induction of inflammation mechanism	The specific inflammation induction mechanism is not completely known. However it is known that TPA, already described above, is extracted from this oil. So, it is assumed that the inflammatory agent is precisely TPA, the mechanism being already stated (Goulart, 18-12-2009).
Laboratorial methodology	Application of 15–25 μl solution of CO in acetone (0.3–3%).to the surface of the animal's ear Acetone is administered in the other ear, serving as control. Inflammation measurements should be performed 0, 6, 24, 48 and 72 h after CO administration. The peak of inflammation is observed 6 h after this administration.[19,46,47]
Inflammation inducing agent	Formaldehyde (FA) 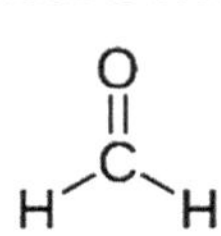

Other information	Formaldehyde (CH_2O) has several industrial applications, being used to obtain resins and as a raw material, as a sterilant, preservative agent for cosmetic and cleaning products, among many others.[45]
Induction of inflammation mechanism	Formaldehyde, such as xylene and toluene, owes its inflammatory characteristics to the ability to stimulate the release of neuropeptides in processes similar to those explained above.[37] Although they have similar inflammation mechanisms, formaldehyde is significantly more irritating than xylene and toluene. In the inflammation induced by formaldehyde, there is an increase of IL-4, brain-derived neurotrophic factor (BDNF), neurotrophin-3 (NT-3) and transient receptor potential vanilloid 1(TRPV-1) in the ear.[37,45]
Laboratorial methodology	25 µl of a solution of FA in acetone (2 to 10%) is applied over the ear of the animal.
	Acetone is administered in the other ear, serving as control.
	Measurement of inflammation should be done 0, 1, 4 and 24 h after AF application.[37,28]

2.2.1.1.2 Paw Edema

Paw edema is obtained by intraplantar administration of the inducing agent, usually allowing the study of inflammation in subcutaneous tissues. In the following table some of the main techniques are presented and organized according to the prevalence of use.

Inflammation inducing agent	Carrageenan 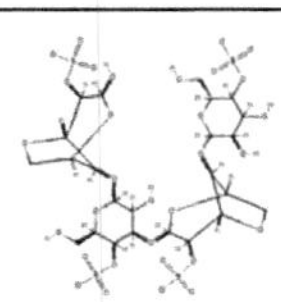
Other information	Carrageenan is the main compound used to induce inflammation in rat paws. Carrageenan refers to a group of complex polysaccharides which repeat galactose monomers in their structure and which may be of different types: *lambda, kappa* and *iota.*[55] Carrageenan is obtained from seaweed extract found in the North Atlantic Ocean and is widely used as a feed additive.[39]
Induction of inflammation mechanism	The use of carrageenan as an inducer of inflammation has been made since Winter et al proposed it in 1962.[54] The inflammation induced by carrageenan occurs because it promotes the release of some biochemical mediators that initiate and amplify the immune response. Among them are histamine and serotonin, followed by prostaglandins which will be the most important in this stimulus.[18]

	Histamine and serotonin are vasoactive amines which cause blood vessel dilation and permeability increase, and also are chemotactic and induce the production of prostaglandin E2. On the other hand, prostaglandin is a derivative of arachidonic acid which promotes vasodilation and is involved in the formation of edema.
Laboratorial methodology	Carrageenan *lambda* is the type commonly used at the experimental level. 100 μl of carrageenan in saline solution (1%) is injected into the rat's paw. Saline solution is administered in the other paw, serving as control. Measurement of inflammation should be done 1, 3, 5, 6, 12 and 24 h after its induction. The peak of inflammation is observed approximately 5 h after injection.[52,55]
Inflammation inducing agent	Histamine 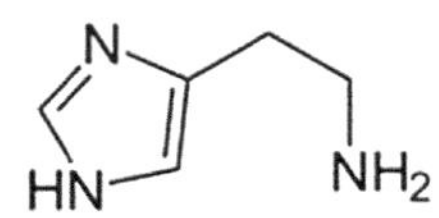 Serotonin 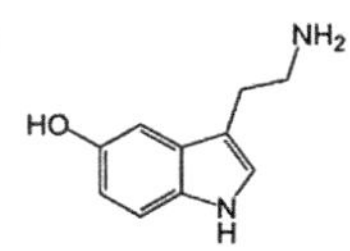
Other information	Histamine and serotonin (5-hydroxytryptamine (5-HT)) are organic molecules that, besides being involved in inflammation, are neurotransmitters.[21] The chemical formula of histamine and serotonin are, respectively, $C_5H_9N_3$ and $N_2OC_{10}H_{12}$.
Induction of inflammation mechanism	Unlike the other molecules referred to so far, histamine and serotonin directly participate in the normal inflammatory reaction of organisms. Histamine originates from basophils, a type of gastric cells and histaminergic nerve endings, while serotonin is stored in platelets and enterochromaffin cells. Under natural inflammation conditions, these vasoactive amines are released when there are physical aggressions, immunological reactions involving immunoglobulin E (IgE), complement system fractions (C3a and C5a), neuropeptides such as substance P, and IL-1 and IL-8 cytokines. When histamine and serotonin are injected in the rat's paw, these chemical mediators of inflammation trigger some of the inflammatory events characteristic of the inflammatory response. One is the formation of edema which can then be analyzed to conclude on the extent of inflammation and the anti-inflammatory activity of test compounds.[21]
Laboratorial methodology	50 μl of saline solution of histamine or serotonin (0.5 μmol) is injected subcutaneously into the paw of the animal. Saline solution is administered in the other paw, serving as control. Analysis of the inflammation should be done several times up to 180 min after administration of the compounds (for example, every 20 min).

	The peak of inflammation is observed approximately 40 min after the induction.[21,28,58]
Inflammation inducing agent	Prostaglandin E2 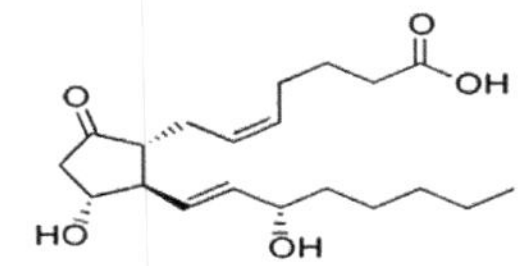
Other information	Prostaglandin E2 ($C_{20}H_{32}O_5$) has several biological actions, including vasodilation, anti-inflammatory and pro-inflammatory activities, modulation of sleep/wake cycles and replication of the HIV virus. Prostaglandin E2 also raises cyclic adenosine monophosphate (AMP) levels, stimulates bone resorption and has thermoregulatory effects. In addition, it regulates sodium excretion and renal hemodynamics.[8]
Induction of inflammation mechanism	Prostaglandin E2, such as histamine and serotonin, is a natural chemical mediator of the inflammatory response. Its action in the formation of edema is due to its ability to bind to a receptor that exists in most plasma membranes of rat cells, EP3, which is part of the G-protein coupled receptor family. In addition to the already known effects of prostaglandin E2 in the inflammatory process—vasodilatation and bronchodilation—it also induces the release of neuropeptides that stimulate inflammation, namely the substance P.[8]
Laboratorial methodology	20 µl of a Phosphate-buffered saline solution with PGE2 (0.1–10 nmol/20 µl) is injected into the rat's paw. Saline solution is administered in the other paw, serving as control. The measurement of inflammation should be done 15, 30, 45 and 60 min after administration of PGE2. The peak of inflammation is observed between 30 and 45 min.[8].
Inflammation inducing agent	Formaldehyde (described above) 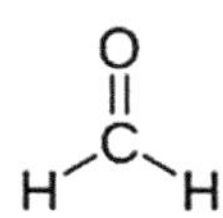
Other information	Formaldehyde (CH_2O) has several industrial applications, being used to obtain resins and as a raw material, sterilant, preservative agent for cosmetic and cleaning products, among many others.[45] In addition to the induction of inflammation in ears presented above, FA may also be used to develop inflammatory response in the intraplantar tissues of rat paws. The mechanism of action responsible is similar to that previously described.[44]
Induction of inflammation mechanism	Formaldehyde, such as xylene and toluene, owes its inflammatory characteristics to the ability to stimulate the release of neuropeptides in processes similar to those explained above.[44]

Induction of inflammation mechanism	0.1–0.2 ml of saline solution with formaldehyde (1–4%) is injected subcutaneously into the rat's paw. Saline solution is administered in the other paw, serving as control. The analysis of the inflammation should be done 0, 3, 6, 24 and 48 h after induction. The peak of inflammation is observed approximately 24 h after formaldehyde administration.[22,44]

2.2.1.1.3 *Peritonitis*

Peritonitis is the inflammation of the lining of the abdominal cavity (peritoneum). The peritoneum is a membrane that lines the internal organs of the abdomen and the inner face of the walls of the abdomen. This inflammation is usually acute and localized and it is triggered by chemical or biological stimuli—for example, internal fluids of the organs released in an injury or presence of microorganisms. The study of therapies that may attenuate this infection are quite pertinent, and for this reason, compounds that induce peritonitis are necessary.[25] Some of them are listed below.

Inflammation inducing agent	Thioglycolate 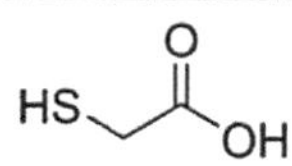
Other information	Thioglycolate—in its most common form of thioglycolic acid ($HSCH_2CO_2H$)—is widely used in the cosmetics and materials industries. On the other hand, sodium glycolate is also very common in bacterial culture media since, by reducing the oxygen, it creates a medium with anaerobic conditions.
Induction of inflammation mechanism	Thioglycolate, when injected intraperitoneally in a guinea pig, will lead to rapid and abundant recruitment of neutrophils. Therefore, it induces the activation of monocytes and infiltration of macrophages at the site with cytokine production and subsequent development of inflammation.[50] In addition to the study of peritonitis, thioglycolate also allows the study of some concrete aspects of inflammation such as proteins released by leukocytes and adhesion molecules to the vascular wall.[4]
Laboratorial methodology	2 ml of a saline solution with thioglycolate (0.3–3%) is injected intraperitoneally into the rat. Rats in identical conditions (age, sex, size, housing conditions) will be injected with saline alone, serving as control. Inflammation assessment is performed at various time periods (3–4 and 72 h) after induction.[3,4]

Inflammation inducing agent	Monosodium urate 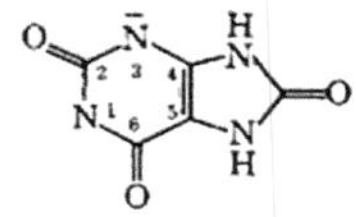 Calcium pyrophosphate 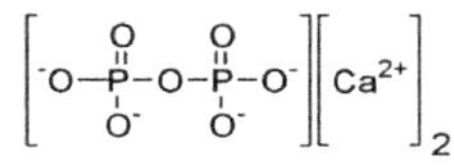
Other information	Monosodium urate (MSU) and calcium pyrophosphate (CPPD) in their crystal forms are well known in the medical community because they accumulate in the joints when high values of uric acid are registered in the blood, inducing inflammation that can become chronic.
Induction of inflammation mechanism	When MSU and CPPD are injected into the peritoneal cavity, they will develop inflammation through a process similar to that of thioglycolate, primarily involving cellular mediators, mainly monocytes. They will also promote the release of IL-1, IL-6, IL-8 and TNF-α.[23]
Laboratorial methodology	MSU crystals are prepared from uric acid and sodium hydroxide and the CPPD crystals are obtained by mixing calcium nitrate and sodium pyrophosphate. Rats in identical conditions (age, sex, size, housing conditions) will be injected with saline alone, serving as control. 3 mg crystals/500 μl saline solution (PBS) are injected intraperitoneally, reaching the peak of inflammation within 27 h after administration.[23,34]

2.2.1.1.4 Acute Joint Inflammation

Acute inflammation in joints is quite common and transversal to all age groups resulting, for example, from injuries acquired in sports.

Inflammation-inducing agent	Monosodium urate 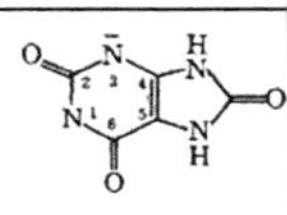 Calcium pyrophosphate 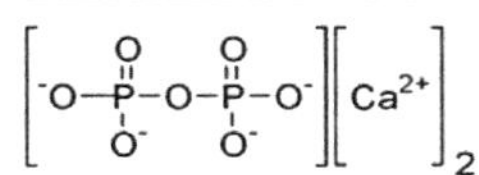
Other information	MSU and CPPD, already referred to in peritonitis, are mainly used to cause inflammatory response in joints. Their inflammatory mechanism results from the interaction between crystals and type B synoviocytes—cells responsible for the production of synovial fluid that lubricates the joints, synovial macrophages and leukocytes that are present in the joint. This interaction leads cells to produce biochemical factors such as IL-8, IL-1b and TNFα that trigger inflammation.[23]
Induction of inflammation mechanism	10 mg MSU or CPPD crystals/5 ml saline solution are injected into the knee joint of the rat. Saline solution is administered, serving as control. The analysis of the inflammation should be done 1, 2, 4, 6, 24 and 48 h after induction.

	In this case, the peak of inflammation becomes variable because there is no concrete parameter—for example, the size of edema—that can be used. Here, the way to assess inflammation is through the quantification of some chemical and cellular mediators present. Of these, there are two major peaks of inflammation located 2 and 24 h after crystal administration.[40]

2.2.1.1.5 Acute Lungs Inflammation

Under natural conditions, acute inflammation in the lungs results mainly from the invasion of this organ by foreign bodies, for example, in cases of allergic inflammation, acute lung injury or acute respiratory distress syndrome.

In order to evaluate inflammation in the lungs, there are no screening methods and the animal must be sacrificed and the lung fluid extracted.

Inflammation inducing agent	Lipopolysaccharide 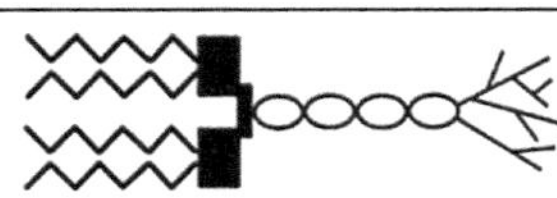
Other information	Lipopolysaccharides are the major components of the cell membrane of gram-negative bacteria. Lipopolysaccharide molecules consist of a lipidic chain—which usually bind to the membrane—and a polysaccharide responsible for cell recognition.[57]
Induction of inflammation mechanism	Lipopolysaccharides used as inflammation inducers are of bacterial origin (e.g., *Pseudomonas aeruginosa* and *Escherichia coli*) and are therefore recognized as foreign molecules to the organism. The presence of lipopolysaccharides leads to the activation of mononuclear phagocytes and the release of mediators such as TNF-α, which in turn, cause tissue damage by the induction of endotoxin release. These endotoxins are associated with various pathologies, namely acute lung injury or acute respiratory distress syndrome. Once the response is triggered by an aggressor stimulus, TNF-α is released, causing the migration of neutrophils that infiltrate the lung space, leading to the formation of pulmonary edema by increasing the typical vascular permeability of the inflammatory response.[57]
Laboratorial methodology	200–250 μg/100 g of animal weight of lipopolysaccharides are administered to the animal. Administration can be performed in two ways: inhalation of aerosolized LPS (250 μg/100 g animal weight) or injected intratracheally (20 μl/10 g animal weight of a saline solution with 2 mg/ml concentration LPS). Rats in identical conditions (age, sex, size, housing conditions) will be injected with saline alone, serving as control. The assessment of inflammation should be performed 6 h after its induction.[57]

2.2.1.2 CHRONIC INFLAMMATION

One of the most immediate and widely used methods for the induction of chronic inflammation is the repeated administration over several days of compounds used for acute inflammation. In this way, the organism is continually subjected to the inducing stimulus and maintains the inflammation. Some of the examples of compounds found in literature that are used in this way are TPA, carrageenan, formaldehyde, and oxazolone. However, given the basis of this methodology, it can be applied to many others.

In many of the experimental models where this principle is used the researchers also create an "air pouch." This bag of air is created at the place where the inflammation is to be studied by injecting a small amount of decontaminated air—5–20 ml. Depending on the time frame of the experiment, it is sometimes necessary to repeat the procedure 3–4 days later. This bag has the advantage of creating a more suitable environment for the development of inflammation when injecting the inducing compound. In addition, this hole houses the exudate resulting from the inflammation that may be later collected and studied, making it much easier to isolate the mediators to be detected in the techniques of analysis of inflammation and avoiding the sacrifice of the animal.[11,31]

2.2.1.2.1 Arthritis

Chronic arthritis is an inflammation that affects a wide range of the population, mainly the elderly. Monosodium urate (MSU), carrageenan and formaldehyde are three of the compounds widely used to induce chronic arthritis. In this case, any of these compounds is administered to the joint under study (usually rat paw joints) repeatedly for several days, which leads to the onset of chronic inflammation.[24,40]

Inflammation inducing agent	Freund's complete adjuvant
Other information	Freund's complete adjuvant (FCA) is constituted by microbial components, namely of *Mycobacterium tuberculosis*, that are identified by the organism as foreign agents, potentiating the inflammation through mechanisms already presented.

Induction of inflammation mechanism	150–750 μl of FCA in the form of emulsion is injected into the joint of the anesthetized rats. The results can be evaluated over time through a perfusion method that avoids animal sacrifice. In this method, when the synovial fluid is to be withdrawn, two needles are injected into the joint of the anesthetized animal and with the aid of an infusion pump, the exudate is replaced with a saline solution.[1]

2.2.1.2.2 Cotton Pellet

This technique involves the subcutaneous introduction, through a minor surgery, of a cotton body (50–100 mg). The introduction is made in the lower back of the animal or near the armpits and remains there for 6–8 days. As a foreign body, the cotton causes an inflammatory response. When the cotton is removed, the exudate can be extracted from it and the molecular and cellular constituents in suspension are studied to evaluate the inflammation. On the other hand it can be dehydrated, its mass calculated and compared to the initial cotton weight. The difference in weight indicates the extent of the inflammation. This procedure is useful for studying inflammation of the fibrinous subtype, for example, if the cotton is embedded in an inducing compound (e.g., carrageenan) before being introduced into the animal, the inflammatory effect is higher.[31,43]

2.2.3 INFLAMMATION EVALUATION

There is a wide variety of ways to analyze and quantify inflammation in laboratorial research. The analysis technique must be chosen considering some factors, the main being the mechanism of action of the inducer of inflammation used and the parameters to be analyzed. Regarding the first one, it is necessary to know what the minor manifested characteristics of inflammation are, for example, which mediators are found in the various phases of the response induced by the specific compound. Concerning the parameters to be analyzed, their definition depends on the method of induction and the action mechanism of the anti-inflammatory agents since these will influence, for example, the mediators more prevalently found.

The methods of analysis of inflammation can be divided into two classes: the screening methods and the biochemical methods. Screening methods allow a more comprehensive assessment of the inflammatory response and are often faster and cheaper. In addition, it is possible to draw conclusions that are qualitative, in which it is observed whether or not the compounds under study have an anti-inflammatory capacity, relating their values to those of the control results and other known compounds. Biochemical methods provide more consistent conclusions about the amplitude of the anti-inflammatory capacity of the compound under study. In addition, they are more rigorous and allow understanding the mechanisms underlying the action of the anti-inflammatory drug. In the studies of new drugs, the two classes of methods are complementary. That is, in the first stage, screening methods should be used to check if there is anti-inflammatory capacity and whether it is worthwhile to continue the study. If so, one should move to the biochemical analysis in the following stage.

2.2.3.1 SCREENING METHODS

The base of the screening methods is to measure the differences in size and weight of edema arising in the inflammatory response and to do the histological analysis of the inflamed tissues.

2.2.3.2 EDEMA SIZE

Measurement of the size of the edema can be done using various instruments, the most common being the micrometer, constant pressure measurement device, and plethysmometer. The latter is mainly used to determine changes in the leg volume and consists of immersing the animal's leg in a volume of water and finding its displacement. This determination should be automatically made in order to avoid operator errors and biased reads.[18,52] The micrometer, on the other hand, is used to measure the volume of any edema and consists of a tool with two ends attached to a common or digital ruler. The edema is placed between the two tips and the ruler shows its size.[7,37] The constant pressure measuring device operates identically to the micrometer, but automatically finds the sample size by pressure variation.[17]

The great advantage is that they are fast, cheap, and simple techniques. On the other hand they have some disadvantages, such as being quite dependent on operator sensitivity and therefore quite susceptible to experimental errors.[39]

2.2.3.3 EDEMA WEIGHT

The measurement of edema weight is a technique used mainly to determine changes in the size of the ears of animals and obliges them to be sacrificed. To perform this, it is necessary that the ear is separated from the cadaver or that a section is removed and that it is weighed, with the other ear of the animal as a negative control.[7,19] Although in a smaller quantity, there are also protocols of experiments in which the feet of animals, with and without edema, are weighted to analyze their evolution.[52]

2.2.3.4 HISTOLOGICAL ANALYSIS

Through histological analysis of the inflamed tissues it is possible to qualitatively examine the infiltration of inflammatory cells and compare them to normal tissues. To do this, sections of the tissues are cut, dehydrated according to standard procedures and then stained—for example, with eosin and hematoxylin—to be analyzed under an optical microscope.[7,37]

2.2.3.5 BIOCHEMICAL METHODS

In the biochemical analysis techniques, the quantities of some biochemical mediators involved in the inflammation are used to determine its extent and can be applied to any type of inflammation, localized or not. The choice of mediators should be made based on the action mechanism of the inducer of the inflammatory response and on what is thought or known to be the mechanism of the anti-inflammatory agent.

Many times the analysis of the mediators implies the sacrifice of the animals so that it is possible to extract tissues and cells from the inflammation site.[37]

The most common mediators in laboratory studies are listed in the following table.

Molecular mediator	Information
Cytokines	Cytokines are biochemical mediators involved in inflammation. Some of them are used to evaluate the inflammation, as IL-4, IL-1B, interferon-gamma (IFN-γ) and TNF-α. The increase in their concentration indicates an increase in inflammatory activity. In laboratorial techniques, the levels of cytokines are measured by the presence of the corresponding mRNA.[37,42] Another more direct technique for determining the concentration of cytokines is through immunological assays, such as those described below for prostaglandins and leukotrienes.[51]
Vascular permeability	As noted before, during the inflammatory process the vascular permeability increases to allow the passage of plasma and immune cells. Thus, the increase in this parameter indicates a more intense inflammatory response. In laboratory, a blood protein called bovine serum albumin (BSA) is used in saline solution which is injected into the tail of the animal prior to the experiment. At its end, the fluid present in the edema is removed and the concentration of BSA is determined, indicating the level of vascular permeability.[6]
Myeloperoxidase (MPO) or peroxidase (PA) assay	Myeloperoxidase is the most abundant protein in neutrophils and is also found in monocytes; therefore, consisting an indicator of the presence of these cells. Its concentration is quite high at inflammation sites because it has the ability to catalyze the reaction between a chloride ion and hydrogen peroxide, leading to the formation of a compound very toxic to bacteria (hypochlorous acid—HClO).[29] In laboratory, the inflamed tissue is withdrawn, homogenized and subjected to a set of steps that promote its purification. Then an artificial electron donor (O-dianisidine dihydrochlofide) and H_2O_2 are added to the solution. The absorbance of the solution is read over a period of time at 460 nm and the faster is the increase in absorbance, the higher is the concentration of myeloperoxidase.[6,51]
Prostaglandins and leukotrienes essays	Prostaglandins and leukotrienes are arachidonic acid metabolites that play a role in the inflammatory response. Thus, their presence at high concentrations indicates the presence of inflammation. In laboratory, the inflamed tissue is extracted and homogenized and a few purification steps are completed. Finally, biomarkers are detected by immunological assays—for example, enzyme-linked immunosorbent assay(ELISA) assay 6, 18—or by liquid chromatography techniques—Mass Spectroscopy (LC-MS) 18. Some of those markers are prostaglandins E2 and F2α, leukotriene C4 and thromboxane B2

Molecular mediator	Information
COX-1 and COX-2	COX-1 and COX-2 enzymes are responsible for metabolizing arachidonic acid, whereby their high concentration indicates inflammation. In the laboratory, the mRNA-encoding COX-1 and COX-2 is detected in a process similar to that described for cytokines.[42] Another of the techniques for its detection are through immunological tests—see the example of prostaglandins and leukotrienes.[18]
^{67}Ga-citrate	^{67}Gallium is a radioactive isotope that can be used to monitor edema evolution and vascular permeability. In laboratory, the Ga-citrate saline solution[67] is injected into the tail of the animal prior to administration of the test compounds and, at different times, the radiation in the edema will be detected with appropriate instruments.[21] This procedure is useful for determining the relative edema size or assessing vascular permeability.
Evans blue probe	Evans blue probe strongly binds to the blood albumin and can therefore be used to study vascular permeability. In laboratory, the probe is injected into the tail of the animal prior to start of the experiment and, with the increased permeability of the vascular endothelium, the probe is released into the edema. Then, the edema-forming fluid is removed and its absorbance at 590 nm is detected.58 This procedure is also useful for determining the relative edema size or assessing vascular permeability.
Direct neutrophils quantification	The MPO assay explained above allows obtaining an indication of the relative amount of neutrophils present in the inflamed tissues. However, if it is desired to make its quantification directly and accurately it will be necessary to resort to more complex techniques, such as flow cytometry. To this end, fluorescence antibodies specific for the animal's Gr-1 are added to the neutrophil-rich solution.[3,50] Gr-1 is a protein present on the surface of granulosa cells, namely neutrophils (MiltenyiBiotec, 5-12-2010). Antibodies, and with them neutrophils, can be quantified by flow cytometry.[3,50] In addition to Gr-1 there is a variety of proteins in the cell membranes of other immune cells involved in inflammation. Thus, the chosen antibodies may be specific for other peptides.

2.2.3.3 *IN VIVO MODEL FOR THE STUDY OF ACUTE INFLAMMATION—THE EXAMPLE OF TTHE RAT PAW EDEMA TEST*

One of the techniques for evaluating anti-inflammatory activity in acute inflammation is the analysis of the rat paw volume variation at which edema is induced by an inflammatory process. One of the possible procedures for this is the injection of carrageenan according to the procedure first proposed by.[54] This procedure is based on the intraplantar injection of 0.1 ml of a 1% κ-carrageenan solution to induce inflammation.[15,36,38,54]

The anti-inflammatory activity is directly related to the evolution of the edema volume that is determined using a plethysmometer (Fig. 2.1-LE7500, Panlab, Barcelona, http://www.cebiolog.com.br/pletismocircmetro.html). This apparatus has two containers filled with a 0.1% NaCl solution and, in one of them, there is an electrode which senses an increase in the level of the solution. There is a channel connecting the two containers that allows the displacement of solution between them. By dipping the animal's foot in the solution, the level of the solution rises and this increment is detected corresponding to the volume of the foot.[39]

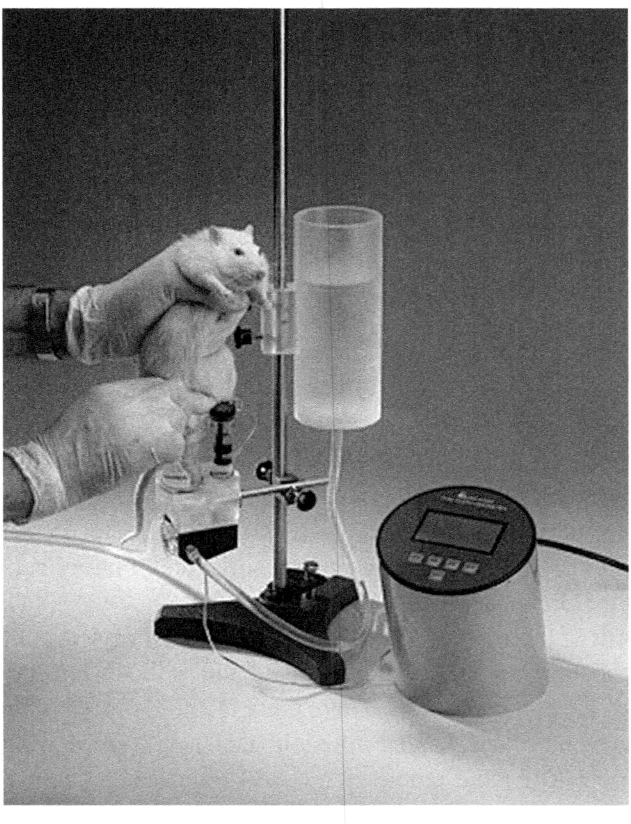

FIGURE 2.1 Plethysmometer.

Carrageenan control and test groups are dosed 1 h before and positive control group is dosed 30 min before a subplantar injection of 0.1 ml of 1% solution of carrageenan in saline, is administered to the right hind footpad of each animal. The paw edema volume is measured with a digital plethysmometer after carrageenan injection and then every hour for 6 h (Refer to Fig. 2.2). The anti-inflammatory effect is calculated 4 h after carrageenan administration and is expressed as percentage of edema inhibition for the treated animals with respect to the carrageenan control group. The volume value at time 0 (before administration) is subtracted from all the following measures, resulting in the edema volume value.[15,36,38]

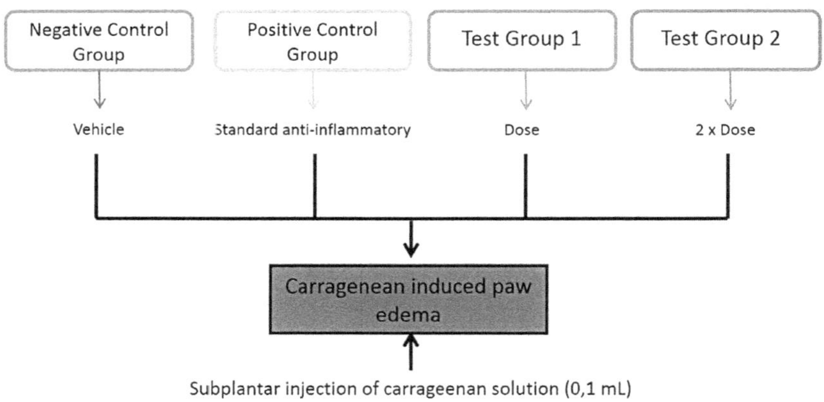

FIGURE 2.2 Experimental design of the rat paw edema test.

Thus, by comparing the mean volume of the edema for the different times or the peak of the edema with those of the negative control group, it is possible to determine the anti-inflammatory capacity of the samples, expressed as inhibition of edema percentage. A greater decrease in the edema's volume corresponds to a greater anti-inflammatory activity of the compound under study.

The result of the anti-inflammatory activity is expressed as the relationship, in percentage, of the volume of the edema of the test group compared to the one of the control groups. Then, there is one example of these results for two fractions of the leaves infusion (CcE) of *Cymbopogon citratus*, flavonoid-rich (CcF) and tannin-rich (CcT) fractions.[14] The evaluation of the anti-inflammatory activity was performed in the carrageenan-induced rat paw edema (Fig. 2.3).

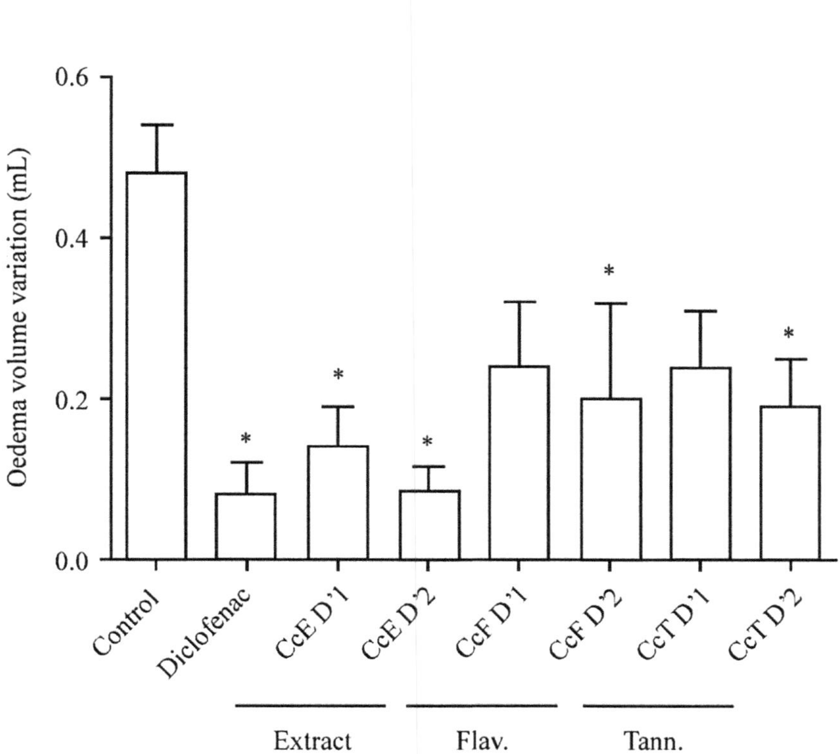

FIGURE 2.3 Results from the carrageenan-induced rat paw edema test were obtained after administering water 0.5 μl gG1 p.o. (negative control), diclofenac sodium 10 mg kgG1 i.p. (positive control) and *Cymbopogon citratus* aqueous extract 34.12 mg kgG1 (CcE D'1) and 68.24 mg kgG1 (CcE D'2), *C. citratus* flavonoid-rich fraction 3.71 mg kgG1 (CcF D'1) and 7.42 mg kgG1 (CcF D'2) and *C. citratus* tannin-rich fraction 2.98 mg kgG1 (CcT D'1) and 5.96 mg kgG1 (CcT D'2) p.o. Each value is the Mean±SEM of six rats. Statistical differences between the treated and the negative control groups were determined by analysis of variance (ANOVA) followed by Bonferroni test. *$p<0.05$ (Based on our current research group communications–2015)

The differences in the edema volume variation between the different experimental groups allow concluding on the anti-inflammatory activity of the tested compounds. It is observable that the group with high activity—lower edema volume—is the positive control group with the commercially available drug. Naturally, the negative control group presents the highest edema because there is no anti-inflammatory agent present. Among the test samples, the one with the higher anti-inflammatory proprieties is the total extract (CcE) whose activity is dependent on the administered dose.[14]

2.2.3.4 IN VIVO MODEL FOR THE STUDY OF CHRONIC INFLAMMATION—THE EXAMPLE OF THE COTTON

A model used to induce chronic inflammation is the cotton pellet test, adapted from Winter and colleagues (1957).[53] This technique induces chronic inflammation in the form of granuloma, allowing the evaluation of the proliferation of the components involved in this process,[31] including anti- and pro-inflammatory mediators. In this test, monocyte/macrophage migration, fibroblast proliferation, apoptosis, and all components associated with chronic inflammation occur in the vicinity of the cotton pellet, wrapping it in granulation tissue.[48]

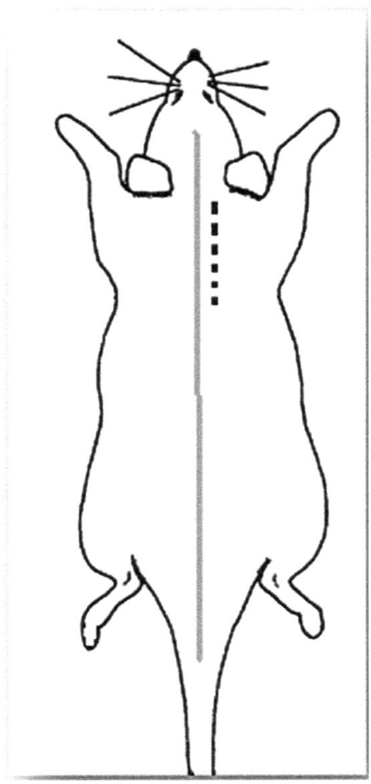

FIGURE 2.4 Graphical representation of the cotton pellet inserted in a rat.

The cotton pellet test is based on the subcutaneous insertion of a sterilized piece of cotton of about 10 mg in the interscapular region of the animals, usually rats[31]—Figure 2.4. Five days after this procedure, the cotton pellets and the surrounding tissue are removed and analyzed.[31,53]

The experimental design (Fig. 2.5) consists of a positive control group (administered with a standard anti-inflammatory agent), a negative group (administered only with a vehicle substance), and the test groups (administered with the compounds to be studied). Each group constitutes of 6–8 animals.

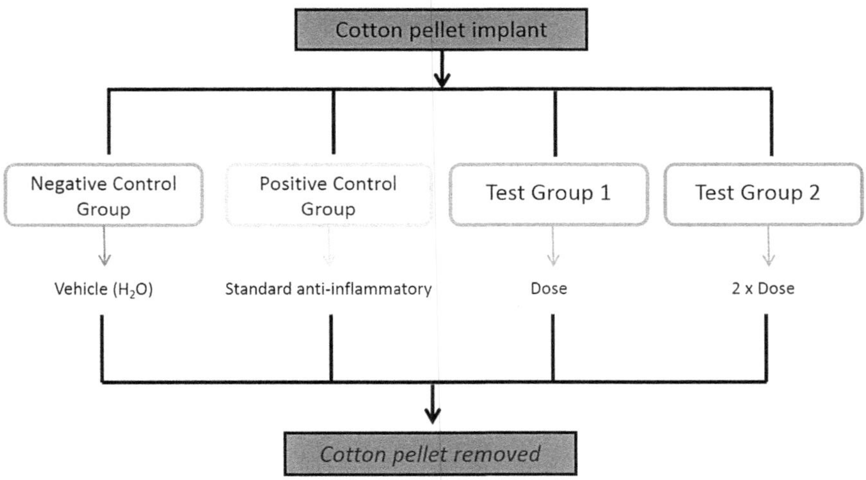

FIGURE 2.5 Experimental design of the cotton pellet-induced granuloma test.

To analyze the anti-inflammatory activity of the studied compounds, there are two techniques that are complementally used: the determination of the cytosine levels at the inflammation site and the body temperature of the animals.

The cotton content and the inflammation surrounding the tissues removed from the animals are solubilized and their content in cytosine is determined through an ELISA test. As explained before, cytokines are molecular mediators of inflammation. Therefore, the existence of lower levels of these molecules in the test groups, when compared to the negative control group, corresponds to an anti-inflammatory activity of the test compounds.

The body temperature is determined by the insertion of a probe in the anesthetized animals' rectum for 2 min. The temperature must be determined 1–2 weeks before the beginning of the experiment and on its last day.[27,49] Fever is a systemic effect strictly associated with the inflammatory processes and provided mainly by the action of cytokines. Therefore, the increase in the animals' temperature during the experiment is associated with the inflammation. Lower body temperatures of the test groups when compared with the negative control group indicate an anti-inflammatory action of the tested compound.

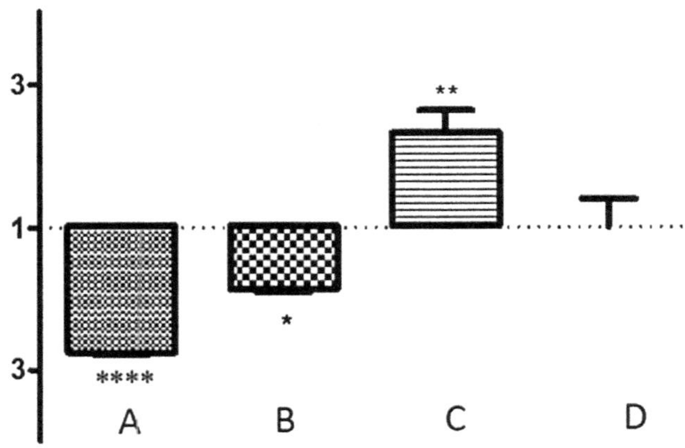

FIGURE 2.6 Levels of IL-2 from the cotton pellet test were obtained after administering (A) *C. citratus* aqueous extract (test compound), (B) Indometacine (positive control group 1), (C) Dexamethasone (positive control group 2) and (D) water (negative control). Each value is the Mean ± SEM of six rats. Statistical differences between the treated and the negative control groups were determined by ANOVA followed by Bonferroni test. *$p<0.05$; ***$p<0.0001$ (Based on our current research group communications—2013).

Figure 2.6 represents the results obtained for the levels of IL-2 with the administration of *C. citratus* as the test compound. It is observable that it clearly reduces the levels of this cytokine, which may be related to the anti-inflammatory action of the test compound. For a conclusive analysis, these results have to be analyzed together with other cytokines.

2.3 EXPERIMENTAL TECHNIQUES FOR PAIN EVALUATION IN VIVO

Pain's perception begins in the receptors present on the skin, muscles, tendons, internal organs, among other, located in the final of the primary afferent neurons. These neurons lead the pain signal to the superficial dorsal horn of the spinal cord.[5] Here, a set of neurotransmitters allow communication with second-order nerves that carry the pain signal to the cortex.

Among the chemical stimuli found in the primary receptors, there are the inflammatory mediators that are at the origin of what can be considered the "inflammatory pain" that is the most significant in the context of this work. In fact, in cases of inflammation, damaged and inflammatory tissue cells—mainly macrophages, mast cells and neutrophils—release the inflammatory mediators that form the "inflammatory soup." Among them, there are some mediators that stimulate the response of the afferent nociceptors, which initiate the propagation of the electrical signal of pain. Among them, the most important are histamine and serotonin (within mast cell origin), bradykinin (kinin system), prostaglandins, interleukins and TNF-α (cytokines), and protons.[41,56]

An analgesic drug may have its activity at the peripheral or central levels. In the first case, the mechanism of action falls on the primary afferent neurons, when the pain signal is generated. In the second case, the analgesic capacity is due to the reduction of the nervous signal of pain in the central nervous system—for example, in the spinal cord when the pain is transmitted to the second-order nerves.

Because of these different central or peripheral activities, when designing experiments to study pain in vivo, it is advisable to include models that allow determining the analgesia mechanism of the test compound. In this document, we present an example of an experiment to test an analgesic active conferred by central mechanisms (hot plate test) and peripheral mechanisms (acetic acid-induced writhing test).

2.3.1 IN VIVO MODEL FOR THE STUDY OF CENTRAL PAIN-THE EXAMPLE OF THE HOT PLATE TEST

The hot plate test (Fig. 2.7) is an experimental alternative for the study of the activity of analgesic compounds with effect on the central nervous system. This test consists of the use of a circumscribed metal plate heated at 55°C where the animals, the mice, are placed individually. The reaction time is recorded when the animals lick their forepaws, shake or jump. The baseline is considered as being the reaction time before treatment.[2] The observer should always be the same to reduce the variability in the appreciation of these signals. For this reason, all these evaluations should be recorded on film for further verification in case of doubt.

FIGURE 2.7 Hot plate.

In a specific experimental design (Fig. 2.8), mice were divided into experimental groups of 6–8 animals, which form the study groups—administered with the study samples. The positive control group was administered with the standard compound which is, in this case, morphine. The negative control group was without any analgesic compound.[39]

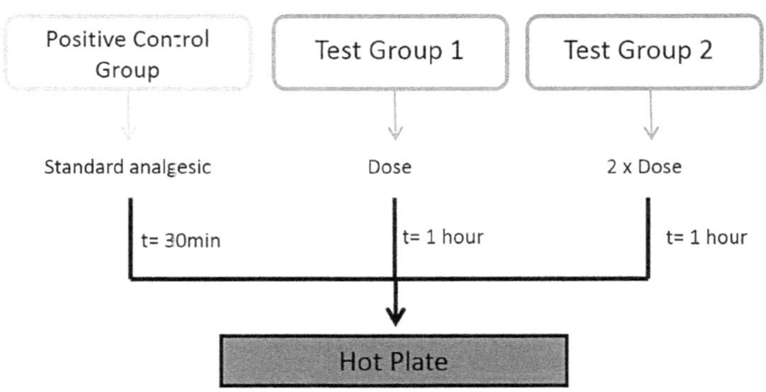

FIGURE 2.8 Experimental design of the hot plate test.

The central analgesic capacity of each sample is expressed as a percentage of the average time that the animals take to show signs of pain when compared to the negative control group. A longer response time directly indicates a greater analgesic activity of the test compound. Figure 2.9 shows an example of the results obtained with this methodology.[14]

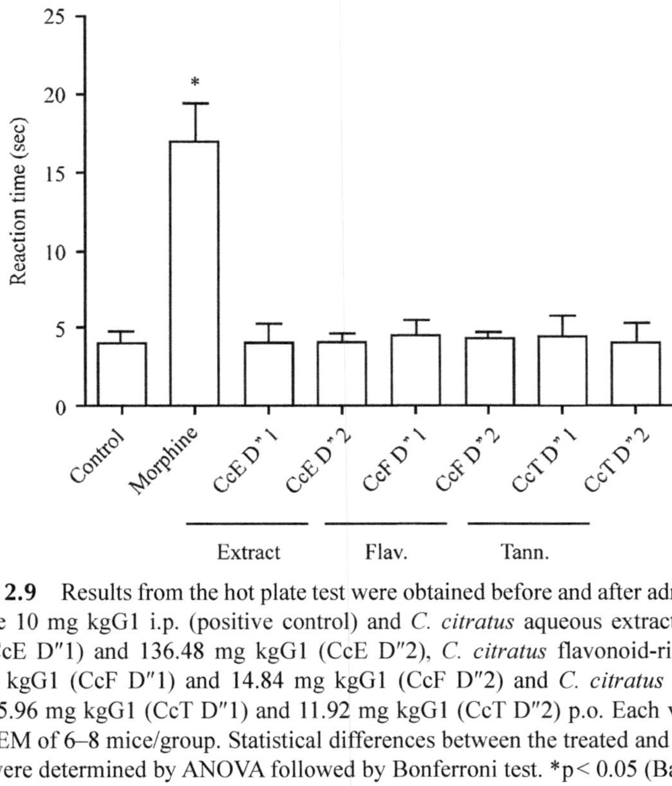

FIGURE 2.9 Results from the hot plate test were obtained before and after administering morphine 10 mg kgG1 i.p. (positive control) and *C. citratus* aqueous extract 68.24 mg kgG1 (CcE D"1) and 136.48 mg kgG1 (CcE D"2), *C. citratus* flavonoid-rich fraction 7.42 mg kgG1 (CcF D"1) and 14.84 mg kgG1 (CcF D"2) and *C. citratus* tannin-rich fraction 5.96 mg kgG1 (CcT D"1) and 11.92 mg kgG1 (CcT D"2) p.o. Each value is the Mean±SEM of 6–8 mice/group. Statistical differences between the treated and the control groups were determined by ANOVA followed by Bonferroni test. *$p < 0.05$ (Based on our current research group communications–2015).

In this example, the only group that shows a statically significant increase on the reaction time is the positive control group (morphine), indicating that the test compounds do not show any central analgesic activity.[14]

2.3.2 IN VIVO MODEL FOR THE STUDY OF PERIPHERAL PAIN-THE EXAMPLE OF THE ACETIC ACID-INDUCED WRITHING TEST

One of the possible models (Figs. 2.10 and 2.11) for the study of the peripheral analgesic activity of a compound is the acetic acid-induced writhing test, also performed with mice. Acetic acid 0.6% in saline is administered i.p., 0.1 mL/10 g b.wt. The number of writhes—a response consisting of abdominal wall contractions followed by hind limb extension—is counted during continuous observation of the animals for 30 min, starting the count 5 min after the injection.[2,38]

FIGURE 2.10 Example of a writhe.

The animals are divided into groups of 6–8 animals: Negative control group does not receive any analgesic compound, the test groups and the positive control group—administered with the standard compound which, in the following case, is diclofenac sodium.

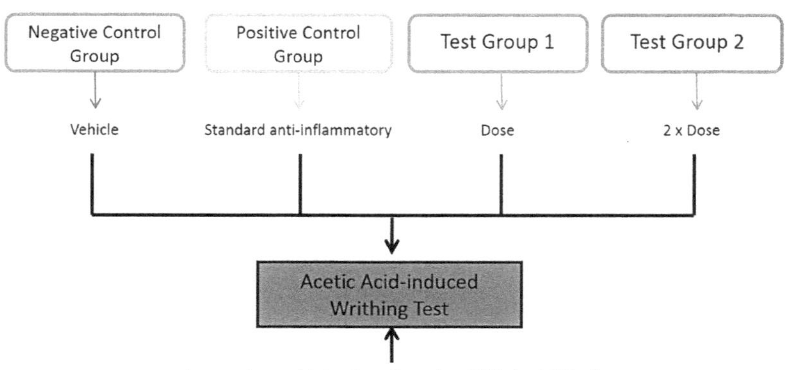

FIGURE 2.11 Experimental design of the acetic acid-induced writhing test.

A significant reduction in the number of writhes by any drug is considered as a positive analgesic response. The peripheral analgesic capacity of each sample was expressed as a percentage as the mean difference in the number of writhes compared to the negative control group. Figure 2.12 shows an example of the results obtained with the Acetic Acid-induced Writhing Test.[14]

In Figure 2.12, it is observed that the group with the higher analgesic group is the positive control one, with diclofenac that is known for its peripheral analgesic activity. All the double doses induce an analgesic activity statistically different from the negative control group. With this information we may conclude that the double doses of the tested compounds have analgesic activity.

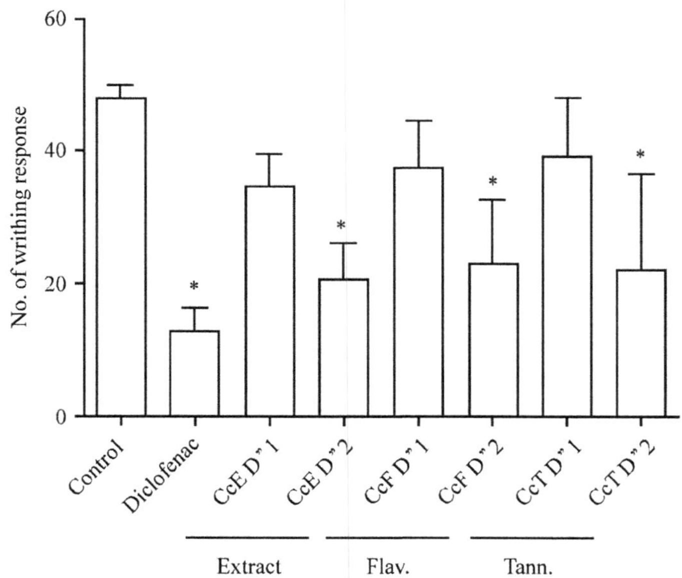

FIGURE 2.12 Results from the acetic acid-induced writhing test were obtained after administering water 0.5 μL gG1 p.o. (negative control), diclofenac sodium 10 mg kgG 1i.p. (positive control) and *C. citratus* aqueous extract 68.24 mg kgG1 (CcE D″1) and 136.48 mg kgG1 (CcE D″2), *C. citratus* flavonoid-rich fraction 7.42 mg kgG1 (CcF D″1) and 14.84 mg kgG1 (CcF D″2) and *C. citratus* tannin-rich fraction 5.96 mg kgG1 (CcT D″1) and 11.92 mg kgG1 (CcT D″2) p.o. Each value is the Mean ± SEM of 6–8 mice/group. Statistical differences between the treated and the control groups were determined by ANOVA followed by Bonferroni test. *p<0.05 (Based on our current research group communications–2015).

Considering that the compounds are the same as in the example from the hot plate test above that did not show central analgesic activity, we conclude that those compounds have their analgesic activity due to peripheral mechanisms.

KEYWORDS

- **biochemical mechanisms**
- **inflammation**
- **pain**
- **experimental techniques**
- **in vivo**
- **in vitro**

REFERENCES

1. Barton, N. Demonstration of a Novel Technique to Quantitatively Assess Inflammatory Mediators and Cells in Rat Knee Joints. *J. Inflamm.* **2007,** *4*, 13.
2. Barua, C. C.; Roy, J. D.; Buragohain, B.; Barua, A. G.; Borah, P.; Lahkar, M. Analgesic and Anti-Nociceptive Activity of Hydroethanolic Extract of *DrymariacordataWilld. Indian J. Pharmacol.* **2011,** *43*(2), 121–125.
3. Borsig, L.; Wang, L.; Cavalcante, M. C.; Cardilo-Reis, L.; Cardilo-Reis, P. L. Selectin Blocking Activity of a Fucosylated Chondroitin Sulfate Glycosaminogly Can from Sea Cucumber. *J. Biol. Chem.* **2007,** *282*, 20, 14, 984–991.
4. Call, D. R.; Nemzek, J. A.; Ebong, S. J.; Bolgos, G. L.; Newcomb, D. E.; Remick, D. G. Ratio of Local to Systemic Chemokine Concentrations Regulates Neutrophil Recruitment. *Am. J. Pathol.* **2001,** *158*(2), 715–721.
5. Carvalho, W. A.; Limônica, L. Mecanismoscelulares e moleculares da dorinflamatória—modulaçãoperiférica e avançosterapêuticos. *Rev. Bras. Anestesiol.* **1998,** *48*(2), 137–158.
6. Chang, J.; Carlson, R. P.; O'neill-Davis, L.; Lamb, B.; Sharma, R. N.; Lewis, A. J. Correlation Between Mouse Skin Inflammation Induced by Arachidonic Acid and Eicosanoid Synthesis. *Inflammation* **1986,** *10*(3), 205–214.
7. Cho, K. H.; Kim, H. D.; Lee, B. W.; Lim, M. K.; Ku, S. K. Effects of Magnetic Infrared Laser on Xylene-Induced Acute Inflammation in Mice. *J. Phys. Ther. Sci.* **2008,** *20*, 255–259.
8. Claudino, R. F.; Kassuya, C. A.; Ferreira, J.; Calixto, J. B. Pharmacological and Molecular Characterization of the Mechanisms Involved in Prostaglandin E2-Induced Mouse Paw Edema. *J. Pharmacol. Exp. Ther.* **2006,** *318*(2), 611–618.

9. Correa, H.; Valenzuela, A. L.; Lucia, L. F.; Duque, C. Anti-Inflammatory Effects of the Gorgonian *Pseudopterogorgia elisabethae* Collected at the Islands of Providencia and San Andrés (SW Caribbean). *J. Inflammation* **2009,** *6*, 5.
10. Craig, A. D. Pain Mechanisms–Labeled Lines Versus Convergence in Central Processing. *Annu. Rev. Neurosci.* **2003,** *26*, 1–30.
11. Davies, D. E.; Stevens, A. J.; Houston, J. B. Use of the Rat Air Pouch Model of Inflammation to Evaluate Regional Drug Delivery. Agents Actions, Special Conference Issue, 1992.
12. Dorsam, R. T.; Gutkind, J. S. G-protein-coupled Receptors and Cancer. *Nat. Rev.* **2007,** *7*, 79–94.
13. Fujii, Y.; Takeuchi, H.; Tanaka, K.; Sakuma, S.; Ohkubo, Y.; Mutoh, S. Effects of fk506 (tacrolimus hydrate) on Chronic Oxazolone-Induced Dermatitis in Rats. *Eur. J. Pharmacol.* **2002,** *452*, 115–121.
14. Garcia, R.; Pinto Ferreira, J.; Costa, G.; Santos, T.; Branco, F.; Caramona, M.; Carvalho, R.; Dinis, A. M.; Batista, M. T.; Castel-Branco, M.; Figueiredo, I. V. Evaluation of Anti-Inflammatory and Analgesic Activities of *Cymbopogoncitratus in Viv*–Polyphenols Contribution. *Res. J. Med. Plants.* **2015,** *9*, 1–13.
15. Garcia, R. Avaliação da actividade anti-inflamatória e analgésica do *Cymbopogon citratus* (DC) *Stapf.*em modelos experimentais *in vivo*. Universidade de Coimbra, 2010, Tese de Mestrado.
16. Goldsby, R. A.; Kindt, T. J.; Kuby, J.; Osborne, B. A. *Immunology*, 6th ed; W. H. Freeman: New York, 2007.
17. Griswold, D.; Webb, E. Arachidonic Acid-Induced Inflammation: Inhibition by Dual Inhibitor of Arachidonic Metabolism. *Inflammation* **1987,** *11*(2), 189–199.
18. Guay, J.; Bateman, K.; Gordon, R.; Mancini, J.; Riendeau, D. Carrageenan-induced paw edema in rat elicits a predominant prostaglandin E2 (PGE2) response in the central nervous system associated with the induction of microsomal PGE2 synthase-1. J. Biol. Chem. 2004, 279:23, 24866–24872.
19. Ismaili, H.; Milella, L., Fkih-Tetouani, S.; Ilidrissi, A.; Camporese, A.; Sosa, S.; et al. *In Vivo* Topical Anti-Inflammatory and in Vitro Antioxidant Activities of Two Extracts of *Thymus Satureioides* Leaves. *J. Ethnopharmacol.* **2004,** *91*, 31–36.
20. Janeway, C. A.; Travers, P.; Walport, M. *Immunobiology—The Immune System in Health and Disease,* 4th ed.; Elsevier Science Ltd./Garland Publishing: N. I., EUA, 2001.
21. Kohno, H.; Sugahara, M.; Ohkubo, Y.; Kubodera, A. Evaluation of Histamine Induced Acute Inflammation. *Eur. J. Nucl. Med.* **1987,** *13*, 371–373.
22. Li, H.; Lu, X.; Zhang, S.; Lu, M.; Liu, H. Anti-Inflammatory Activity of Polysaccharide from *Pholiotanameko. Biochemistry (Moscow)* **2008,** *73*(6), 669–675.
23. Liu-Bryan, R.; Lioté, F. Monosodium Urate and Calcium Pyrophosphate Dihydrate (cppd) Crystals, Inflammation, and Cellular Signaling. *Joint Bone Spine* **2005,** *72*, 295–302.
24. Maruyama, N.; Ishibashi, H.; Hu, W.; Morofuji, S.; Inouye, S.; Yamaguchi, H.; et al. Suppression of Carrageenan- and Collagen ii-Induced Inflammation in Mice by *Geranium Oil. Mediators Inflammation* **2006,** 1–7.
25. Merck. (30-12-2009). Peritonite. http://www.manualmerck.net/artigos/imprime.asp?id=138&cn=1135.

26. Moodley, I.; Sotsios, Y.; Bertinc, B. Modulation of Oxazolone-Induced Hypersensitivity in Mice by Selective PDE Inhibitors. *Mediat. Inflamm.* **1995,** *4*, 112–116.
27. Mythreyi R.; Sasikala E.; Geetha A.; Madhavan, V. Antipyretic Activity of Leaves of *Cadabafruticosa* (L.) Druce. *Pharmacology online* **2008,** *3*, 136–142.
28. Northover, B. J.; Subramanian, G. Some Inhibitors of Histamine-Induced and Formaldehyde-Induced Inflammation in Mice. *Brit. J. Pharmacol.* **1961,** *16*, 163–169.
29. Northwest Life Science. (05-01-2010). www.nwlifescience.com.
30. Oliveira, L. F. Actualizaçãoemmecanismos e fisiopatologia da dor. Primer Simposio Virtual de Dolor, Medicinapaliativa e avances en farmacologiadel Dolor, 2001.
31. Paschapur, M.; Patil, M.; Kumar, R.; Patil, S. Evaluation of Anti-Inflammatory Activity of Ethanolic Extract of *Borassus flabellifer* l. Male Flowers (inflorescences) in Experimental Animals. *J. Med. Plants. Res.* **2009,** *3*(2), 49–54.
32. Pereira, R. Avaliação da actividade anti-inflamatória de *Cymbopogoncitratus* (DC) *Stapf.*emmodelos *in vivo* de inflamaçãocrónica. Universidade de Coimbra, 2013, Tese de Mestrado.
33. Pinto, A. M. *RespostaInflamatória in Fisiopatologia—Fundamentos e Aplicações,* 1st Ed.; LIDEL: Coimbra, 2007; pp 185–279.
34. Pouliot, M.; James, M. J.; McColl, S. R.; Naccache, P. H.; Cleland, L. G. Monosodium Urate Microcrystals Induce Cyclooxygenase-2 in Human Monocytes. *Blood J.* **1998,** *91*, 1769–1776.
35. Rennecke, J.; Rehberger, P. A.; Furstenberger, G.; Johannes, F. J.; Stohr, M. Protein-kinase-cμ Expression Correlates with Enhanced Keratinocyte Proliferation in Normal and Neoplastic Mouse Epidermis and in Cell Culture. *Int. J. Cancer.* **1999,** *80*, 98–103.
36. Sadeghi, H.; Hajhashemi, V.; Minaiyan, M.; Movahedian, A.; Talebi, A. A Study on the Mechanisms Involving the Anti-Inflammatory Effect of Amitriptyline in Carrageenan-Induced Paw Edema in Rats. *Eur. J. Pharmacol.* **2011,** *667*(1–3), 396–401.
37. Saito, A.; Tanaka, H.; Usuda, H.; Shibata, T.; Higashi, S. Characterization of Skin Inflammation Induced by Repeated Exposure of Toluene, Xylene, and Formaldehyde in Mice. *Environ. Toxicolo.* **2009,** *26*(3), 224–232.
38. Santos, E. N. Anti-Inflammatory, Antinociceptive, and Antipyretic Effects of Methanol Extract of *Carinianarubra* Stem Bark in Animal Models. *Anais da Academia Brasileira de Ciências* **2011,** *83*(2), 557–566.
39. Santos, T. N.; Costa, G.; Ferreira, J. P; Liberal, J.; Francisco, V.; Paranhos, A.; Cruz, M. T.; Castel-Branco, M.; Figueiredo, I. V.; Batista, M. T. Antioxidant, Anti-Inflammatory, and Analgesic Activities of *Agrimonia eupatoria* L. Infusion. *Evid. Based Complement Alternat. Med.* **2017.** Article ID 8,309,894, 13 pages.
40. Schiltz, C.; Liote, F.; Prudhommeaux, F.; Meunier, A.; Champy, R.; Callebert, J.; et al. Monosodium Urate Monohydrate Crystal–Induced Inflammation in Vivo–Quantitative Histomorphometric Analysis of Cellular Events. *Arthritis Rheumat.* **2002,** *46*(6), 1643–1650.
41. Scholz, J.; Woolf, C. J. Can We Conquer Pain? *Neuroscience* **2002,** *5*, 1062–1067.
42. Shin, Y. W.; Bae, E. A.; Kim, S. S.; Lee, Y. C.; Kim, D. H. Effect of Ginsenoside Rb1 and Compound K in Chronic Oxazolone-Induced Mouse Dermatitis. *Int. Immunopharmacol.* **2005,** *5*, 1183–1191.

43. Srinivas, K.; Rao, M.; Rao, S. Anti-Inflammatory Activity of *Heliotropium indicum* Linn. And *Leucas aspera* spreng. in Albino Rats. *Indian J. Pharmacol.* **2000,** *32*, 37–38.
44. Suleyman, H.; Buyukokuroglu, M. E. The Effects of Newly Synthesized Pyrazole Derivatives on Formaldehyde-, Carrageenan-, and Dextran-Induced Acute Paw Edema in Rats. *Biol. Pharm. Bull.* **2001,** *24*(1)0, 1133–1136.
45. Takano, N.; Sakurai, T.; Kurachi, M. Effects of Anti-Nerve Growth Factor Antibody on Symptoms in the NC/Nga Mouse, an Atopic Dermatitis Model. *J. Pharmacol. Sci.* **2005,** *99*, 277–286.
46. Towbin, H.; Pignat, W.; Wiesenberg, I. Time-Dependent Cytokine Production in the Croton Oil-Induced Mouse Ear Oedema and Inhibition by Prednisolone. *Inflamm. Res.* **1995,** *44*, 160–161.
47. Tubaro, A.; Dri, P.; Delbello, G.; Zilli, C.; Loggia, R. D. The Croton Oil Ear Test Revisited. *Agents Actions* **1985,** *17*, 347–349.
48. Uzkeser, H.; Cadirci E.; Halici Z.; Odabasoglu F.; Polat B.; Yuksel T.; Ozaltin S.; Atalay, F. Anti-Inflammatory and Antinociceptive Effects of Salbutamol on Acute and Chronic Models of Inflammation in Rats: Involvement of an Antioxidant Mechanism. *Mediat. Inflamm.* **2012,** *2012*, 438912.
49. Victoria, S.; Das S.; Lalhlenmawia H.; Phuco L.; Shantabi L. Study of Analgesic, Antipyretic and Anti-Inflammatory Activities of the Leaves of *Thunbergia coccinea* Wall. *Int. Multidiscip. Res. J.* **2002,** *2*, 83–88.
50. Wan, H.; Coppens, J. M.; Helden-Meeuwsen, C. G.; Leenen, P. J.; Rooijen, N. V.; Khan, N. A.; et al. Chorionic Gonadotropin Alleviates Thioglycollate-Induced Peritonitis by Affecting Macrophage Function. *J. Leukocyte Biol.* **2009,** 86.
51. Webb, E. F.; N., Newsholme, S. J.; Griswold, D. E. Intralesional Cytokines in Chronic Oxazolone-Induced Contact Sensitivity Suggest Roles for Tumor Necrosis Factor Alpha and Interleukin-4. *J. Invest. Dermatol.* **1998,** *111*, 86–92.
52. Whiteley, P. E.; Dalrymple, S. A. Models of Inflammation: Carrageenan-Induced Paw Edema in the Rat. *Curr. Prot. Pharmacol.* **1998,** *0*(1), 5–6.
53. Winter, C. A.; Porter C. Effect of Alterations in Side Chain upon Anti-Inflammatory and Liver Glycogen Activities of Hydrocortisone Esters. *J. Am. Pharm. Assoc.* **1957,** *46*(9), 515–519.
54. Winter, C. A.; Risley, E. A.; Nuss G. W. Carrageenin-Induced Edema in Hind Paw of the Rat as an Assay for Anti-Inflammatory Drugs. *Proc. Soc. Exp. Biol. Med.* **1962,** *111*, 544–547.
55. Winyard, P. G.; Willoughby, D. A. *Biology Methods in Molecular Biology*; Humana Press: Totowa, NJ, 2001; vol. 225, pp 115–121.
56. Woolf, C. J.; Max, M. B. Mechanism-Based Pain Diagnosis. *Anesthesiology* **2001,** *95*, 241–249.
57. Xie, Y. C.; Dong, X. W.; Wu, X. M.; Yan, X. F.; Xie, Q. M. Inhibitory Effects of Flavonoids Extracted from Licorice on Lypopolysaccharide-Induced Acute Pulmonary Inflammation in Mice. *Int. Immunopharmacol.* **2009,** *9*, 194–200.
58. Yesilada, E.; Kupeli, E. *Berberis crataegina* DC root Exhibits Potent Anti-Inflammatory, Analgesic and Febrifuge Effects in Mice and Rats. *J. Ethnopharmacol.* **2002,** *79*, 237–248.

59. Young, L. D.; Kheifets, J.; Ballaron, S.; Young, J. Edema and Cell Infiltration in the Phorbol Ester-Treated Mouse Ear are Temporally Separate and can be Differentially Modulated by Pharmacologic Agents. *Agents Actions* **1989,** *26*(3–4), 335–341.

CHAPTER 3

REVEALING ENERGY LANDSCAPES OF ATOMIC CLUSTERS BY APPLYING ADAPTIVE BIOINSPIRED ALGORITHMS

J. M. C. MARQUES[1,*], W. S. JESUS[1,2], F. V. PRUDENTE[2], F. B. PEREIRA[3,4], and N. LOURENÇO[4,5]

[1]*CQC, Department of Chemistry, University of Coimbra, 3004-535 Coimbra, Portugal*

[2]*Instituto de Física, Universidade Federal da Bahia, 40, 170-115 Salvador, Brazil*

[3]*Instituto Superior de Engenharia de Coimbra, Quinta da Nora, 3030-199 Coimbra, Portugal*

[4]*Centro de Informática e Sistemas da Universidade de Coimbra (CISUC), 3030-290 Coimbra, Portugal*

[5]*Department of Informatics Engineering, University of Coimbra, 3030-290 Coimbra, Portugal*

[*]*Corresponding author. E-mail: qtmarque@ci.uc.pt*

CONTENTS

ABSTRACT

We review our work on the development of evolutionary algorithms (EAs) for revealing low-energy structures of atomic and molecular clusters. The application of EAs on the study of the microsolvation of alkali-metal ions with argon and assessing the chemical ordering of binary clusters of transition-metal elements is discussed. Additionally, we discuss the application of novel self-adaptive bioinspired algorithms to model cluster systems. Several adaptive strategies dealing both with control parameters and algorithmic components will be presented and some preliminary results are described and analyzed.

3.1 INTRODUCTION

The structure of a chemical compound (e.g., a drug) in biological media is intrinsically related to its function. In particular, proteins are formed by chains of amino acids that fold to acquire the specific native structure. Such an aggregate of coils (chains) may lead to several available minima, but there is just one structure that is effective to carry out the biological function (role) and, indeed, the sequence of amino acids forming a protein is able to fold into the corresponding native state.[1] Extra complexity in the folding process can be observed when competition between two native states is manifest.[2,3] This connection between protein folding and its biological function has been well-established theoretically through the analysis of the energy landscapes,[4] whose framework benefits of employing global geometry optimization methods. Indeed, several algorithms that search for low-energy configurations of atomic and molecular systems have been applied to the study of the structure of molecules with biological interest.[5–15]

Among global optimization techniques,[16–21] evolutionary algorithms (EAs)[22] are state-of-the-art methods that have shown to be very successful for discovering low-energy structures of atomic and molecular aggregates. In this work, we review the development of bioinspired algorithms carried out by our group over the past 10 years, including their application, for the first time, to discover putative global minima of some relevant aggregates (see refs[23,24] and references therein). We also propose a novel self-adaptive approach[25] that aims to enhance optimization effectiveness when exploring the energy landscape of atomic clusters with unknown properties. There are several reports in the literature that advocate the development of adaptive bioinspired approaches, spanning from the early evolutionary strategies[26] efforts to the recent area of hyper-heuristics.[27] They all share the common goal of developing an optimization framework that autonomously adapts to the problem being addressed. This adaptation simplifies the task of the practitioner that does not have to rely on expert knowledge to tailor an algorithm to a specific situation and enhances the ability of the method to effectively sample the search space of the problem under study. Self-adaptation may occur at the parameters level and/or at the algorithmic components.[27,28] Moreover, it may consider the selection of a subset of existing components or it may foster the discovery of a novel optimization strategy. Here, we focus on a framework that selects EA components (transformation operators and diversity measures) and favorable settings. Preliminary results presented in Section 3.5 suggest that the self-adaptive EA is able to autonomously identify the best components that enhance the likelihood of discovering high-quality solutions. Our expectation is that, in the near future, such novel methods can ultimately be applied to the study of biochemical systems, such as protein folding or drug design.

The application of global optimization methods may be carried out together with the calculation of the interaction energy at either the ab initio or density functional theory (DFT) levels of theory. This approach has been recently reviewed by Heiles and Johnston.[29] A less computationally demanding alternative for calculating the interaction energy consists in the use of semiempirical potential energy functions. We follow here the latter approach. Thus, some well-established potential models were employed: (i) sum of simple pair-potential functions for the clusters resulting from the microsolvation of alkali-metal ions with argon and (ii) the usual Gupta potential for the transition-metal alloy, with the parameters taken from Cleri and Rosato.[30] Specifically, we have studied $Li^+(Ar)_N$, $Na^+(Ar)_N$, and $K^+(Ar)_N$ clusters as well as bimetallic $(RhCu)_N$ aggregates. These are relevant chemical systems that can also be appropriate to show the flexibility and robustness of the EAs. Accordingly, the main features of our bioinspired algorithms are described in Section 3.2, while examples of their application to the microsolvation of alkali-metal ions and $(RhCu)_N$ clusters are reported in Sections 3.3 and 3.4, respectively. In Section 3.5, we present a self-adaptive framework that can autonomously identify the best components for an effective cluster geometry optimization EA. Finally, the main conclusions are given in Section 3.6.

3.2 EVOLUTIONARY CLUSTER OPTIMIZATION

The first work describing the application of EAs to cluster geometry optimization was published in 1993. Hartke pioneered this area by applying what can be considered a naif binary-encoded algorithm for the optimization of Si_4 atomic clusters.[31] These first attempts were quite inefficient and were unable to compare with existing state-of-the-art techniques, which were mainly based on stochastic single-point search methods (e.g., simulated annealing). In 1995, a breakthrough paper from Deaven and Ho completely changed the area of evolutionary cluster optimization and set the foundations for modern algorithms.[32] They describe the application of an EA to carbon clusters and propose several novel components that clearly enhanced the effectiveness of this class of methods: (i) the coordinates of the atoms composing the aggregate are codified as real values, thus departing from the binary encoding (this same modification was simultaneously proposed by Zeiri[33]; (ii) genetic operators are applied directly in the 3D space, thus preserving some semantic properties of the solutions. The proposed crossover is considered the first cut-and-splice (C&S)-like operator; (iii) conjugate gradient minimization is applied to all solutions generated by the EA, moving them to the nearest local optimum; (iv) a simple mechanism prevents identical individuals from belonging to the population. An energy-based similarity is considered. Over the last 20 years, the proposals of Deaven and Ho[32] have been improved with the aim of further enhancing the effectiveness of EAs for cluster geometry optimization.[19,34–36] Two key issues have been considered by researchers: the development of genetic operators that are aware of the properties of the solutions when transforming them[32,37–39] and the definition of enhanced distance measures that better estimate the existence of duplicates/identical solutions in the population.[19,36,40–42]

Our group has been developing cutting-edge EAs for cluster optimization for over a decade.[36,43–45] The most distinguished feature of our approaches is the unbiased nature of the search performed by the algorithm. Although this might compromise absolute effectiveness, it allows for an increased robustness, thus fostering its application to novel chemical systems for which specific knowledge about the general organization of the global minimum is not available. Considering existing global optimization approaches, our algorithms comprise two key enhancements: improved set of genetic operators[36,43] and a suite of distance measures that effectively maintain diversity throughout an optimization run.[36] The representation of solutions builds upon the proposals of Zeiri[33] and Deaven and Ho.[32] In accordance, to represent an atomic aggregate with N particles, the chromosome is composed of 3N real values, codifying the Cartesian coordinates of each particle. In turn, when representing molecular clusters, another set of 3 real values is added to each particle in order to encode the Euler angles that describe the orientation of the corresponding molecule in the 3D space.[23] All genes have a predetermined minimum and maximum bound: for the coordinates, they define a 3D box in which the particles must be placed, whereas one of the orientation angles varies in the interval $[0, \pi]$ and the other two range between 0 and 2π. The EA explores the search space defined by these settings, trying to identify the cluster with the global minimum energy.

To enhance search efficiency, all solutions are locally optimized before evaluation and therefore the solution pool is composed just by local optima. The limited-memory Broyden-Fletcher-Goldfarb-Shanno (L-BFGS) method is the quasi-Newton standard local search procedure adopted to hybridize evolutionary cluster geometry optimization, as it cleverly combines modest memory requirements with fast convergence.[46] Several works describe modified local search procedures to speed up convergence and to increase the likelihood of discovering promising solutions. The so-called two-stage local search considers a modified biased potential that changes the potential energy surface (PES) and enhances the likelihood of discovering nonicosahedral global minima.[18,37] This modified potential comprises a set of parameters that favor the appearance of specific shapes (e.g., ellipsoidal vs. spherical). As these settings must be specified prior to the optimization, this local search strategy cannot be used in situations where the shape of the global optimum is not known. In the algorithm applied in this work, only the standard L-BFGS is considered.

Selection of parents is performed with the tournament operator, a method that allows for an effective adjustment of the selective pressure.[22] In a previous work, we proposed a structural elite operator that was able to keep the best solutions in the population, together with the promotion of clusters with diverse structural shapes.[47] In our most recent publications, we have been proposing a more efficient steady-state approach: it relies on a fitness-based replacement strategy, in which the new solutions generated by the EA replace the worst individuals from the population, providing that they have better quality (i.e., lower potential).[36] To prevent premature convergence, a diversity mechanism forbids the coexistence of similar solutions in the pool. The similarity between individuals can be accessed either by fitness or by structural properties. Results obtained with several chemical systems demonstrate the advantage of relying on structural distance measures.[36]

C&S crossover operators are clearly the best option to combine subclusters of parents when creating novel solutions.[32,35,38,39,43] Their distinctive feature is the operation in the 3D space, aiming at preserving some semantic properties of the parents. Different flavors exist in the literature, depending on the constraints considered when applying the operator (e.g., selection based on a cutting plane vs. selection based on minimal Euclidean distance) and on the type of cluster being optimized (atomic/molecular; homogeneous/heterogeneous). As for mutation, two classes of operators are usually considered: random mutation takes a particle and chooses a new arbitrary location for it inside the 3D box defining the search space, and sigma mutation slightly perturbs the coordinates of a particle with a value sampled from a normal distribution.[36]

3.3 MICROSOLVATION OF ALKALI-METAL IONS

The theoretical investigation on the microsolvation of alkali-metal ions by water and other molecules has been recently reviewed by us.[23] The substitution of the solvent molecules by atoms reduces the complexity of the theoretical problem, both in modeling the potential energy interaction and during the search of the global minimum structure (which, now, does not depend on orientational degrees of freedom). Because of this, we have chosen the microsolvation study of alkali-metal ions with argon atoms to illustrate the application of our global optimization methodology.

The study of the microsolvation of alkali-metal ions with argon constitutes a good starting point for understanding the solvation phenomena involving these ionic species in the more complex biological environment. To establish the interaction potential model, we have employed two-body functions. While the Ar–Ar pair-potential is represented by the Rydberg–London function of Cahill and Persegian,[48] the interaction between the alkali-cation and argon was modeled as:

$$V(R) = aR^c \exp(-bR) - \chi_{pol}(d_{pol}, R)\frac{C_4}{R^4} - \chi_{disp}(d_{disp}, R)\frac{C_6}{R^6} \quad (3.1)$$

where a, b, c, C_4, C_6 are fitting parameters, while the damping functions for both polarization and dispersion components have the general form:

$$\chi(d,R)=\begin{Bmatrix} 1 & R>d \\ \exp\left[-\left(d/R-1\right)^2\right] & R\leq d \end{Bmatrix} \tag{3.2}$$

We should note that d refers to d_{pol} (d_{disp}) in the calculation of polarization (dispersion) component. The parameters have been obtained by fitting the function to ab initio data calculated at the coupled-cluster with singles, doubles and estimated-triples [CCSD(T)] framework by using GAMESS package[49] and a basis set with quadruple-zeta quality for Li^+–Ar and Na^+–Ar (correlation-consistent, polarized valence, quadruple-zeta, i.e., cc-pVQZ, for lithium[50] and augmented correlation-consistent, polarized valence, quadruple-zeta, i.e., aug-cc-pVQZ, for argon[51]) and with triple-zeta quality for K^+–Ar (augmented triple-zeta plus polarization functions, i.e., ATZP basis set[52,53]). The ab initio electronic energy was corrected for the basis set superposition error with the counterpoise method,[54] and during the calculations, 7 and 10 orbitals were frozen for Na^+–Ar and K^+–Ar systems, respectively. The values of the parameters are given in Table 3.1, according to the fit performed using 25 and 26 ab initio points distributed in the intervals of 2.3 up to 10.0 Å and 2.5 up to 10.0 Å for Na^+–Ar and K^+–Ar, respectively, while the corresponding ones for Li^+–Ar have been published in ref.[55] All ab initio points are available from the authors upon request. It is apparent from Figure 3.1 that the potential curves for Na^+–Ar and K^+–Ar represent very well the corresponding ab initio data.

TABLE 3.1 Parameters of the Na^+–Ar and K^+–Ar Potentials that were Obtained by Fitting Equations 3.1 and 3.2 to CCSD(T) Ab Initio Energies. (See the Text.)

Potential parameters	Systems	
	Na^+–Ar	**K^+–Ar**
a/eV	5040.47	4523.69
b/Å^{-1}	4.25	3.478
ϵc	1.815	0.25
C_4/eV Å^{-4}	8.34	11.078
C_6/eV Å^{-6}	181.207	79.51
d_{pol}/Å	7.95	3.415
d_{disp}/Å	2.60	3.669

This pair-potential approach appears to be reasonable, since we are mainly concerned with the structural features of the microsolvation clusters and the three-body interactions are not expected to affect too much the geometries arising from models based on two-body functions. Actually, we have recently investigated the microsolvation of Li^+ by argon[55] and we showed that the inclusion of three-body interactions to model the $Li^+(Ar)_N$ clusters is relevant for describing energetic features, but it is less important to establish the global minimum geometry. In this reference,[55] we have shown that, in general, the global minima of the potential with three-body interactions have always higher energy and, usually, lower symmetry than the corresponding ones for the PES that includes only two-body terms. Indeed, the $Li^+(Ar)_N$ clusters obtained from a potential based on pair-wise interactions failed the main energetic features up to $N=10$, and the $Li^+(Ar)_2$, and $Li^+(Ar)_3$ structures become wrong when compared with the ones from the MP2 optimization and the single-point CCSD(T) calculation for the MP2 geometries. On the other hand, the structures obtained using the PES including three-body terms had a good agreement with the corresponding ones optimized at the ab initio level up to $N=8$. For larger clusters, additionally, the comparison between potentials with and without three-body forces showed significant energetic and some structural differences for several of the cluster sizes. However, the results obtained with both surfaces were able to explain the high-stability peaks for structures with "magic numbers" $N=4$, 6, 14, 16, and 34 that arise from the experimental mass spectral intensities.[56]

In the remaining of this section, we report new results for $Na^+(Ar)_N$ and $K^+(Ar)_N$ (with $N=2$–40) and compare them with previous studies on similar systems. In Figure 3.2, we show the putative global minimum structures for $Na^+(Ar)_N$. This system has been previously studied by Rhouma et al.[57] by employing two potential models that include three-body interactions. In spite of some discrepancies for small-size clusters, where three-body effects are expected to be more apparent, the present results give similar global minimum structures up to $N=10$; exceptions arise only for $Na^+(Ar)_2$ and $Na^+(Ar)_3$. As for $Li^+(Ar)_N$, the clusters grow up by surrounding the Na^+ by an increasing number of argon atoms. A similar pattern is observed in Figure 3.3 for $K^+(Ar)_N$. Nonetheless, several structural differences arise for the solvation of Li^+, Na^+, and K^+ with argon. While the maximum coordination number is 6 for the $Li^+(Ar)_N$ clusters,[55] it increases to 8 (12) for the solvation of $Na^+(K^+)$, as one can observe from Figure 3.2 (Fig. 3.3). This may be attributed to the increasing size of the ions, which is apparent from the diatomic curves displayed in Figure 3.1: the equilibrium geometries are 2.37, 2.85, and 3.33 Å for Li^+–Ar, Na^+–Ar, and K^+–Ar, respectively. Although such a behavior with the increasing size of alkali-metal ion may be expected for other solvents, exceptions have been observed when there is a certain degree of competition between ion–solvent and solvent–solvent interactions (see, e.g., refs 58–61 for the microhydration case).

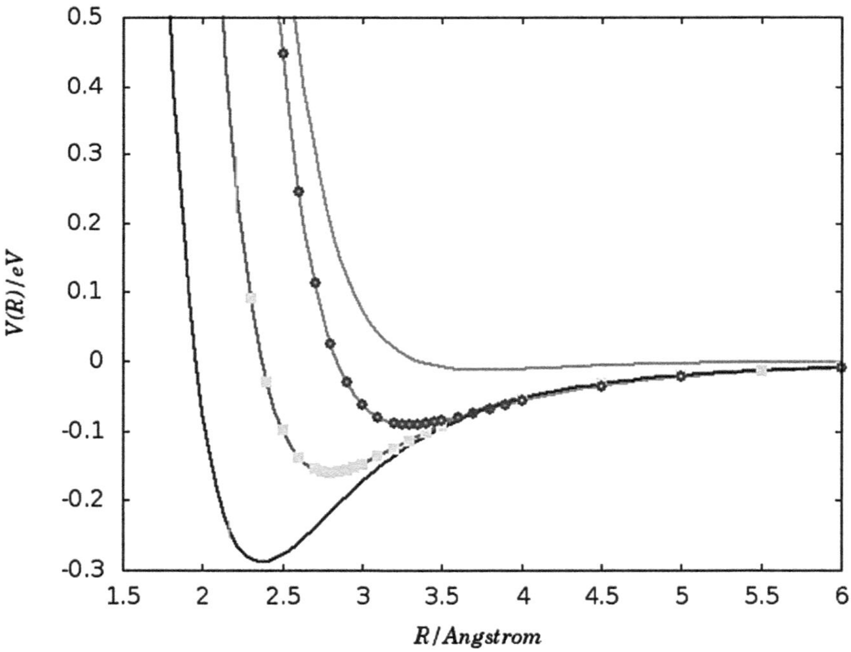

FIGURE 3.1 **(See color insert.)** Potential energy curves for Li^+–Ar (magenta), Na^+–Ar (green), K^+–Ar (blue), and Ar–Ar (orange) interactions. The Ar–Ar curve is from ref 48, while the ion Ar ones correspond to fits of Equation 3.1 to ab initio points (also represented for Na^+–Ar and K^+–Ar).

We show in Figure 3.4 that, for most cases, the ion is displaced off the center of the cluster. In fact, during the completion of the first solvation shell, the ion occupies the center of the structure only for the clusters

$Li^+(Ar)_4$, $Li^+(Ar)_6$, $Na^+(Ar)_6$, $Na^+(Ar)_8$, $K^+(Ar)_9$, and $K^+(Ar)_{12}$. It is worth noting that such clusters correspond to maxima of the second energy difference, that is,

$$\Delta_2 E = -2E_N + E_{N-1} + E_{N+1} \tag{3.3}$$

where E_N, E_{N-1}, and E_{N+1} are the energies of clusters with N, $N-1$, and $N+1$ argon atoms, respectively; the $\Delta_2 E$ curves are represented in Figure 3.5 and the maxima are usually designated as "magic numbers." For the three alkali-ions, the strongest "magic number" is associated with the closure of the corresponding first solvation shell. While the $Li^+(Ar)_6$ cluster has O_h symmetry, the $K^+(Ar)_{12}$ structure is an icosahedron. In the case of Na^+, though the first solvation shell closes at $N=8$ (in agreement with the work of Rhouma et al.,[57] which uses a potential model with explicit three-body terms), another strong peak occurs for the $Na^+(Ar)_{10}$ cluster; we note from Figure 3.2 that both $Na^+(Ar)_8$ and $Na^+(Ar)_{10}$ structures belong to the D_{4d} point group of symmetry, but two of the argon atoms of the latter are already in the second solvation.

Concerning larger clusters, the $K^+(Ar)_N$ global minimum structures tend to be more symmetric than the $Na^+(Ar)_N$ ones. As for Ar_N clusters, the $K^+(Ar)_N$ global minima are icosahedric-type structures, which is against the global optimization result achieved for a pair-potential model by Hernández-Rojas and Wales,[62] and that predicts icosahedral packing only beyond $N=49$. Indeed, the equilibrium distance and the well depth of the K^+–Ar bond are both more similar to the corresponding values in the Ar–Ar interaction than the Li^+–Ar and Na^+–Ar ones. It is also clear from Figure 3.4 that, in the case of large clusters, the Li^+ ion tends to occupy a more central position than Na^+ or K^+, which may be explained by the fact that the Li^+–Ar bond is the strongest one (see Figure 3.1). In turn, only very small peaks of the curves are observed for the larger clusters in Figure 3.5, which emphasizes the small contribution of each Ar–Ar interaction in the second solvation shell for the total energy of the aggregate. Nonetheless, it is worth noting that we can reproduce the mass-spectra "magic numbers" observed in time-of-flight experiments[63] for $K^+(Ar)_{12}$, $K^+(Ar)_{18}$, and $K^+(Ar)_{22}$, while a previous global optimization study by Hernández-Rojas and Wales[62] estimated a completely different series of clusters with enhanced stability.

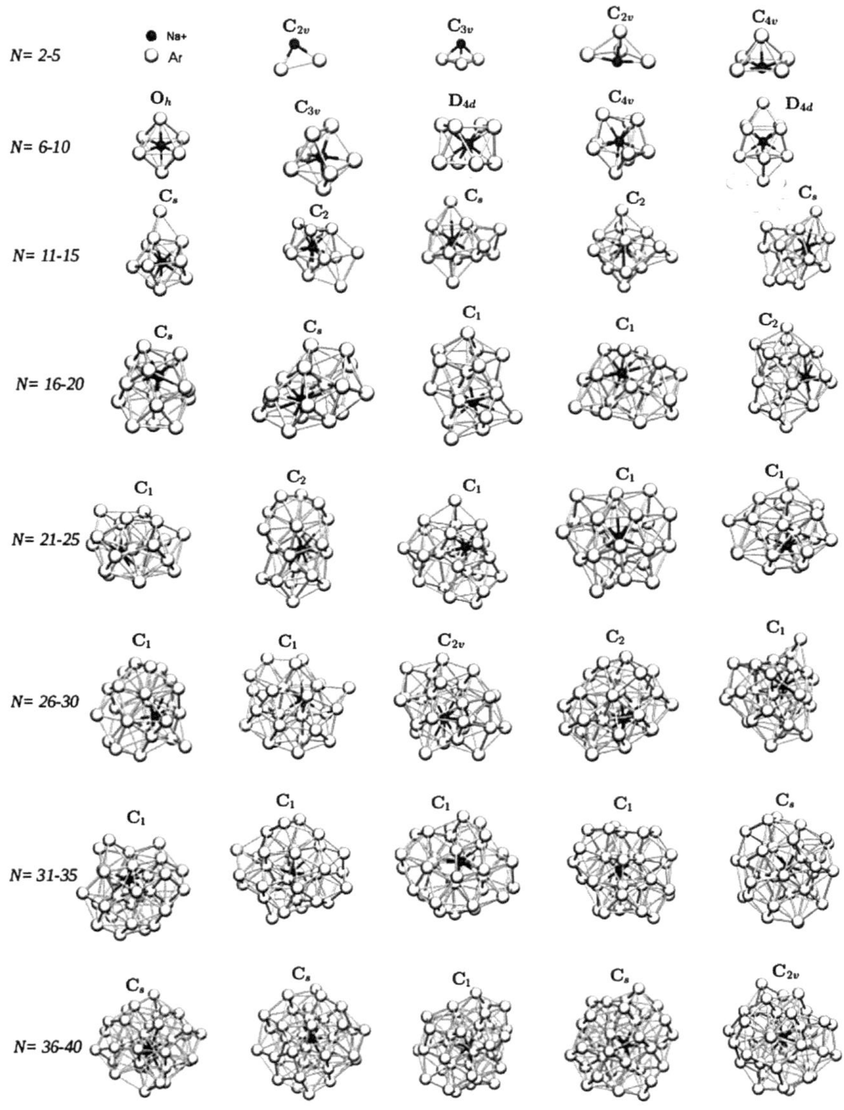

FIGURE 3.2 Putative global minimum structures for the $Na^{+}(Ar)_N$ clusters. The Cartesian coordinates are available from https://apps.uc.pt/mypage/faculty/qtmarque/en/clusters.

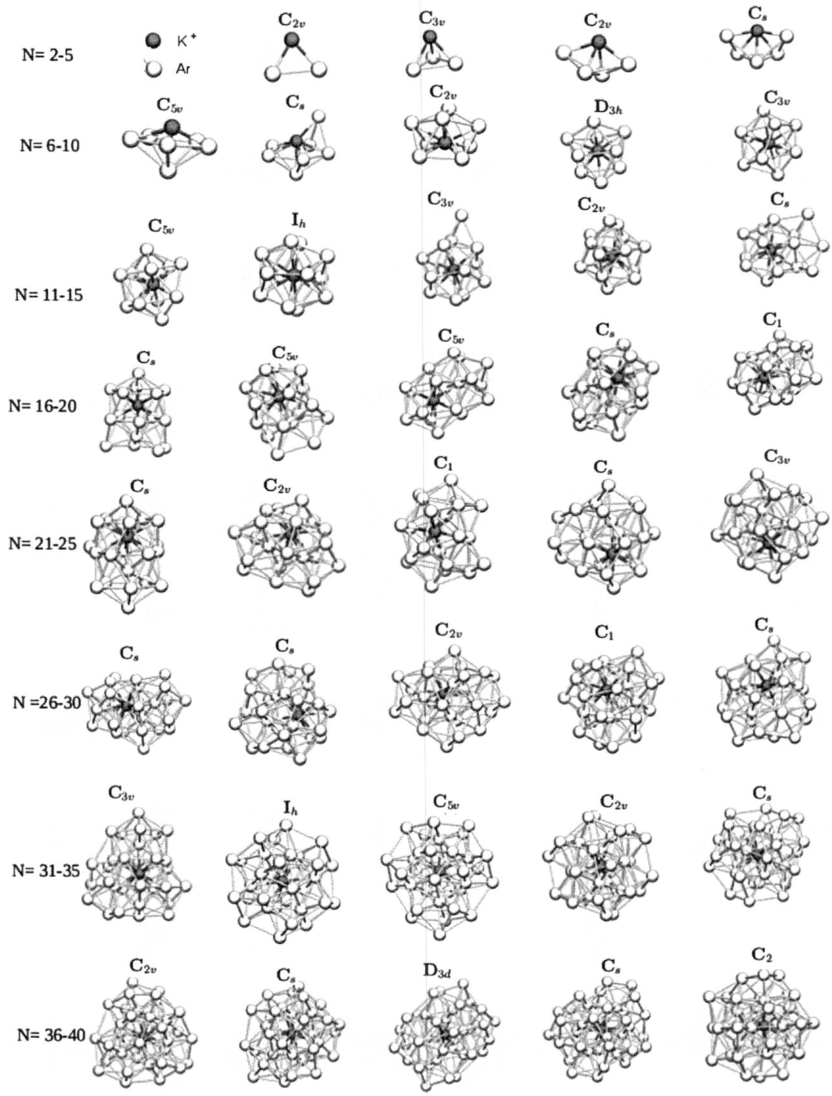

FIGURE 3.3 Putative global minimum structures for the $K^+(Ar)_N$ clusters. The Cartesian coordinates are available from https://apps.uc.pt/mypage/faculty/qtmarque/en/clusters.

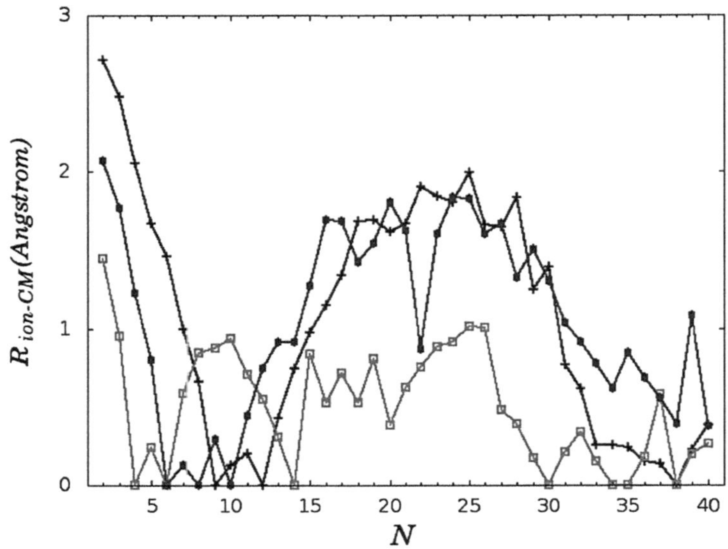

FIGURE 3.4 (See color insert.) Distance separating the ion from the center of mass of the cluster: orange line and open squares, $Li^+(Ar)_N$; red line and solid squares, $Na^+(Ar)_N$; magenta line and crosses, $K^+(Ar)_N$.

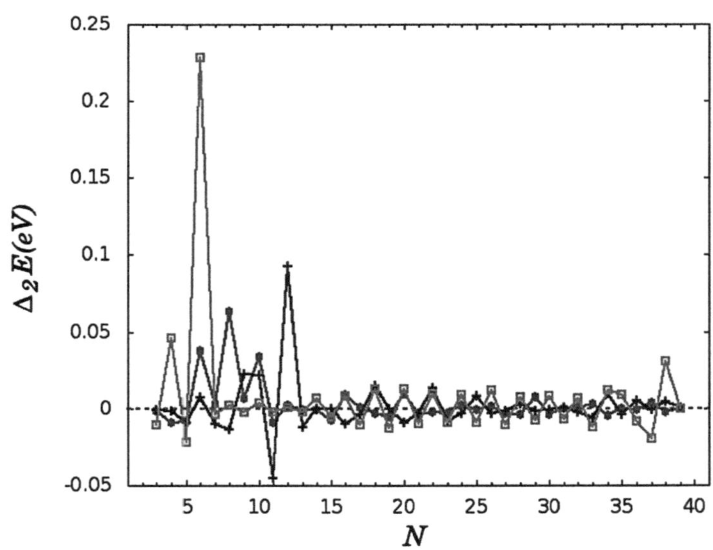

FIGURE 3.5 (See color insert.) Second energy difference of the global minimum structures: orange line and open squares, $Li^+(Ar)_N$; red line and solid squares, $Na^+(Ar)_N$; magenta line and crosses, $K^+(Ar)_N$.

3.4 BINARY TRANSITION METAL CLUSTERS

Metal alloys constitute a stringent test for global optimization algorithms since there are several minima resulting from swapping different types of atoms within the same structure (i.e., the so-called homotops[64]). Our EA has been applied to study several two-component systems, including binary Lennard-Jones clusters,[43] mixtures of rare-gas atoms,[44] Zn–Cd alloy,[65] and colloidal aggregates.[45,66] In this section, we analyze the global minimum structures of binary clusters of the Cu–Rh alloy that ~~has~~have been searched with our EA. Figure 3.6 shows the putative global minimum structures of the $(RhCu)_N$ (N = 2–21) clusters. In spite of having atoms of two different species, there are some symmetric structures, especially for small-size clusters, and even the $(RhCu)_{13}$ global minimum has C_{3v} symmetry. Owing to the difference in the cohesion energies of rhodium and copper (they are[67] 5.75 and 3.49 eV/atom, respectively), the tendency of the growing clusters is to keep the Rh atoms next to each other, while increasing the Rh–Cu nearest-neighbor interactions as much as possible. Thus, it is apparent for larger clusters that the Cu atoms segregate on the surface of the structure, whereas rhodium occupies preferentially more central positions. Since the number of atoms of each type is the same and the radius of Rh is larger than the Cu one, we cannot observe a perfect core–shell structure. Nonetheless, we may say that the global minimum structures display a clear core–shell type of segregation with the copper atoms on the surface. In contrast, we recall here our recent work on $(ZnCd)_N$ clusters,[65] where a perfect icosahedral core–shell structure is observed for most sizes.

It is interesting to observe in Figure 3.7 that most of $(RhCu)_N$ (N = 2–21) clusters are not stable in comparison with the neighbor size ones. Indeed, the strongest "magic number" arises for N= 19, though two other small peaks occur for N= 13 and N= 17. In particular, the corresponding N= 19 homogeneous cluster resembles a very organized structure that essentially has O_h symmetry, that is, it displays a truncated octahedron motif (cf. Figure 3.6).

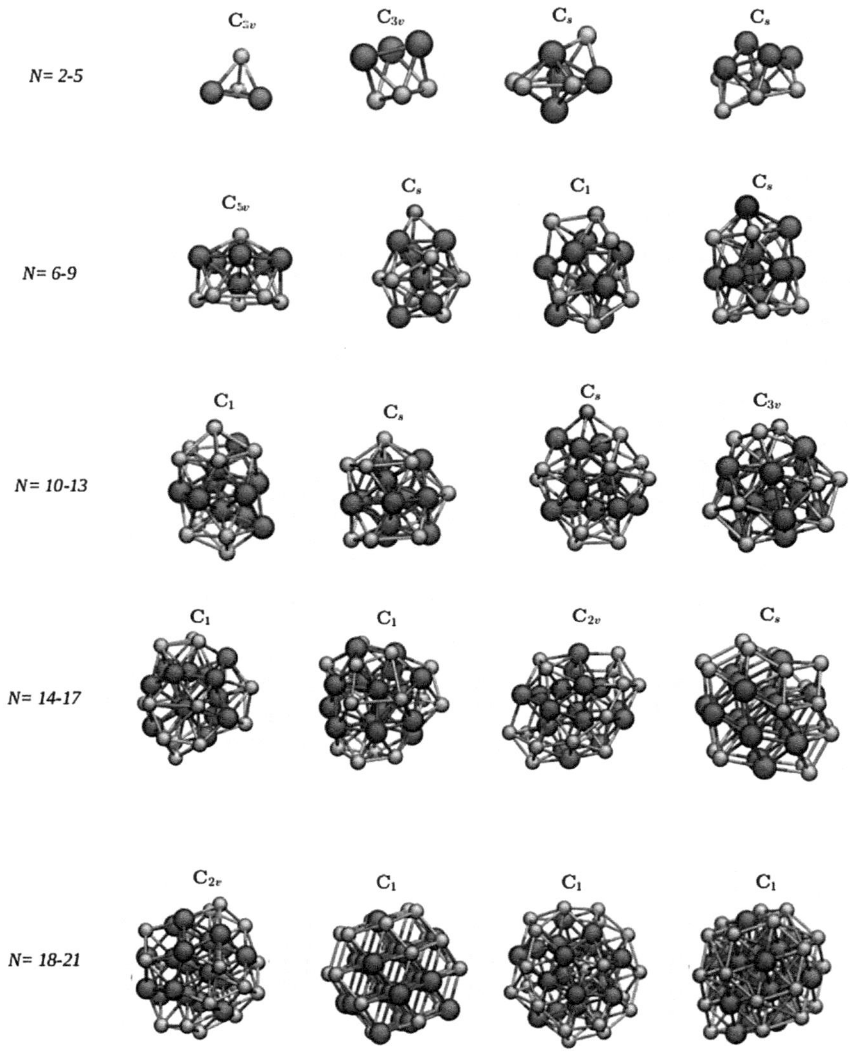

FIGURE 3.6 Putative global minimum structures for the $(RhCu)_N$ clusters as modeled by the Gupta potential (see ref 30 for the parameters of the potential): dark gray (light grey) spheres represent Rh (Cu) atoms. The corresponding point groups of symmetry are also shown. The Cartesian coordinates are available from https://apps.uc.pt/mypage/faculty/qtmarque/en/clusters.

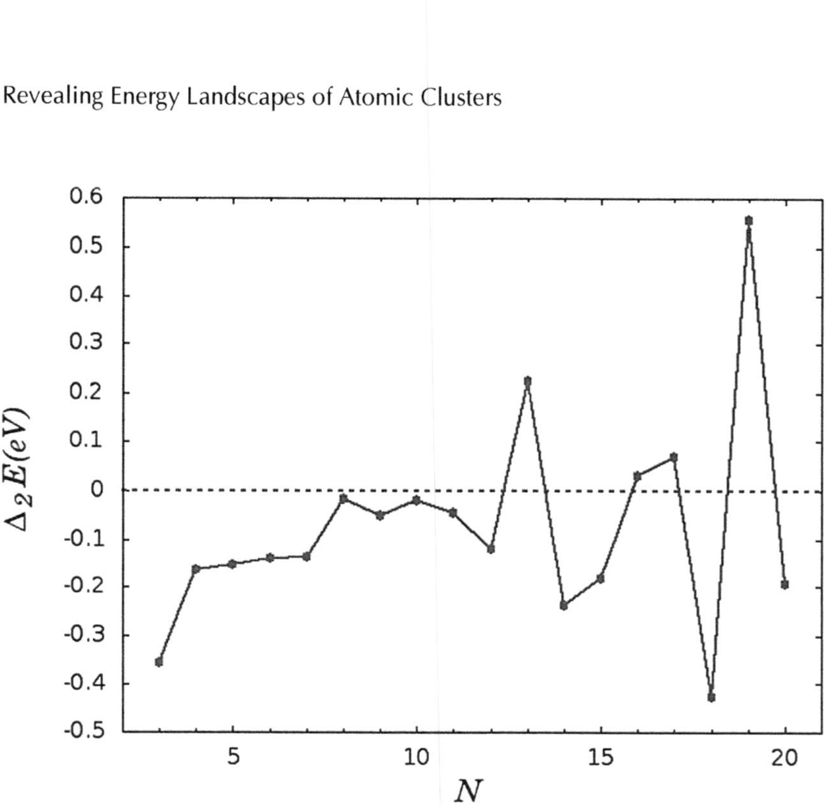

FIGURE 3.7 Second energy difference of the global minimum structures for the $(RhCu)_N$ clusters.

3.5 NEW SELF-STRATEGIES FOR GLOBAL OPTIMIZATION: APPLICATION TO MORSE CLUSTERS

Current research in the evolutionary cluster area mainly focuses on enhancing specific components aiming at a better efficiency/effectiveness tradeoff. Remarkable examples are the definition of simple recipes for maintaining diversity or the proposal of novel genetic operators.[68]

Such efforts correspond to an exploitation of existing frameworks and we believe that these proposals are not enough to foster the appearance of innovative evolutionary approaches. On the contrary, we consider that truly original ideas should be tested, in order to allow for a relevant enhancement of the behavior of EAs for cluster geometry optimization. Das and Wales recently proposed one interesting idea.[69] They presented a machine learning algorithm to estimate the potential energy level of a given configuration. Results are reported for a simple triatomic molecule and the system is able to make a reliable prediction about the final local minimum, taking a sequence of configurations as input. Despite being tested only in simple systems, the accuracy of the classification algorithm suggests that it might be used to advise a hybrid EA to only perform the costly local minimization step in promising departing configurations.

In this work, we propose a novel self-adaptive EA for cluster geometry optimization. The proposal is directly related to the area of multimeme memetic algorithms,[70] which comprises self-adaptive memetic methods that simultaneously coevolve solutions and settings encoded in each individual. It also draws inspiration from hyper-heuristics,[27,28,71] a field that deals with the development of computational frameworks whose goal is to automatize the design of methods to solve hard global optimization problems. In concrete, our approach allows individuals to autonomously choose several settings and algorithmic components, thus defining their own search strategy.

A brief perusal of the cluster evolutionary optimization literature confirms that many different components can be selected to create a hybrid EA for this task, for example, in what concerns the distance measure used to estimate similarity or the transformation operators that are applied to selected parents. Moreover, a considerable number of settings need to be defined. These are important decisions, as they will clearly impact the behavior of the algorithm. In addition, particularly when dealing with chemical systems with unknown properties, it might be difficult to correctly identify which components are better for that specific task. Finally, an optimization run comprises several stages and different components/settings might be better adapted to specific periods.

The EA proposed in this section has a general behavior similar to the one presented in Section 3.2. However, each solution from the population pool will comprise two components: a regular representation of the coordinates of the particles that compose the aggregate and individual settings that specify how this solution will be processed by the algorithm. We consider the following individual information pertaining to the definition of the self-optimization strategy:

- **Crossover type:** Individuals can adopt either generalized C&S or standard one-point crossover.
- **Mutation type:** Individuals can adopt either sigma mutation or random mutation. If sigma mutation is active, then the individual also encodes its own standard deviation, which can take one of the following values: 0.01, 0.05, 0.1, or 0.2.

- **Distance measure:** Individuals can estimate similarity to other solutions using fitness distance, structural distance, or center of mass distance.[36]

Figure 3.8 highlights an example of an individual chromosome. The first part of the solution encodes the Cartesian coordinates of the atoms that belong to the cluster being optimized. The second part contains the selection of components and settings that help define its own search strategy.

Cluster coordinates	Self-strategy
$atom_1$ *(x,y,z)*, $atom_2$ *(x,y,z)*, ..., $atom_N$ *(x,y,z)*	Cx_type, Mut_type, $Distance_M$

FIGURE 3.8 General organization of an individual chromosome: each individual comprises the coordinates of the atoms that belong to the cluster and its self-strategy.

The individual settings encoded in a solution are inherited by its descendants. When generating a descendant, encoded self-strategies are mutated at a predetermined rate (0.05 in this study) to prevent stagnation of the optimization behavior. All encoded settings are specific to a single solution, with the exception of the crossover type. When two parents engaged in a crossover operation have different crossover types, one of them is randomly selected.

The self-strategy encoded in solutions is straightforward and several other components could be considered. Moreover, evolution and cooperation/competition between existing strategies might be performed using other enhanced rules. Here, our goal is to propose a simple approach that can act as a proof of concept for the advantage of relying on self-adapted EAs for cluster geometry optimization. In accordance, we applied the self-adaptive EA to 14 hard Morse instances identified in ref,[36] corresponding to the following cluster sizes: {41, 43, 46, 47, 53, 59, 60, 61, 62, 68, 70, 73, 74, and 75}. We do not seek for an enhanced effectiveness in this specific situation, as the EA success rate reported in the abovementioned reference results from a carefully tuned algorithm to the potential function being addressed. Following our research hypothesis, our main goal is to verify whether the self-adaptive framework is able to successfully identify the components/settings that foster the discovery of the global minima for the instances considered in this study. The EA global settings are the same as reported in the abovementioned study: number of runs: 30; population size: 100; tournament size: 5; crossover rate: 0.7; mutation rate: 0.05; evaluations: 1×10^7; maximum local search length: 1000.

As a rule, we verified that the self-adaptive EA was able to find the global minimum in several runs, for all the instances considered in the study. The success rate ranged between 3 and 33% and, for each instance, tended to be slightly below that reported for the fine-tuned EA.[36] As mentioned before, this was an expected result, as the optimization algorithm is solving two problems simultaneously: it seeks for the best optimization strategy, at the same time it is trying to find optimal configurations for the clusters being optimized. Given the limited computational budget, we could anticipate that the final absolute optimization performance would be inferior when compared to carefully tuned algorithms that only need to seek for a cluster with minimum potential. The most important outcome here is to verify if self-adaptation is identifying the most promising components and results confirm that this is indeed happening. The three panels from Figure 3.9 show the prevalence of C&S crossover and sigma mutation throughout the optimization run: panel (a) refers to the Morse cluster with 43 atoms, panel (b) for the instance with 68 atoms, and panel (c) for the instance with 74 atoms. The same trend is visible in other instances. Since the beginning of the run, the percentage of individuals in the solution pool encoding C&S crossover and sigma mutation steadily increases and reaches over 90% for this mutation operator and around 80% for the crossover. The self-adaptive EA is thus able to identify and promote the application of the suite of operators that the literature in this area recognizes as the most effective ones. A detailed inspection of the strategy configurations encoded in global optimal solutions discovered by the hybrid EA reveals that not a single one was obtained with random mutation. This outcome confirms that sigma mutation is mandatory to perform the final adjustments that lead to the discovery of clusters with minimum potential energy. As for the crossover, around 75% of global optimal solutions encode the application of C&S crossover, corroborating that semantic-aware operators are better equipped to explore the search space. It is interesting to notice that when taking the largest instances considered in this study ($N \geq 70$), the advantage of C&S vanishes. For these instances, the percentage of global minima encoding this operator drops to about 50%, the same percentage of standard one-point crossover. This suggests that C&S-like crossovers might have some difficulties when dealing with clusters above a given number of particles. This is an effect that was not previously reported in the literature and that might be another factor that helps justify the decrease in the EA success rate when clusters grow in size. We will return to this topic in a future work.

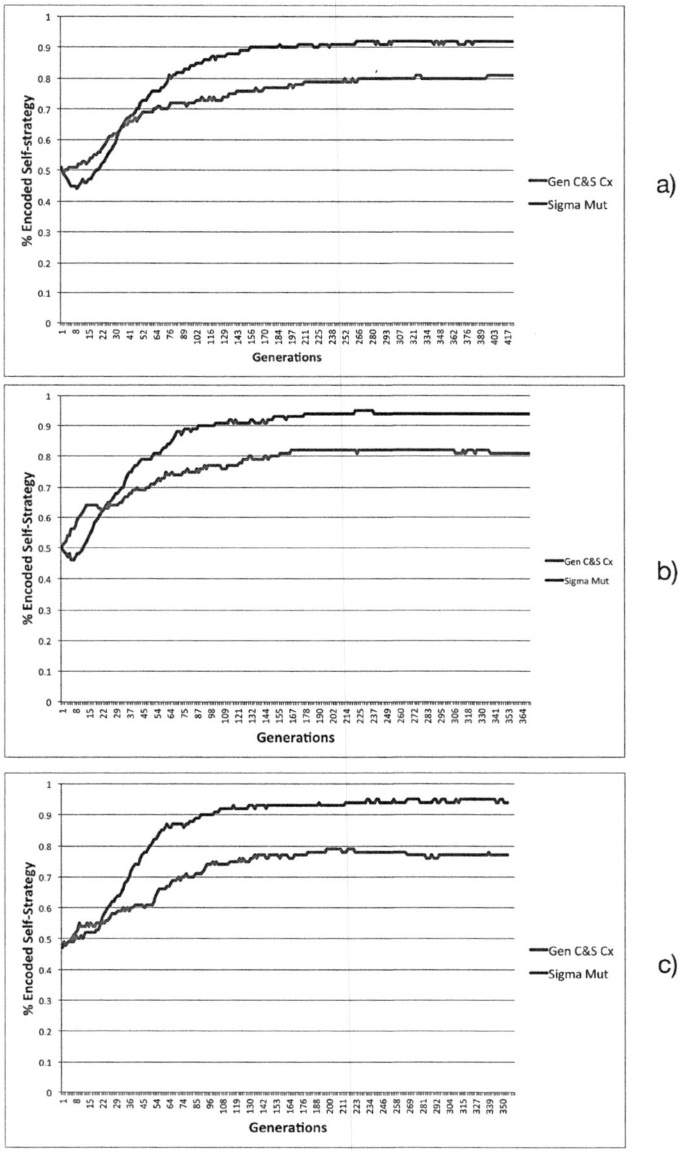

FIGURE 3.9 (See color insert.) Evolution of the percentage of individuals encoding generalized C&S crossover and sigma mutation in the optimization of three Morse instances: (a) 43 atoms; (b) 68 atoms; and (c) 74 atoms.

To complete the analysis of the self-adaptive EA, in Figure 3.10, we present the evolution of the three distance measures encoded in the solutions belonging to the solution pool. The three-panel report results obtained with the three instances from the previous figure. The outcomes

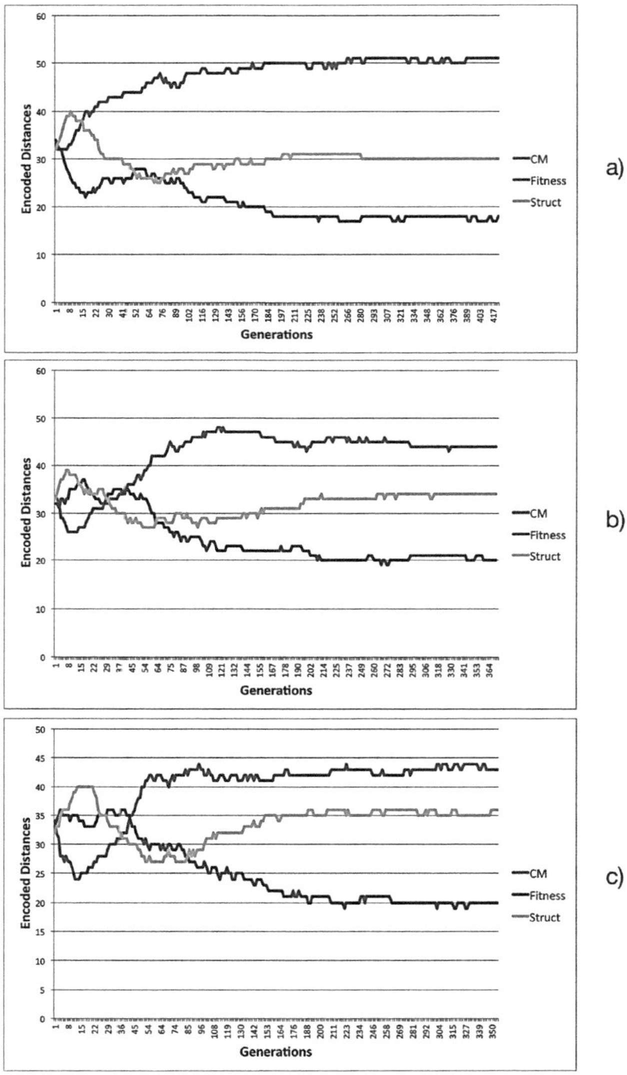

FIGURE 3.10 (See color insert.) Evolution of the percentage of individuals encoding each of the distance measures in the optimization of three Morse instances: (a) 43 atoms; (b) 68 atoms; and (c) 74 atoms.

are similar for instances of different size and confirm that fitness-based distances are not suitable for accessing the similarity between atomic clusters. The two structural-based distances are predominant in the solution pool and among the two, the center of mass distance obtains the higher percentage. This is in line with previous studies that suggest this distance measure as the best option to estimate cluster similarity[19,36] and support the hypothesis that a straightforward self-adaptive strategy is able to correctly identify components that enhance the optimization ability of an EA for cluster optimization.

3.6 CONCLUSIONS

We reviewed the work of our group on the development of EAs for searching low-energy structures of atomic clusters. To illustrate the potentialities of this methodology, we have applied our EAs to the study of alkali-ion microsolvation with argon and for discovering global minimum structures of Rh–Cu binary clusters. The results for $Li^+(Ar)_N$, $Na^+(Ar)_N$, and $K^+(Ar)_N$ show that strong "magic numbers" are associated with the closure of the first solvation shell. Among these three systems, the global minimum structures for $K^+(Ar)_N$ clusters are, in general, the most symmetric ones. It is also worth noting that simple two-body potential employed for $K^+(Ar)_N$ clusters allowed to reproduce the experimental "magic numbers" observed[63] for $K^+(Ar)_{12}$, $K^+(Ar)_{18}$, and $K^+(Ar)_{22}$. Concerning the bimetallic clusters, we have shown that the global minimum structures present segregation of Cu atoms on the surface of the aggregates, which is compatible with the experimental cohesion energies of rhodium and copper.

Finally, we should emphasize that EAs are state-of-the-art methods for geometry optimization and, in the past 20 years, they have been successfully applied to chemical systems with distinct properties. In addition, the preliminary results presented in this work suggest that self-adaptation is a promising avenue for future research. We proposed a novel self-adaptive framework that is able to simultaneously discover good quality solutions for a given cluster optimization problem, while it crafts a method specifically adapted to the search space being explored. We hypothesize that this framework is particularly advantageous in situations where knowledge about the structural properties of the system being optimized is not available and our future research will address this important endeavor.

ACKNOWLEDGMENTS

We acknowledge the Fundação para a Ciência e a Tecnologia/Coordenação de Aperfeiçoamento de Pessoal de Nível Superior bilateral project (Ref: 2984/DRI and 88887.125439/2016–00/CAPES). J. M. C. M. acknowledges the support from the Coimbra Chemistry Centre, which is financed by the Portuguese "FCT" through the Project No 7630 UID/QUI/00313/2013, cofunded by COMPETE2020-UE. J. M. C. M. is grateful to the COST Action CM1405 "Molecules in Motion (MOLIM)" for support. F. V. P. is grateful to "Conselho Nacional de Desenvolvimento Científico e Tecnológico" for grants (311093/2016–7). We are also grateful for the provision of computational time in the supercomputer resources hosted at Laboratório de Computação Avançada, Universidade de Coimbra.

KEYWORDS

- **microsolvation**
- **Rh-Cu alloys**
- **global optimization**
- **self-adaptation**
- **Morse clusters**

REFERENCES

1. Anfinsen, C. B. Principles that Govern the Folding of Protein Chains. *Science* **1973,** *181*, 223–230.
2. Schug, A.; Whitford, P. C.; Levy, Y.; Onuchic, J. N. Mutations as Trapdoors to Two Competing Native Conformations of the Rop-Dimer. *Proc. Natl. Acad. Sci. U. S. A.* **2007,** *104*, 17674–17679.
3. Levy, Y.; Cho, S. S.; Shen, T.; Onuchic, J. N.; Wolynes, P. G. Symmetry and Frustration in Protein Energy Landscapes: a Near Degeneracy Resolves the Rop Dimer-Folding Mystery. *Proc. Natl. Acad. Sci. U. S. A.* **2005,** *102*, 2373–2378.
4. Schug, A.; Onuchic, J. N. From Protein Folding to Protein Function and Biomolecular Binding by Energy Landscape Theory. *Curr. Opin. Pharmacol.* **2010,** *10*, 709–714.
5. Miller, M. A.; Wales, D. J. Energy Landscape of a Model Protein. *J. Chem. Phys.* **1999,** *111*, 6610–6616.

6. Cox, G. A.; Mortimer-Jones, T. V. R.; Taylor, P.; Johnston, R. L. Development and Optimisation of a Novel Genetic Algorithm for Studying Model Protein Folding. *Theor. Chem. Acc.* **2004,** *112*, 163–178.
7. Cox, G. A.; Johnston, R. L. Analyzing Energy Landscapes for Folding Model Proteins. *J. Chem. Phys.* **2006,** *124*, 204714.
8. Bennett, A. J.; Johnston, R. L.; Turpin, E.; He, J. Q. Analysis of an Immune Algorithm for Protein Structure Prediction. *Inform* **2008,** *32*, 245–251.
9. Prentiss, M. C.; Wales, D. J.; Wolynes, P. G. Protein Structure Prediction Using Basin-Hopping. *J. Chem. Phys.* **2008,** *128*, 225106.
10. Prentiss, M. C.; Wales, D. J.; Wolynes, P. G. The Energy Landscape, Folding Pathways and the Kinetics of a Knotted Protein. *PLoS Comput. Biol.* **2010,** *6*, e1000835.
11. Klenin, K.; Strodel, B.; Wales, D. J.; Wenzel, W. Modelling Proteins: Conformational Sampling and Reconstruction of Folding Kinetics. *Biochim. Biophys. Acta.* **2011,** *1814*, 977–1000.
12. Oakley, M. T.; Wales, D. J.; Johnston, R. L. Energy Landscape and Global Optimisation for a Frustrated Model Protein. *J. Phys. Chem. B* **2011,** *115*, 11525–11529.
13. Oakley, M. T.; Wales, D. J.; Johnston, R. L. The Effect of Non-Native Interactions on the Energy Landscapes of Frustrated Model Proteins. *J. Atom. Molec. Opt. Phys.* **2012,** *2012*, 192613.
14. Wales, D. J.; Head-Gordon, T. Evolution of the Potential Energy Landscape with Static Pulling Force for Two Model Proteins. *J. Phys. Chem. B* **2012,** *136*, 8394–8411.
15. Oakley, M. T.; Richardson, E. G.; Carr, H.; Johnston, R. L. Protein Structure Optimisation with a "Lamarckian" Ant Colony Algorithm. *IEEE/ACM Trans. Comput. Biol. Bioinf.* **2013,** *10*, 1548–1552.
16. Wales, D. J.; Doye, J. P. K. Global Optimization by Basin-Hopping and the Lowest Energy Structures of Lennard-Jones Clusters Containing Up to 110 Atoms. *J. Phys. Chem. A* **1997,** *101*, 5111–5116.
17. Wales, D. J.; Scheraga, H. A. Global Optimization of Clusters, Crystals, and Biomolecules. *Science* **1999,** *285*, 1368–1372.
18. Locatelli, M.; Schoen, F. Efficient Algorithms for Large Scale Global Optimization: Lennard-Jones Clusters. *Comput. Optim. Appl.* **2003,** *26*, 173–190.
19. Grosso, A.; Schoen, M. F. A Population-Based Approach for Hard Global Optimization Problems Based on Dissimilarity Measures. *Math. Program. Ser. A* **2007,** *110*, 373–404.
20. Shao, X.; Cheng, L.; Cai, W. A Dynamic Lattice Searching Method for Fast Optimization of Lennard-Jones Clusters. *J. Comput. Chem.* **2004,** *25*, 1693–1698.
21. Takeuchi, H. Clever and Efficient Method for Searching Optimal Geometries of Lennard-Jones Clusters. *J. Chem. Inf. Model.* **2006,** *46*, 2066–2070.
22. Eiben, A. E.; Smith, J. E. *Introduction to Evolutionary Computing*, 2nd ed.; Springer: Berlin Heidelberg, 2015.
23. Marques, J. M. C.; Pereira, F. B.; Llanio-Trujillo, J. L.; Abreu, P. E.; Albertí, M.; Aguilar, A.; Pirani, F.; Bartolomei, M. A Global Optimization Perspective on Molecular Clusters. *Philos. Trans. R Soc. A* **2017,** *375*, 20160198.
24. Marques, J. M. C.; Pereira, F. B. Colloidal Clusters from a Global Optimization Perspective. *J. Mol. Liq.* **2015,** *210*:51–63.

25. Meyer-Nieberg, S.; Beyer, H.G. Self-adaptation in Evolutionary Algorithms. In *Parameter Setting in Evolutionary Algorithms*; Lobo, F. G., Lima, C. F., Michalewicz, Z., Eds.; Springer: Berlin, Heidelberg, 2007, pp 47–75.
26. Beyer, H. G.; Schwefel, H. P. Evolution Strategies—A Comprehensive Introduction. *Nat. Comput.* **2002,** *1*(1), 3–52.
27. Burke, E. K.; Hyde, M.; Kendall, G.; Ochoa, G., Özcan, E.; Woodward, J. R. A Classification of Hyper-heuristic Approaches. In *Handbook of Metaheuristics*; Gendreau, M.; Potvin, J. Y., Eds; Springer: USA; 2010, pp 449–468.
28. Pappa, G.; Ochoa, G.; Hyde, M.; Freitas, A.; Woodward, J.; Swan, J. Contrasting Meta-Learning and Hyper-Heuristic Research: the Role of Evolutionary Algorithms. *Genet. Program. Evol. M.* **2014,** *15*(1), 3–35.
29. Heiles, S.; Johnston, R. L. Global Optimization of Clusters Using Electronic Structure Methods. *Int. J. Quantum. Chem.* **2013,** *113*, 2091–2109.
30. Cleri, F.; Rosato, V. Tight-Binding Potentials for Transition Metals and Alloys. *Phys. Rev. B: Condens. Matter* **1993,** *48*, 22–33.
31. Hartke, B. Global Geometry Optimization of Clusters Using Genetic Algorithms. *J. Phys. Chem.* **1993,** *97*, 9973–9976.
32. Deaven, D. M.; Ho, K. Molecular Geometry Optimization with a Genetic Algorithm. *Phys. Rev. Lett.* **1995,** *75*(2), 288–291.
33. Zeiri, Y. Prediction of the Lowest Energy Structure of Clusters Using a Genetic Algorithm. *Phys. Rev.* **1995,** *51*, 2769–2772.
34. Hartke, B. Application of Evolutionary Algorithms to Global Cluster Geometry Optimization. In *Applications of Evolutionary Computation in Chemistry*; Johnston, R. L., Ed.; Springer Verlag: Berlin Heidelberg, 2004; vol. 110, pp 33–53.
35. Roberts, C.; Johnston, R. L.; Wilson, N. T. A Genetic Algorithm for the Structural Optimization of Morse Clusters. *Theor. Chem. Acc.* **2000,** *104*, 123–130.
36. Pereira, F. B.; Marques, J. M. C. A Study on Diversity for Cluster Geometry Optimization. *Evol. Intel.* **2009,** *2*(3), 121–140.
37. Pereira, F. B.; Marques, J. M. C. A Self-Adaptive Evolutionary Algorithm for Cluster Geometry Optimization, In *Proceedings of the Eight International Conference on Hybrid Intelligent Systems (HIS 2008)*; IEEE Press: Washington, DC, 2008; pp 678–683.
38. Pereira, F. B.; Marques, J. M. C. Analysis of Crossover Operators for Cluster Geometry Optimization. In *Computational Intelligence for Engineering Systems: Emergent Applications, Intelligent Systems, Control and Automation, Science and Engineering Series*; Madureira, A. et al., Eds.; Springer-Verlag: Netherlands, 2010; pp 77–89.
39. Chen, Z.; Xiangwei, J.; Li, J.; Shushen, L. A Sphere-Cut-Splice Crossover for the Evolution of Cluster Structures. *J. Chem. Phys.* **2013,** *138*, 214303.
40. Cheng, L.; Cai, W.; Shao, X. A Connectivity Table for Cluster Similarity Checking in the Evolutionary Optimization Method. *Chem. Phys. Lett.* **2004,** *389*, 309–314.
41. Cassioli, A.; Locatelli, M.; Schoen, F. Dissimilarity Measures for Population-Based Global Optimization Algorithms. *Comput. Optim. Appl.* **2010,** *45*(2), 257–281.
42. Marques, J. M. C.; Llanio-Trujillo, J. L.; Abreu, P. E.; Pereira, F. B. How Different are Two Chemical Structures? *J. Chem. Inf. Model.* **2010,** *50*, 2129–2140.

43. Marques, J. M. C.; Pereira, F. B. An Evolutionary Algorithm for Global Minimum Search of Binary Atomic Clusters. *Chem. Phys. Lett.* **2010,** *4485*, 211–216.
44. Marques, J. M. C.; Pereira, F. B. A Detailed Investigation on the Global Minimum Structures of Mixed Rare-Gas Clusters: Geometry, Energetics and Site Occupancy. *J. Comput. Chem.* **2013,** *34*(6), 505–522.
45. Cruz, S. M. A.; Marques, J. M. C.; Pereira, F. B. Improved Evolutionary Algorithm for the Global Optimization of Clusters with Competing Attractive and Repulsive Interactions. *J. Chem. Phys.* **2016,** *145*, 154109.
46. Liu, D. C.; Nocedal, J. On the Limited Memory Method for Large Scale Optimization. *Math. Program. B* **1989,** *45*, 503–528.
47. Pereira, F. B.; Marques, J. M. C. A Hybrid Evolutionary Algorithm for Cluster Geometry Optimization: the Importance of Structural Elitism, In *Proceedings of the Eight International Conference on Hybrid Intelligent Systems (HIS-2008)*, IEEE Press; 2008, pp 911–914.
48. Cahill, K.; Persegian, V. A. Rydberg-London Potential for Diatomic Molecules and Unbonded Atom Pairs. *J. Chem. Phys.* **2004,** *121*, 10839–10842.
49. Gordon, M. S.; Schmidt, M. W. Advances in Electronic Structure Theory: GAMESS a Decade Later. In *Theory and Applications of Computational Chemistry: the First Forty Years*; Dykstra, C. E., Frenking, G., Kim, K. S, Scuseria, G. E., Eds; Elsevier: Amsterdam, 2005; pp 1167–1189.
50. Dunning, T. H Jr. Gaussian Basis Sets for Use in Correlated Molecular Calculations. I. the Atoms Boron Through Neon and Hydrogen. *J. Chem. Phys.* **1989,** *90*. 1007–1023.
51. Woon, D. E.; Dunning, T. H J. Gaussian Basis Sets for Use in Correlated Molecular Calculations. Iii. the Atoms Aluminum Through Argon. *J. Chem. Phys.* **1993,** *98*, 1358–1371.
52. Fantin, P. A.; Barbieri, P. L.; Canal Neto, A.; Jorge, F. E. Augmented Gaussian Basis Sets of Triple and Quadruple Zeta Valence Quality for the Atoms H and from Li to Ar: Applications in HF, MP2, and DFT Calculations of Molecular Dipole Moment and Dipole (Hyper)Polarizability. *J. Mol. Struct. (THEOCHEM)* **2007,** *810*, 103–111.
53. Camiletti, G. G.; Canal Neto, A.; Jorge, F. E.; Machado, S. F. Augmented Gaussian Basis Sets of Double and Triple Zeta Valence Qualities for the Atoms K and Sc–Kr: Applications in HF, MP2, and DFT Calculations of Molecular Electric Properties. *J. Mol. Struct. (THEOCHEM)* **2009,** *910*, 122–125.
54. Boys, S. F.; Bernardi, F. The Calculation of Small Molecular Interactions by the Differences of Separate Total Energies. Some Procedures with Reduced Errors. *Mol. Phys.* **1970,** *19*, 553–566.
55. Prudente, F. V.; Marques, J. M. C.; Pereira, F. B. Solvation of Li^+ by Argon: How Important are Three-Body Forces? *Phys. Chem. Chem. Phys.* **2017**, *19*, 25707–25716.
56. Froudakis, G. E.; Farantos, S. C.; Velegrakis, M. Mass Spectra and Theoretical Modeling of Li^+Ne_n, Li^+Ar_n and Li^+Kr_n Clusters. *Chem. Phys.* **2000,** *258*, 13–20.
57. Rhouma, M. B. E. H.; Calvo, F.; Spiegelman, F. Solvation of Na^+ in Argon Clusters. *J. Phys. Chem. A* **2006,** *110*, 5010–5016.
58. Schulz, F.; Hartke, B. Dodecahedral Clathrate Structures and Magic Numbers in Alkali Cation Microhydration Clusters. *Chem. Phys. Chem.* **2002,** *3*, 91–106.
59. Schulz, F.; Hartke, B. A New Proposal for the Reason of Magic Numbers in Alkali Cation Microhydration Clusters. *Theor. Chem. Acc.* **2005,** *114*, 357–379.

60. González, B. S.; Hernández-Rojas, J.; Wales, D. J. Global Minima and Energetics of $Li^+(H_2O)_n$ and $Ca^{2+}(H_2O)_n$ Clusters for $n \leq 20$. *Chem. Phys. Lett.* **2005,** *412*, 23–28.
61. Llanio-Trujillo, J. L.; Marques, J. M. C.; Pereira, F. B. New Insights on Lithium-Cation Microsolvation by Solvents Forming Hydrogen-Bonds: Water Versus Methanol. *Comput. Theor. Chem.* **2013,** *1021*, 124–134.
62. Hernández-Rojas, J.; Wales, D. J. Global Minima for Rare Gas Clusters Containing One Alkali Metal Ion. *J. Chem. Phys.* **2003,** *119*, 7800–7804.
63. Lüder, C.; Prekas, D.; Velagrakis, M. Ion-Size Effects in the Growth Sequences of Metal-Ion-Doped Noble Gas Clusters. *Laser Chem.* **1997,** *17*, 109–122.
64. Jellinek, J.; Krissinel, E. B. Ninalm Alloy Clusters: Analysis of Structural Forms and Their Energy Ordering. *Chem. Phys. Lett.* **1996,** *258*, 283–292.
65. Zanvettor, C. M. A.; Marques, J. M. C. On the Lowest-Energy Structure of Binary Zn-Cd Nanoparticles: Size and Composition. *Chem. Phys. Lett.* **2014,** *608*, 373–379.
66. Cruz, S. M. A.; Marques, J. M. C. Low-Energy Structures of Clusters Modeled with Competing Repulsive and Either Long- or Moderate Short-Range Attractive Interactions. *Comput. Theor. Chem.* **2017,** *1107*, 82–93.
67. Kittel, C. *Introduction to Solid State Physics,* 8th ed.; John Wiley & Sons, Inc.: Hoboken-NJ, 2005.
68. Dittner, M.; Hartke, B. Conquering the Hard Cases of Lennard-Jones Clusters with Simple Recipes. *Comput. Theor. Chem.* **2017,** *1107*, 7–13.
69. Das, R.; Wales, D. Machine Learning Prediction for Classification of Outcomes in Local Minimisation. *Chem. Phys. Lett.* **2017,** *667*, 158–164.
70. Krasnogor, N.; Smith, J. Emergence of Profitable Search Strategies Based on a Simple Inheritance Mechanism. In *Proceedings of the 3rd Annual Conference on Genetic and Evolutionary Computation*; Morgan Kaufmann Publishers Inc: San Francisco-CA, 2001, pp 432–439.
71. Burke, E. K.; Hyde, M.; Kendall, G.; Woodward, J. A Genetic Programming Hyper-Heuristic Approach for Evolving 2-D Strip Packing Heuristics. *IEEE Trans. Evolut. Comput.* **2010,** *14*(6), 942–958.

CHAPTER 4

ANTIOXIDATIVE ACTIVITY OF *AZADIRACHTA INDICA, OCIMUM TENUIFLORUM*, AND *WITHANIA SOMNIFERA*

KATARÍNA VALACHOVÁ[1,*], DOMINIKA TOPOL'SKÁ[1], MILAN NAGY[2], IVO JURÁNEK[1], and LADISLAV ŠOLTÉS[1]

[1]*Institute of Experimental Pharmacology and Toxicology, SK 84104 Bratislava, Slovakia*

[2]*Department of Pharmacognosy and Botany, Faculty of Pharmacy, Comenius University in Bratislava, SK 83232 Bratislava, Slovakia*

[*]*Corresponding author. E-mail: Katarina.Valachova@savba.sk*

CONTENTS

ABSTRACT

The antioxidative activity of aqueous and methanolic extracts of Ayurvedic plants, *Azadirachta indica* A. Juss, *Ocimum tenuiflorum* L., and *Withania somnifera* (L.) Dunal were assessed by four in vitro methods. The hydrogen atom-donating property of the aqueous extracts of plants was investigated in accordance with their influence on the free radical-mediated degradation of high-molar-mass hyaluronan. Kinetics measurements to assess the electron donor capacity of the aqueous extracts were performed by the ABTS assay. Inhibitory activity of the aqueous extracts was assessed by means of site-specific hydroxyl radical-mediated 2-deoxy-D-ribose degradation (SRD) and standard spectrocolorimetric ABTS assay. The spectrocolorimetric DPPH assay was exploited to accomplish the determination of the IC_{50} values of methanol extracts of the three Ayurvedic plants.

By comparing the hydrogen atom-donating efficacy of all extracts at their lowest concentration, it can be stated that the aqueous extract of *A. indica* inhibited hyaluronan degradation more markedly compared to the one of *O. tenuiflorum*; the aqueous extract of *W. somnifera* was found to be the least effective. These findings were in accordance with the results obtained from SRD assay. In particular, the IC_{50} values of *A. indica, O. tenuiflorum*, and *W. somnifera* extracts were on average 15, 91, and 155 μg/ml, respectively. The mean IC_{50} values determined by ABTS assay for *A. indica* and *O. tenuiflorum* extracts were 29 and 18 μg/ml, respectively. On the other hand, using their methanol extracts, the mean IC_{50} values determined by DPPH assay were 378 and 345 μg/ml, respectively. As concerns the methanol extract of *W. somnifera*, neither by ABTS nor DPPH assay the IC_{50} values could be determined. The kinetics measurements showed that the methanol extract of *O. tenuiflorum* reduced $ABTS^{\bullet+}$ cation radicals more effectively than that of *A. indica*, and that of *W. somnifera* exerted no electron donor properties at all.

4.1 INTRODUCTION

Ayurveda is one of the oldest health systems in the world with fundamental principles and theory-based practices. The meaning of *Ayu* is life and *Veda* is knowledge or science; therefore, Ayurveda is also generally translated as the "science of life."[23] Ayurveda physicians use a wide variety of preparations originated from medicinal plants such as fresh leaves, dried powder, tincture, mash, and tea. Many studies regarding pharmacognosy, chemistry, pharmacology, and clinical therapeutics with a large variety of Ayurvedic medicinal plants were performed, and major pharmaceutical companies have renewed their strategies in favor of natural products drug discovery.[24]

Ayurveda uses a holistic approach to healthcare, which is accepted by the World Health Organization. The major aim involved in Ayurveda is prevention of diseases and treatment of chronic diseases. It concerns mainly with lifestyle improvement and techniques to reduce stress; in addition, Ayurveda utilizes a wide range of medicinal plants, including *Azadirachta indica* and *Ocimum tenuiflorum*. The extracts of *A. indica* and *O. tenuiflorum* as antioxidants have been in focus of numerous studies (e.g., refs 1, 2, 8, 9, 15, 30, 32, 33, 36).

4.1.1 AZADIRACHTA INDICA

A. indica A. Juss or neem is a tree, whose parts are used in traditional medicine in India, and contains active substances with multiple medicinal properties. The preparations originated from neem tree are popular due to their low toxicity to organisms, along with low cost and easy availability.[28] Due to their antibacterial properties, preparations from leaves of neem tree are used for controlling airborne bacterial contamination in residential areas.[38]

More than 135 compounds with diverse structure have been isolated from various parts of the neem tree, yet just few of them have been studied for their pharmacological and biological actions. They comprise a vast array of biologically active compounds that are chemically diverse and structurally complex, for example, isoprenoids, such as diterpenoids, triterpenoids, tetranortriterpenoids, and non-isoprenoids, such as polyphenolics, carbohydrates, proteins, hydrocarbons, flavonoids, tannins, sulfurous compounds, fatty acids, and so forth.[3,5,10] Studies on neem extracts have revealed anti-inflammatory, antipyretic, antitumor, antidiabetic, immunomodulatory, anticarcinogenic effects and also positive effects on central nervous system.[13,14,18,19,25,26,27,37,43]

Clinical studies showed that the lotion prepared from neem leaf, when locally applied, can heal eczema and scabies within 3–4 days in acute stage or a fortnight in chronic case. Indian Ayurvedic physicians prescribe neem leaf extract for oral use when treating malaria to lower hyperlipidemia, particularly cholesterol, in malaria patients. Some papers reported on the use of neem extracts in treating patients with various cancers[5].

4.1.2 OCIMUM TENUIFLORUM

O. tenuiflorum L., also known as *Ocimum sanctum* L, Holy basil, or *tulsi*, is an erect softy hairy aromatic herb or undershrub found throughout India. This plant is noted as an "elixir of life," used not only as medicine but also as spice in cuisine. The *tulsi* plants are traditionally used as treatment for diverse conditions such as infections, skin diseases, hepatic disorders, kidney stones, heart disorder, and as an antidote for snake bite and scorpion sting.[6,33] The plant's essential oil contains eugenol; β-caryophyllene; β-*trans*-guaiene; (E)-α-bergamotene; caryophyllene oxide; 1,8-cineole; 1,10-di-epi-cubenol and *trans*-β-farnesene methyl chavicol; eugenol linalool; camphor; and methyl cinnamate as its main constituents.[6,34] The oil of *tulsi* is prepared by steam distillation from its leaves, and flowering tops are used in food, dental, and oral products and in fragrances and medicines and during traditional rituals. Extracted oil's constituents have been shown to possess insecticidal, nematicidal, and fungicidal activity. The leaf extract of *O. tenuiflorum* containing flavonoids orientin and vicenin has been reported for a significant in vitro antioxidative activity and in vivo anti-lipid peroxidation effect, thus indicating a free radical scavenging mechanism of *O. tenuiflorum* action in preventing cellular damage and tumor formation[7].

Patients suffering from gastric and hepatic disorders are administered with aqueous decoction of *tulsi* leaves. Herbal preparations containing this plant were shown to accelerate process of healing, mitigate clinical symptoms, and improve biochemical parameters in patients with viral hepatitis. The leaf juice of *O. tenuiflorum* is a popular remedy for cold; it is also used in Ayurvedic eye drop preparations used for the treatment of glaucoma, cataract, and chronic conjunctivitis. This juice is also given to patients when treating chronic fever, dysentery, hemorrhage, and dyspepsia.

4.1.3 WITHANIA SOMNIFERA

Another plant used in Ayurveda is *W. somnifera* (L.) Dunal, which is also called as ashwagandha. It is commonly known as "Indian winter cherry" or "Indian ginseng." It is a branched shrub, cultivated widely in Central and Western India and North America. The shrub has traditionally been used for relieving weakness, nervous exhaustion, and arthritis; it helps the body adapt to physical stress. In addition, it has mildly sedative, anti-inflammatory, antipyretic, and analgetic properties. Ashwagandha roots are used as a tonic for brain and nervous system and also in preventive healthcare. *W. somnifera* contains alkaloids such as aswagandhine, anahygrine, cuscohygrine, and steroidal compounds including ergostane-type steroidal lactones, withaferin A, and withanolides. The withanolide class of phenolic compounds acts to prevent or reduce oxidative stress by scavenging free radicals.[16,17,20,21,31,35] Widodo et al.[42] found out that ashwagandha leaf extract selectively killed cancer cells by induction of reactive oxygen species (ROS)-signaling and thus it can be considered as a potential reagent for ROS-mediated cancer chemotherapy. Studies using *W. somnifera*-derived preparations have been performed revealing their moderate analgesic, anti-inflammatory, and disease-modifying activity in experimental animals and arthritis patients.[22]

The aim of the present study was to assess effects of the aqueous and methanol extracts of *O. tenuiflorum*, *A. indica*, and *W. somnifera* on scavenging free radicals by the ABTS, DPPH, and site-specific hydroxyl radical-mediated 2-deoxy-D-ribose degradation (SRD) assays and on their potential to inhibit oxidative degradation of high-molar-mass hyaluronan using rotational viscometry.

4.2 MATERIAL AND METHODS

4.2.1 CHEMICALS

Hyaluronan samples (sodium salt) of the average molar mass 662.4 kDa coded P0207–1B and HA15M-5 of the average molar mass 1.93 MDa were purchased from Lifecore Biomedical Inc., Chaska, MN, USA. Analytical purity grade NaCl and $CuCl_2 \cdot 2H_2O$ were purchased from Slavus Ltd., Bratislava, Slovakia. *A. indica, O. tenuiflorum* (powder from leaves) were obtained from Himalaya, Bangalore, India. *W. somnifera* (powder from leaves) was purchased from Organics India, India. L-Ascorbic acid and potassium persulfate ($K_2S_2O_8$ p.a. purity, max. 0.001% nitrogen) were the products of Merck KGaA, Darmstadt, Germany. 2,2′-Azinobis (3-ethylbenzothiazoline-6-sulfonic acid) diammonium salt (ABTS (purum, >99%) was from Fluka, Germany. 2,2-Diphenyl-1-picrylhydrazyl

(DPPH; 95%), D-deoxyribose, trichloracetic acid, thiobarbituric acid, hydrogen peroxide, $FeCl_3$, hide powder, sodium carbonate, pyrogallol, Folin–Ciocalteu's phenol reagent were from Sigma-Aldrich, Germany. Methanol was the product of MikroChem, Pezinok, Slovakia. Redistilled deionized high-quality grade water, with conductivity of <0.055 μS/cm, was produced using the TKA water purification system from Water Purification Systems GmbH, Niederelbert, Germany.

4.3 EXPERIMENTAL

4.3.1 PREPARATION OF STOCK AND WORKING SOLUTIONS

Hyaluronan P0207–1B (24 mg) was dissolved in 0.15 mmol/l aqueous NaCl solution for 24 h in the dark. Hyaluronan solutions were prepared in two steps. First, 4.0 ml, and after 6 h, 3.9 ml of 0.15 mmol/l NaCl was added in the absence of the plant extracts, 3.8 or 2.9 ml in the presence of the *A. indica* extract, and 3.8 or 3.4 ml in the presence of the *O. tenuiflorum* extract. Second, solutions of ascorbate (16 mmol/l), and cupric chloride solution (320 μmol/l) were prepared also in 0.15 mol/l aqueous NaCl.

Hyaluronan HA15M-5 (14 mg) was dissolved in 0.15 mmol/l aqueous NaCl solution for 24 h in the dark. Hyaluronan sample solutions were prepared in two steps. First, 4.0 ml, and after 6 h, 3.9 of 0.15 mmol/l NaCl was added in the absence of the plant extract, 3.8 or 3.4 ml of 0.15 mmol/l NaCl was added in the presence of the *W. somnifera* extract. Second, solutions of ascorbate (16 mmol/l) and cupric chloride solution (160 μmol/l) were prepared also in 0.15 mol/l aqueous NaCl.

All aqueous extracts were prepared as follows: the powder of leaves (50 mg) was leached for 15 min in boiled distilled water (40 ml). Then, the extracts were cooled down, filtered, and completed with phosphate buffer (5.0 mmol/l, pH 7.4) to the volume 50 ml. Concentrations of *A. indica*, *O. tenuiflorum*, and *W. somnifera* extracts were 2, 4, and 0.4 mg/ml, respectively.

4.3.2 SITE-SPECIFIC HYDROXYL RADICAL-MEDIATED 2-DEOXY-D-RIBOSE DEGRADATION (SRD)

For SRD method, the stock aqueous solutions of the plant extracts *A. indica* (100 mg/ml), *W. somnifera* (100 mg/ml), and *O. tenuiflorum*

(90 mg/ml) were sonicated for 15 min and dried in a desiccator to reach a final concentration of solutions 14, 14, and 29 mg/ml, respectively.

SRD method was performed according to Halliwell et al.[11] with slight adaptations. Unlike 2-deoxy-D-ribose, all reagents were prepared in distilled water. The reagents were added in order: 200 μl of the mixture of $FeCl_3$ solution (100 μmol/l) with 5.0 mmol/l phosphate buffer (pH 7.4) in ratio 1:1, H_2O_2 (100 μl, 1.0 mmol/l), ascorbic acid (100 μl, 1.0 mmol/l), samples (100 μl, 29–14 mg/ml), followed by 2-deoxy-D-ribose (500 μl, 5.6 mmol/l in 5.0 mmol/l phosphate buffer, pH 7.4), vortexed and incubated at 50°C for 30 min. Further, thiobarbituric acid (1 ml, 1.0%) was added, vortexed, and heated at 95°C for 40 min, followed by the addition of CCl_3COOH (1 ml, 2.8%). The absorbance of the samples was measured in UV-Star, 96U-well microplates (Greiner-BioOne GmbH, Germany) at 532 nm using TECAN Infinite M200 (TECAN AG, Austria). Measurements were performed in quadruplicate. From dose–response relationship, IC_{50} was calculated.

4.3.3 DETERMINATION OF CONTENT OF TANNINS

The aqueous extracts in the volume of 25 ml (1.0, 1.0, and 0.7 mg/ml) of *O. tenuiflorum, A. indica, and W. somnifera*, respectively, were prepared.

The content of total phenols was determined as follows: 2 ml of each aqueous extract was mixed with 1 ml of Folin–Ciocalteu's phenol reagent and 10 ml of distilled water. The reaction mixture was made up to the total volume 25 ml with 29% sodium carbonate in water. Absorbance (A_1) at 760 nm of the samples was recorded after 30 min.

The content of polyphenols not adsorbed by hide powder was determined as follows: 10 ml of each aqueous extract was mixed with 0.1 g of hide powder and left on a shaker for 60 min. Then, the solution was filtered and 5 ml of the filtered solution was mixed with 1 ml of Folin–Ciocalteu's phenol reagent and 10 ml of distilled water. The reaction mixture was made up to total volume 25 ml with 29% sodium carbonate in water. Absorbance (A_2) at 760 nm of the samples was recorded after 30 min.

The content of tannins (%) in the extracts was calculated according to the following equation:

$$\% = 62.5 \times (A_1 - A_2) \times m_2 / A_3 \times m_1 \quad (4.1)$$

A_1—absorbance of the measured solution when determining the total polyphenol content
A_2—absorbance of the measured solution when determining tannins, that is, the polyphenol content not absorbed by hide powder
A_3—absorbance of pyrogallol as a reference ($A_3 = 0.35$)
m_1—weight of the sample (g)
m_2—weight of pyrogallol (g)

4.3.4 ROTATIONAL VISCOMETRY(UNINHIBITED/INHIBITED HYALURONAN DEGRADATION)

The procedure was performed as described by Baňasová et al.[4], Valachova et al.[39,41], and Hrabarova et al.[12] Degradation of high-molar-mass hyaluronan was induced in vitro by Weissberger's biogenic oxidative system comprising 100 µmol/l ascorbate and 1 or 2 µmol/l $CuCl_2$, applied under aerobic conditions. The reaction mixture was transferred into the Teflon® cup reservoir of a Brookfield LVDV-II+PRO digital rotational viscometer (Brookfield Engineering Labs., Inc., Middleboro, MA, USA) and changes in dynamic viscosity of hyaluronan were recorded at 25.0±0.1°C in 3-min intervals for 5 h. Two experimental regimes were applied for assessing the influence of the plant extracts on hyaluronan degradation in vitro. First, each drug was added to the reaction mixture 30 s before the addition of ascorbic acid, which initiates oxidative degradation of hyaluronan by producing •OH radicals. Second, each extract was added to the reaction mixture 1 h later, when production of alkoxy/peroxy-type radicals prevails.

4.3.5 2ABTS ASSAY: KINETICS OF SCAVENGING $ABTS^{•+}$

The first step of the standard ABTS assay was preparation of the aqueous solution of $ABTS^{•+}$ cation radical. $ABTS^{•+}$ was prepared 24 h before the measurements at room temperature as follows: ABTS aqueous stock solution (7 mmol/l) was mixed with $K_2S_2O_8$ aqueous solution (2.45 mmol/l) in the equivolume ratio. On the next day, 1 ml of the resulting solution was diluted with distilled water to the final volume of 60 ml.[29,40]

Stock solutions of the extract of *A. indica* at concentrations 1, 1.5, and 1.8 μg/ml and of the extract of *O. sanctum* at concentrations 0.2, 2.0, and 2.6 μg/ml in the volume of 50 μl was added to 2 ml of the $ABTS^{\bullet+}$ solution. Kinetics of scavenging $ABTS^{\bullet+}$ was performed in triplicate at the wavelength 734 nm. Measurements of reduction of $ABTS^{\bullet+}$ using each extract were performed in time interval 2–20 min using ultraviolet-visible (UV-Vis) 1800 spectrophotometer (Shimadzu, Japan).

4.3.6 ABTS ASSAY: DETERMINATION OF IC_{50} VALUES

$ABTS^{\bullet+}$ cation radical was prepared as mentioned above. The aqueous reagent in the volume of 225 μl was added to 25 μl of the aqueous solutions of the plant extracts. Absorbance at 734 nm of samples was recorded within 6 min using UV-Vis 1800 spectrophotometer (Shimadzu, Japan).

The aqueous solutions of all plant extracts were prepared as follows: the plant extracts (8 mg/ml) were dissolved in distilled water (25 ml). The solutions were sonicated for 15 min and dried in a desiccator to reach a final concentration of 2.4, 5.2, and 1.8 mg/ml for the extracts of *O. tenuiflorum*, *A. indica*, and *W. somnifera*, respectively.

4.3.7 DPPH ASSAY: DETERMINATION OF IC_{50} VALUES

The first step of a standard DPPH assay was preparation of $DPPH^{\bullet}$ radical as follows: 2,2-diphenyl-1-picrylhydrazyl (1.1 mg) was dissolved in 50 ml of distilled methanol to generate $DPPH^{\bullet}$. The $DPPH^{\bullet}$ solution in the volume of 225 μl was added to 25 μl of the methanol solution of the plant extracts. Absorbance (517 nm) of samples was recorded after 30 min.

The methanolic solutions of all plant extracts were prepared as follows: The plant extracts (8 mg/ml) were dissolved in methanol (25 ml). The solutions were sonicated for 15 min and dried in a desiccator to reach a final concentration of 2.4, 0.6, and 1.5 mg/ml for the extracts of *O. tenuiflorum*, *A. indica*, and *W. somnifera*, respectively.

In both assays, the measurements were performed in the quadruplicate in Greiner UV-Star 96U-well microplates (Greiner-BioOne GmbH, Germany) by using the Tecan Infinite M 200 reader (Tecan AG, Austria).

4.4 RESULTS AND DISCUSSION

Potential antioxidative effects of the aqueous extracts of *A. indica*, *O. tenuiflorum*, and *W. somnifera* have been tested against oxidative degradation of high-molar-mass hyaluronan, whose results are depicted in Figures 4.1–4.3. At first, hyaluronan was exposed to oxidative degradation by cupric ions in the presence of ascorbate, which led to a decrease in dynamic viscosity of the hyaluronan solution from 11.5 to 6.07 mPa·s (black curve). To achieve the inhibition of hyaluronan degradation induced by •OH radicals, the plant extracts (*A. indica:* 5, 25, and 125 μg/ml; *O. tenuiflorum:* 5, 50, and 500 μg/ml; and *W. somnifera*: 5 and 25 μg/ml) were added to the reaction system before the application of ascorbic acid. Time- and dose-dependent changes in dynamic viscosity of the hyaluronan solutions were monitored in the time interval from 0 up to 300 min. Figure 4.1 (left panel) indicates that the extract of *A. indica* effectively scavenged •OH radicals in a dose-dependent manner. The decreases in dynamic viscosity of the hyaluronan solutions were 2.86, 2.07, and 1.11 mPa·s at the concentrations of the extract 5, 25, and 125 μg/ml, respectively. The extract at concentrations 25 and 125 μg/ml, added into the reaction system at propagation phase, when alkoxy- and peroxy-type radicals are generated, was effective in inhibiting hyaluronan degradation (Fig. 4.1 [right panel]).

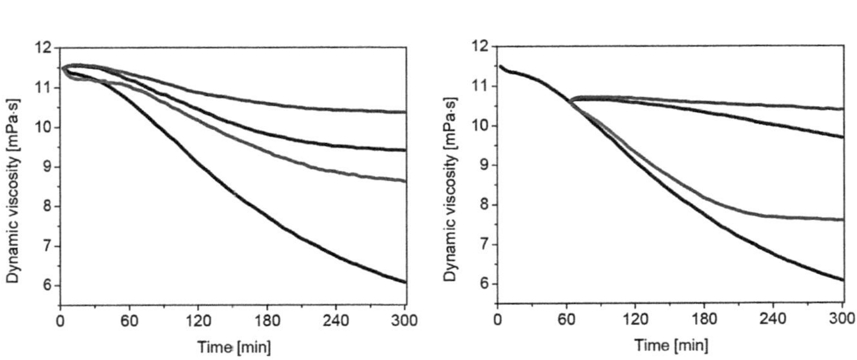

FIGURE 4.1 (See color insert.) Time-dependent changes in dynamic viscosity of the hyaluronan solution in the presence of *A. indica* extract added to the reaction mixture just before the initiation of hyaluronan degradation (left panel) or 1 h later (right panel) in μg/ml: 0 (black), 5 (red), 25 (blue), and 125 (green).

As depicted in Figure 4.2 (left panel), the extract of *O. tenuiflorum* effectively inhibited ˙OH radical-induced hyaluronan degradation as seen by decreasing in dynamic viscosity of the hyaluronan solution in 5-h period by 3.27, 2.7, and 0.9 mPa·s for concentrations 5, 50, and 500 μg/ml, respectively. Similar effect was observed when adding the extract to the reaction system 1 h after the initiation of hyaluronan degradation (Fig. 4.2 [right panel]).

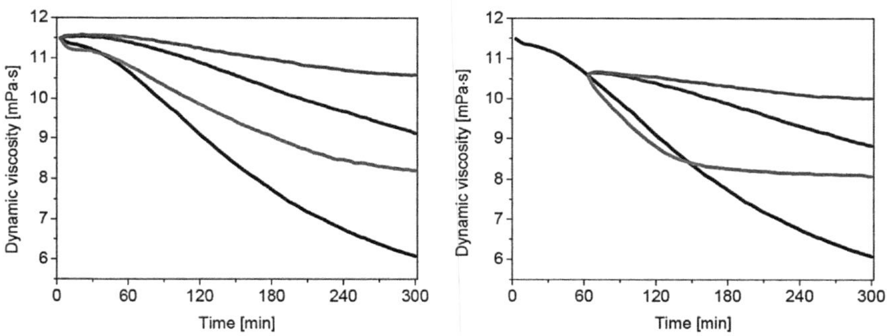

FIGURE 4.2 (See color insert.) Time-dependent changes in dynamic viscosity of the hyaluronan solution in the presence of the extract of *O. tenuiflorum* added to the reaction mixture just before the initiation of hyaluronan degradation (left panel) or 1 h later (right panel) in μg/ml: 0 (black), 5 (red), 50 (blue), and 500 (green).

Figure 4.3 illustrates the results of the addition of the extract of *W. somnifera* into the hyaluronan degradation system. As seen, this extract dose dependently inhibited ˙OH radical-induced degradation of hyaluronan (left panel). Within 300 min, the decrease in dynamic viscosity of the hyaluronan solution was 3.52 and 1.78 mPa·s for the extract at concentration 5 and 25 μmol/l, respectively. Similarly, positive results were observed when the extract was added to the reaction system 1 h after the initiation of hyaluronan degradation, where alkoxy/peroxy-type radicals were predominately present (Fig. 4.3 [right panel]). Using the extract at concentration 5 and 25 μmol/l, the decrease in dynamic viscosity of the hyaluronan solution was 3.06 and 0.66 mPa·s, respectively. The effect of the extract at the latter concentration resembles chain-breaking antioxidant action (blue curve).

Protective effects of the extracts against hyaluronan degradation can be related to the content of tannins, which was 72.4, 61.3, and 17.0% for *A. indica*, *O. tenuiflorum*, and *W. somnifera*, respectively.

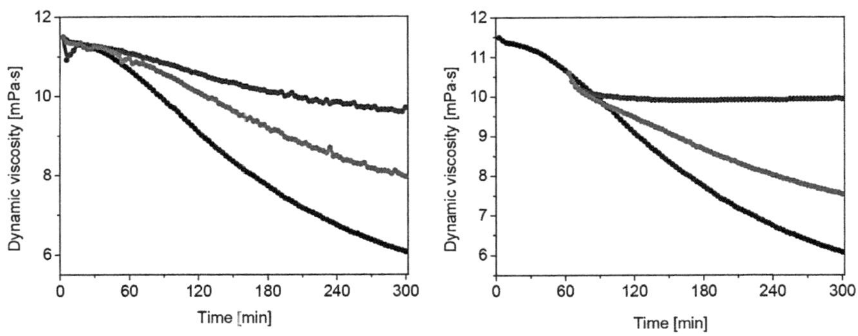

FIGURE 4.3 (See color insert.) Time-dependent changes in dynamic viscosity of the hyaluronan solution in the presence of the extract of *W. somnifera* added to the reaction mixture just before the initiation of hyaluronan degradation (left panel) or 1 h later (right panel) in μg/ml: 0 (black), 5 (blue), and 25 (red).

Next approach to test scavenging capacity of the extracts against $^{\bullet}OH$ radicals was the SRD assay. We found the IC_{50} value for the extract of *A. indica* to be 15±0.9 μg/ml, which was four times less than that of *O. tenuiflorum* (IC_{50}=90.8±0.9 μg/ml). It corresponds to the results of rotational viscometry, which showed a higher efficacy also for the extract *A. indica*. The IC_{50} value of *W. somnifera* using the SRD assay was 155.4±0.9 μg/ml, which also corresponds to the effect of this extract in hyaluronan degradation.

As depicted in Figure 4.4, dose-dependent kinetics of radical scavenging by the extracts *A. indica* and *O. tenuiflorum* was observed using the standard ABTS decolorization assay. In the 20th min of the measurement, the amount of nonreduced $ABTS^{\bullet+}$ was 32%, when using the extract of the *O. tenuiflorum* at the concentration 125 μg/ml. The value for the extract of *A. indica* was 20%, whereas *W. somnifera* did not reduce $ABTS^{\bullet+}$. The percentage of nonreduced $ABTS^{\bullet+}$ was about 96% independently of the concentration used (20−50 μg/ml; results not shown).

Table 4.1 displays the IC_{50} values of the extracts using ABTS, DPPH, and SRD assays. As depicted, the extract of *O. tenuiflorum* reduced $ABTS^{\bullet+}$ and $DPPH^{\bullet}$ more efficiently than that of *A. indica*. Furthermore, both extracts were effective in reducing $ABTS^{\bullet+}$ rather than that of $DPPH^{\bullet}$. On the other hand, in the extract of *W. somnifera*, we were not able to obtain IC_{50} values.

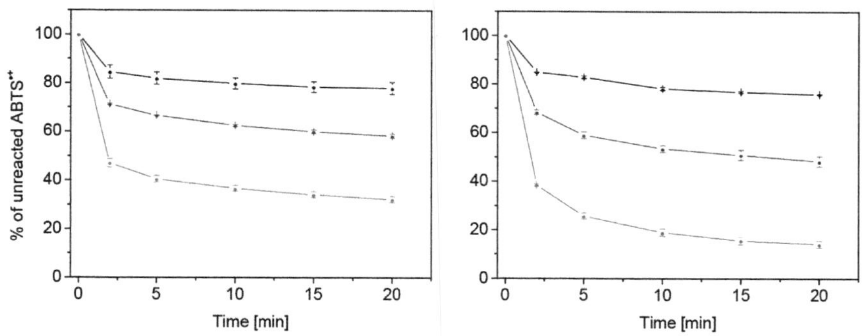

FIGURE 4.4 (See color insert.) Percentage of unreacted ABTS•+ after addition of the *O. tenuiflorum* extract (left panel) and the *A. indica* extract (right panel). Concentrations of the *O. tenuiflorum* extract in ABTS•+ solution were: 5 (black), 50 (red), and 65 μg/ml (green). Concentrations of the *A. indica* extract in ABTS•+ solution were: 25 (black), 35 (red), and 45 (green) μg/ml.

TABLE 4.1 IC_{50} Values of the Extracts Tested Determined by the ABTS and DPPH Assays.

Method	Extract	IC_{50} (μg/ml)
ABTS	*Azadirachta indica*	28.5 ± 0.9
	Ocimum tenuiflorum	19.7 ± 0.8
	Withania somnifera	Not determined
DPPH	*A. indica*	378 ± 0.8
	O. tenuiflorum	345 ± 0.8
	W. somnifera	Not determined

Different assays of antioxidant activity of the plant extracts are implemented and used to date; each of them has advantages and disadvantages. By the method of rotational viscometry, the ability of the substances to donate H• atom is determined. The ABTS and DPPH assays serve for determination of electron donor properties of substances, which was shown in both extracts examined.

Hyaluronan was subjected to oxidative degradation by Cu(II) ions in the presence of ascorbate under aerobic conditions (Scheme 4.1). As seen, during the initiation cycle, just 1 μmol of ascorbate reacts from its total concentration 100 μmol/l. The formed amount of H_2O_2 is therefore just 1 μmol because in the reactive system, just 1 μmol/l of Cu(II) ions is present. The produced amount of hydrogen peroxide is dependent on the presence of the intermediate complex Cu(I), which is immediately decomposed to form maximally 1 μmol/l of •OH radicals. However, released Cu(II) ions with the remaining unreacted ascorbate (99 μmol/l) promptly react to form other 1 μmol/l of hydrogen peroxide followed by production of next 1 μmol of •OH radicals.

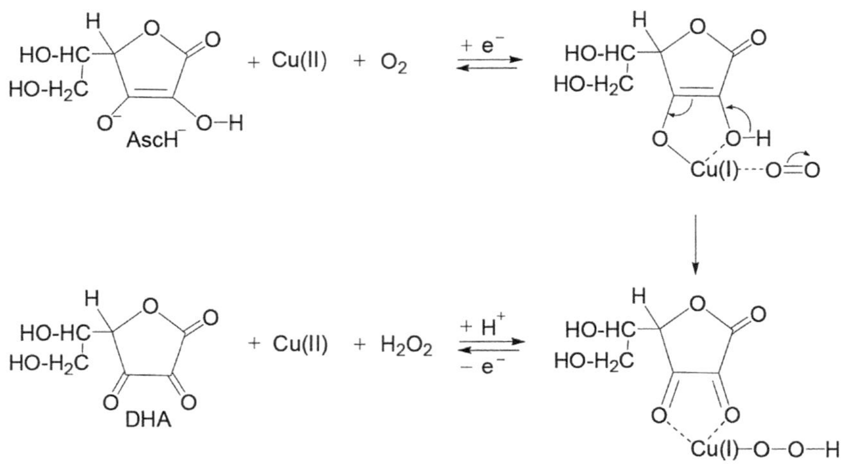

$$H_2O_2 + Cu(I)\text{---complex} \rightarrow {}^{\bullet}OH + Cu(II) + OH^-$$

SCHEME 4.1 The Weissberger biogenic oxidative system.

4.5 CONCLUDING REMARKS

1. The results of rotational viscometry showed that the extracts tested dose dependently inhibited the degradation of hyaluronan mediated by hydroxyl or by the alkoxy- and peroxy-type radicals, indicating that the extracts were effective H• donors.
2. The SRD assay, which determines the capacity of substances to scavenge hydroxyl radicals, was supportive to the results of rotational viscometry.
3. Based on the IC_{50} values in the ABTS and DPPH assays, a bit higher radical scavenging capacity was observed for the *O. tenuiflorum* extract than that of *A. indica*. In contrast, in kinetics measurements, a moderately higher efficacy in the reduction of

the radicals was observed for the *A. indica* extract than that of *O. sanctum*, because lower concentrations were needed to reduce ABTS•+ cation radical, which points to its capability to donor electrons. On the other hand, the extract of *W. somnifera* was not effective in reducing both ABTS•+ cation radicals and DPPH• radicals; its IC_{50} value could not be determined. The IC_{50} values of SRD indicated that the extract of *A. indica* scavenged •OH radicals more effectively than did the extract of *O. sanctum*; though the extract of *W. somnifera* was not effective in this assay.

Each extract, that is, aqueous, alcoholic, or the other one of any Ayurvedic natural supplement is a multicomponent mixture. If we can simply accept a thesis that besides ballast components an investigated mixture contains also two different "active species," which give data when measured, for example, by an X-meter, then it may mean that one species is active and the other one is not active. Therefore, the final result measured with an X-meter will be reported as no effect. Putatively, we can think that both species will be active. Then, the X-meter allows an equal response for both components or even we can indicate a synergic effect of individual species. As mentioned above that any extract consists of many components, and thus the result of measurements using the X-meter is a "black result."

When administered a two-component Ayurvedic mixture, it is distributed in an organism and an observed positive effect can be attributed to the effect of the active substance. In contrast, if a positive effect of the component was not observed, it means that the active species was not distributed to the site or besides the active species also the antagonist species penetrates to the particular site. We can state that antioxidant effects of Ayurvedic preparations have been evaluated for many years and that is why these preparations can be administered regardless of whether a positive, negative, or no response was measured "in a tube" or by using a measuring apparatus.

ACKNOWLEDGMENT

The study was supported by the grants of VEGA (2/0065/15 and 2/0155/16) and APVV-15–0308.

KEYWORDS

- **Ayurveda**
- **glycosaminoglycans**
- **natural supplements**
- **radical scavenging capacity**
- **rotational viscometry**

REFERENCES

1. Alzohairy, M. A. Therapeutics Role of Azadirachta indica (Neem) and Their Active Constituents in Diseases Prevention and Treatment. *Evid. Based Complement. Altern. Med.* **2016,** *2016* Article ID 7382506.
2. Arakaki, J.; Suzui, M.; Morioka, T.; Kinjo, T.; Kaneshiro, T.; Inamine, M.; Sunagawa, N.; Nishimaki, T.; Yoshimi, N. Antioxidative and Modifying Effects of a Tropical Plant *Azadirachta indica* (Neem) on Azoxymethane-Induced Preneoplastic Lesions in the Rat Colon. *Asian Pacific J. Cancer Prev.* **2006,** *7*(3), 467–471.
3. Bandyopadhyay, U.; Biswas, K.; Chatterjee, R.; Bandyopadhyay, D.; Chattopadhyay, I.; Ganguly, C. K.; Chakraborty, T.; Bhattacharya, K.; Banerjee, R. K. Gastroprotective Effect of Neem (*Azadirachta indica*) Bark Extract: Possible Involvement of H^+-K^+ -ATPase Inhibition and Scavenging of Hydroxyl Radical. *Life Sci.* **2002,** *71*, 2845–2865.
4. Baňasová, M.; Sasinková, V.; Mendichi, R.; Perečko, T.; Valachová, K.; Juránek, I.; Šoltés, L. Free-Radical Degradation of High-Molar-Mass Hyaluronan Induced by Weissberger's Oxidative System: Potential Antioxidative Effect of Bucillamine. *Neuroendocrinol. Lett.* **2012,** *33*(3), 151–154.
5. Biswas, K.; Chattopadhyay, I.; Banerjee, R. K.; Bandyopadhyay, U. Biological Activities and Medicinal Properties of Neem (*Azadirachta indica*). *Curr. Sci.* **2002,** *82*, 1336–1345.
6. Chandra S.; Dwivedi P.; Arti K. M.; Shinde L. P. An Industrial Cultivation of Tulsi (*Ocimum sanctum*) for Medicinal use and Rural Empowerment. *J. Med. Plants Stud.* **2016,** *4*(6), 213–218.
7. Devi, U. M. Radioprotective, Anticarcinogenic and Antioxidant Properties of the Indian Holy Basil, *Ocimum Sanctum* (Tulasi). Indian *J. Exp. Biol.* **2001,** *39*, 185–190.
8. Dkhil, M. A.; Al-Quraishy, S.; Aref, A. M.; Othman, M. S.; El-Deib, K. M.; Abdel Moneim, A. E. The Potential Role of *Azadirachta indica* Treatment on Cisplatin-Induced Hepatotoxicity and Oxidative Stress in Female Rats. *Oxid. Med. Cell. Longev.* **2013,** *2013*, Article ID 741817, 9 pages.
9. Farahna, M.; Bedri, S.; Khalid, S.; Idris, M.; Pillai, C. R.; Khalil, E. A. Anti-Plasmodial Effects of Azadirachta indica in Experimental Cerebral Malaria: Apoptosis of

Cerebellar Purkinje Cells of Mice as a Marker. *North Am. J. Med. Sci.* **2010,** *2*(11), 518–525.

10. Ganguli, S. Neem: A Therapeutic for All Seasons. *Curr. Sci.* **2002,** *82*, 1304.
11. Halliwell, B.; Gutteridge, J. M. C.; Aruoma, O. The Deoxyribose Method: A Simple "test tube" Assay for Determination of Rate Constants for Reactions of Hydroxyl Radicals. *Anal. Biochem.* **1987,** *165*, 215–219.
12. Hrabarova, E.; Valachova, K.; Rychly, J.; Rapta, P.; Sasinkova, V.; Malikova, M., L.; Soltes L. High-Molar-Mass Hyaluronan Degradation by Weissberger's System: Pro- and Anti-Oxidative Effects of Some Thiol Compounds. *Polym. Degrad. Stab.* **2009,** *94*, 1867–1875.
13. Kurin, E.; Mučaji, P.; Nagy, M. In Vitro Antioxidant Activities of Three Red Wine Polyphenols and Their Mixtures: an Interaction Study. *Molecules*, **2012,** *17*, 14336–14348. DOI: 10.3390/molecules171214336.
14. Manikandan, P. P.; Letchoumy, P. V.; Gopalakrishnan, M.; Nagini, S. Evaluation of *Azadirachta indica* Leaf Fractions for in Vitro Antioxidant Potential and in Vivo Modulation of Biomarkers of Chemoprevention in the Hamster Buccal Pouch Carcinogenesis Model. *Food Chem. Toxicol.* **2008,** *46*, 2332–2343.
15. Minhas, U.; Minz, R.; Das, P.; Bhatnagar, A. Therapeutic Effect of *Withania Somnifera* on Pristane-Induced Model of SLE. *Inflammopharmacology* **2012,** *20*(4), 195–205.
16. Mirjalili, M. H.; Moyano, E.; Bonfill, M.; Cusido, R. M.; Palazón, J. Steroidal Lactones from Withania Somnifera, an Ancient Plant for Novel Medicine. *Molecules* **2009,** *14*(7), 2373–2393.
17. Mishra, S. K.; Trikamji B. A Clinical Trial with *Withania Somnifera* (*Solanaceae*) Extract in the Management of Sarcopenia (Muscle Aging). *J. Org. Biomol. Chem.* **2013,** *1*, 187–194.
18. Mukherjee, N.; Mukherjee, S.; Saini, P.; Roy, P.; Babu, S. P. S. Antifilarial Effects of Polyphenol Rich Ethanolic Extract from the Leaves of *Azadirachta Indica* Through Molecular and Biochemical Approaches Describing Reactive Oxygen Species (ROS)-Mediated Apoptosis of Setaria Cervi. *Exp. Parasitol.* **2014,** *136*, 41–58.
19. Okpanyi, S. N.; Ezeukwu, G. C. Anti-inflammatory and Antipyretic Activities of *Azadirachta indica. Planta Medica* **1981,** *41*, 34–39.
20. Palash, M.; Mitali G.; Kumar, M. T.; Prasad, D. A. Pharmacognostic and Free-Radical Scavenging Activity in the Different Parts of Ashwagandha *Withania somnifera (L. Dunal). Int. J. Drug Dev. Res.* **2010,** *2*(4), 830–843.
21. Patel, D. S.; Shah, P. B.; Managoli, N. B. Evaluation of in Vitro Anti-Oxidant and Free Radical Scavenging Activities of *Withania Somnifera* and *Aloe vera. Asian J. Pharm. Technol.* **2012,** *2*(4), 143–147.
22. Patwardhan, B. Ethnopharmacology and Drug Discovery. *J. Ethnopharmacol.* **2005,** *100*, 50–52.
23. Patwardhan, B.; Vaidya, A. D. B. Ayurveda: Scientific Research and Publications. *Curr. Sci.* **2009,** *97*, 1117–1121.
24. Patwardhan, B.; Vaidya, A. D. B.; Chorghade, M. Ayurveda and Natural Products Drug Discovery. *Curr. Sci.* **2004,** *86*, 789–799.
25. Ponnusamy, S.; Haldar, S.; Mulani, F.; Zinjarde, S.; Thulasiram, H.; Ravi Kumar, A. Gedunin and Azadiradione: Human Pancreatic Alpha-Amylase Inhibiting Limonoids from Neem (*Azadirachta indica*) as Anti-Diabetic Agents. *PLoS One* **2015,** *10*, 1–19.

26. Prakash, P.; Gupta, N. Therapeutic Uses of *Ocimum Sanctum Linn* (Tulsi) with a Note of Eugenol and its Pharmacological Actions: A Short Review. *Indian J. Physiol. Pharmacol.* **2005,** *49*(2), 125–131.
27. Raghavendra, M.; Rituparna, M.; Kumar, S.; Acharya, S. B. Role of Aqueous Extract of *Azadirachta Indica* Leaves in an Experimental Model of Alzheimer's Disease in Rats. *Int. J. App. Basic Med. Res.* **2013,** *3*, 37–47.
28. Raut, R. R.; Sawant, A. R.; Jamge, B. R. Antimicrobial Activity of *Azadirachta indica* (Neem) Against Pathogenic Microorganisms. *J. Acad. Ind. Res.* **2014,** *3*, 327–329.
29. Re, R.; Pellegrini, N.; Proteggente, A.; Pannala, A.; Yang, M.; Rice-Evans, C. Antioxidant Activity Applying an Improved ABTS Radical Cation Decolorization Assay. *Free Radic. Biol. Med.* **1999,** *26*(9–10), 1231–1237.
30. Schumacher, M.; Cerella, C.; Reuter, S.; Dicato, M. Anti-Inflammatory, Pro-Apoptotic, and Anti-Proliferative Effects of a Methanolic Neem (Azadirachta indica) Leaf Extract are Mediated Via Modulation of the Nuclear Factor-κB Pathway. *Genes* Nutr. **2011,** *6*(2), 149–160.
31. Senthil, K.; Thirugnanasambantham, P.; Oh, T. J.; Kim, S. H.; Choi, H. K. Free Radical Scavenging Activity and Comparative Metabolic Profiling of in Vitro Cultured and Field Grown *Withania somnifera* Roots. *PLoS One* **2015,** *10*(4), e0123360.
32. Sethi, J.; Sood, S.; Seth, S.; Talwar A. Evaluation of Hypoglycemic and Antioxidant Effect of *Ocimum Sanctum. Indian J. Clin. Biochem.* **2004,** *19*(2), 152–155.
33. Shetty, S.; Udupa, S.; Udupa, L. Evaluation of Antioxidant and Wound Healing Effects of Alcoholic and Aaqueous Extract of *Ocimum sanctum Linn.* in Rats. *Evid. Based Complement. Altern. Med.* **2008,** *5*, 95–101.
34. Sims, C. A.; Juliani, H. R.; Mentreddy, S. R.; Simon, J. E. Essential Oils in Holy Basil (*Ocimum tenuiflorum L.*) as Influenced by Planting Dates and Harvest Times in North Alabama. *J. Med. Act. Plants* **2014,** *2*, 33–41.
35. Singh, N.; Bhalla, M.; de Jager, P.; Gilca, M. An Overview on Ashwagandha: A Rasayana (rejuvenator) of Ayurveda. *Afr. J. Tradit. Complement. Altern. Med.* **2011,** *8*(S), 208–213.
36. Sithisarn, P.; Supabphol, R.; Gritsanapan, W. Antioxidant Activity of Siamese Neem Tree (VPI209). *J. Ethnopharmacol.* **2005,** *99*, 109–112.
37. Somsak, V.; Chachiyo, S.; Jaihan, U.; Nakinchat, S. Protective Effect of Aqueous Crude Extract of Neem (*Azadirachta indica*) Leaves on *Plasmodium berghei*-Induced Renal Damage in Mice. *J. Trop. Med.* **2015,** *2015*, Article ID 961,205, 5 pages.
38. Thanigaivel, S.; Vijayakumar, S.; Gopinath, S.; Mukherjee, A.; Chandrasekaran, N.; Thomas, J. In Vivo and in Vitro Antimicrobial Activity of *Azadirachta indica (Lin)* Against *Citrobacter freundii* Isolated from Naturally Infected Tilapia (*Oreochromis mossambicus*). *Aquaculture* **2015,** *437*, 252–255.
39. Valachová, K.; Baňasová, M.; Topoľská, D.; Sasinková, V.; Juránek, I.; Collins, M. N.; Šoltés, L. Influence of Tiopronin: Captopril and Levamisole Therapeutics on the Oxidative Degradation of Hyaluronan. *Carbohydr. Polym.* **2015,** *134*, 516–523.
40. Valachová, K.; Hrabárová, E.; Dráfi, F.; Juránek, I.; Bauerová, K.; Priesolová, E.; Nagy, M.; Šoltés, L. Ascorbate and Cu(II)-Induced Oxidative Degradation of High-Molar-Mass Hyaluronan, pro- and Antioxidative Effects of Some Thiols. *Neuroendocrinol. Lett.* **2010,** *31*(2), 101–104.

41. Valachova, K.; Rapta, P.; Kogan, G.; Hrabarova, E.; Gemeiner, P.; Soltes, L. Degradation of High-Molar-Mass Hyaluronan by Ascorbate Plus Cupric Ions: Effects of D-Penicillamine Addition. *Chem. Biodivers.* **2009,** *6*, 389–395.
42. Widodo, N.; Priyandoko, D.; Shah, N.; Wadhwa, R.; Kaulfound, S. C. Selective Killing of Cancer Cells by Ashwagandha Leaf Extract and Its Component Withanone Involves ROS Signaling. *PLoS One* **2010,** *5*(10), e13536.
43. Yadav, N.; Kumar, S.; Kumar, R.; Srivastava, P.; Sun, L.; Rapali, P.; Marlowe, T.; Schneider, A.; Inigo, J. R.; O'Malley, J.; Londonkar, R.; Gogada, R.; Chaudhary, A. K.; Yadava, N.; Chandra, N. Mechanism of Neem Limonoids-Induced Cell Death in Cancer: Role of Oxidative Phosphorylation. *Free Radic. Biol. Med.* **2016,** 90, 261–271.

PART II

Nanoscience and Technology

CHAPTER 5

ZnO NANOSTRUCTURE–POLYMER DIELECTRIC FILMS FOR ELECTRONIC AND OPTOELECTRONIC DEVICE APPLICATIONS

SUVRA PRAKASH MONDAL*, SAURAB DHAR, and TANMOY MAJUMDER

Department of Physics, National Institute of Technology, Agartala, Tripura 799046, India

**Corresponding author. E-mail: suvraphy@gmail.com, suvra.phy@nita.ac.in*

CONTENTS

ABSTRACT

ZnO is an important II–VI semiconductor, which have been extensively studied in photovoltaics, photoelectrochemical cells, piezoelectric nanogenerators (NGs), ultraviolet detectors, lasers, light-emitting diodes, and gas sensors, due to its unique optical and electrical properties including wide bandgap, higher excitonic binding energy, superior electron mobility, low cost, and environmental friendliness. ZnO–polymer nanocomposites are especially interesting due to their control particle size and high stability. The influence of polymer matrix plays a major role in the optical and electrical properties of the composite materials. In recent years, significant progress has been made in the synthesis of various types of ZnO–polymer nanocomposites for the fabrication of electronic and optoelectronic devices. Particularly, ZnO nanostructures embedded or encapsulated in the insulating polymer matrix are attractive for high dielectric materials, memory devices, gas sensors, piezoelectric NGs, and photosensing applications. Here, we discuss the growth and characteristics of ZnO nanostructure-insulating polymer dielectric thin films and their device applications.

5.1 INTRODUCTION

ZnO is a very promising and attractive material for various types of applications. ZnO has several interesting properties such as direct and wide bandgap in the near-ultraviolet (UV) spectral region.[1,2] As it has large free exciton binding energy (~60 meV), room temperature excitonic emission can be observed.[3,4] ZnO crystallizes in the hexagonal wurtzite structure and large bulk single crystals can be formed easily compared to GaN or other wurtzite structure crystals.[5–7] From the early days, electrical, optical, and optoelectronic properties of ZnO have been studied in thin films or bulk forms. Recently, ZnO nanostructures have been studied in several field of research such as photovoltaics,[8–10] photoelectrochemical cells,[11,12] piezoelectric nanogenerators (NGs), photodetectors,[13–16] UV lasers,[17–19] light-emitting diodes (LED),[20–23] and gas sensors.[24,25] The major advantage of ZnO is that it can be easily used in synthesis of various kinds of nanostructures such as nanowires,[26,27] nanorods,[11–13] nanotubes,[28–30] nanoflowers,[31,32] nanobelts,[33,34] nanorings,[35,36] nanoterapods-tripods,[37] nanoforest,[38,39] and so forth due to its unique crystal structure. Semiconductor–polymer hybrid nanomaterials are composed of semiconductor nanostructures embedded in the polymer matrix. Recently, ZnO–polymer nanocomposites have been studied due to their various applications such as solar cell,[40,41] piezoelectric NGs,[42] UV detectors,[13] LED,[20–23] and gas sensors.[43,44] Such composite materials are especially important because it carries advantages of both semiconductors as well as polymers. More importantly, they can be used in flexible electronics. Several research works have been carried out with ZnO nanostructure-conducting polymer composites such as poly(3-hexylthiophene-2,5-diyl), poly[2-methoxy-5-(3′,7′-dimethyloctyloxy)-1,4-phenylenevinylene], poly(3,4-ethylenedioxythiophene)-poly(styrene sulfonate), poly(p-phenylene vinylene) and so forth for solar cell, photodetectors, resistive switching memory, and light-emitting device applications.[13,45–50] On the other hand, combination of ZnO nanostructures with insulating polymer such as polyvinyl alcohol (PVA), poly(methyl methacrylate) (PMMA), polydimethylsiloxane (PDMS), and polyvinylidene fluoride (PVDF) have several applications such as charge storage devices, piezoelectric NGs, high dielectric material applications, and so forth.[51–56] ZnO-conducting polymer nanocomposites have been intensively studied by several researchers for various electronic and optoelectronic device applications.[57–59] However, the study of ZnO-insulating dielectric polymer composites is also important for some specific applications. Here, we report a brief review on ZnO nanostructure-insulating polymer nanocomposites for some important electronic and optoelectronic device applications.

5.2 ZnO NANOSTRUCTURES

ZnO nanostructured materials have been attracted much attention due to their distinguished performance in electronics, optics, and photonics. From the early 1960s, ZnO thin films have been studied intensively because of their applications in sensors,[43,44] photodetectors,[13–16,47] solar cells,[8–12] transducers,[60] and catalysts.[61] ZnO is a wurtzite hexagonal structure (space group C6mc) with lattice parameters $a = b = 0.3296$ and $c = 0.52065$ nm. It is a wide bandgap (3.37 eV) II–VI compound semiconductor with high exciton binding energy (60 meV).[62,63] In ZnO crystals, tetrahedrally coordinated O^{2-} and Zn^{2+} ions are stacked alternately in planes along the c-axis.

Figure 5.1a and b shows the schematic representation of a typical ZnO unit cell and different crystal planes. The most important property of such unit cell is the existences of polar surfaces.[64] The positively charged Zn-terminated faces at (0001) and negatively charged O-terminated surfaces at $(000\bar{1})$ create a normal dipole moment and spontaneous polarization along the c-axis. ZnO creates various kinds of nanostructures due to their unique crystal structure. ZnO has three majors of fast growth directions such as $(2\bar{1}\ \bar{1}\ 0)$, $(01\ \bar{1}\ 0)$, and (00 01). These crystal planes play major roles to form various nanostructures such as nanorods, nanowires, nanotubes, nanobelts, nanorings, nanocombs, nanosprings, and nanoflowers, and so forth.

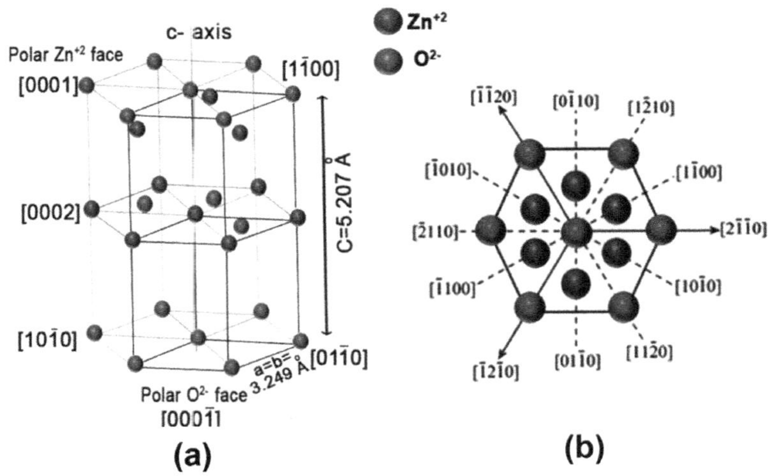

FIGURE 5.1 **(See color insert.)** (a) Unit cell of hexagonal wurtzite structure of zinc oxide (ZnO) and (b) various crystal planes of ZnO.

One-dimensional (1D) ZnO nanostructures such as nanorods, nanowires can be grown by vapor–liquid–solid (VLS) growth process.[65,66] Figure 5.2a shows the schematic representation of ZnO nanorods/nanowires grown by VLS process. In VLS process, metal nanoparticles (NPs) such as Au and Fe are used as catalyst and ZnO vapors are used as the reactant gases. At sufficiently high temperature (~650°C), ZnO diffuses into Au and eutectic liquid droplets are formed. When more and more ZnO vapors are absorbed by Au droplet, ZnO precipitate downward to the catalyst and nanowires/nanorods growth is initiated. The diameter of the nanowires highly depends on the size of the catalysts. Figure 5.2b represents the typical scanning electron microscopy (SEM) micrograph of ZnO nanowires grown by VLS method. On the other hand, ZnO nanorods of comparable shorter length can be grown chemically at very low temperature (65–90°C) using a hydrothermal process.[11–14]

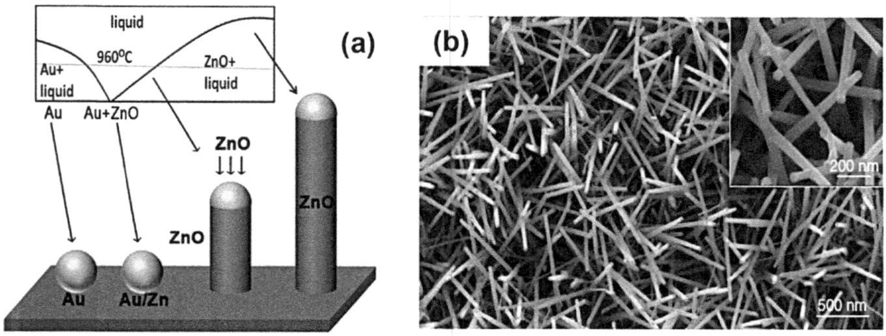

FIGURE 5.2 (a) Vapor–liquid–solid growth (VLS) mechanism of ZnO nanowires and (b) scanning electron microscopy (SEM) micrograph of ZnO nanowires. (b: Reproduced with permission from reference 65. © 2008 Elsevier B.V)

Figure 5.3a shows the schematic representation of ZnO nanorod growth by hydrothermal process. At first, 0.1 M of zinc acetate dihydrate was dissolved in 50 ml isopropyl alcohol at 65°C. Afterward, a transparent solution was achieved by adding 0.02 M diethanolamine in the previous solution. A thin ZnO nanoparticle seed layer was deposited from the previous solution on substrates by spin coating, followed by heating at 200°C for 1 h. For the growth of nanorods, ZnO-seeded substrates were dipped inside a hydrothermal reactor, containing an equimolar solution of zinc nitrate and hexamethylenetetramine. The temperature of the solution was maintained at 90°C for several hours. The length and diameter of the chemically grown nanorods are typically 1–5 μm and 50–200 nm, respectively. Nanobelts are 1D nanostructures which have a well-defined geometrical shape and side surfaces. ZnO nanobelts were synthesized and studied by several researches by sublimation of ZnO powder without introducing any catalyst.[67–69]

Figure 5.4a shows the transmission electron microscopy (TEM) micrographs of ZnO nanobelts of widths in the range of 50–300 nm and thicknesses 10–30 nm. The absence of Au nanoparticle at the end nanobelts confirmed that the growth mechanism was dominated by vapor–solid process. ZnO nanobelts usually grow along $(01\bar{1}0)$ direction. The growth direction of top and bottom flat surfaces of the nanobelts is along $(2\bar{1}\bar{1}0)$

plane and side surfaces along (0001) plane. Figure 5.4b shows the SEM image of ZnO nanobelts with the Y-shaped morphology of widths several hundred of nanometers and lengths up to 10 μm. Modification of the composition of source materials and geometry of the substrates can create complex oxide nanostructures.

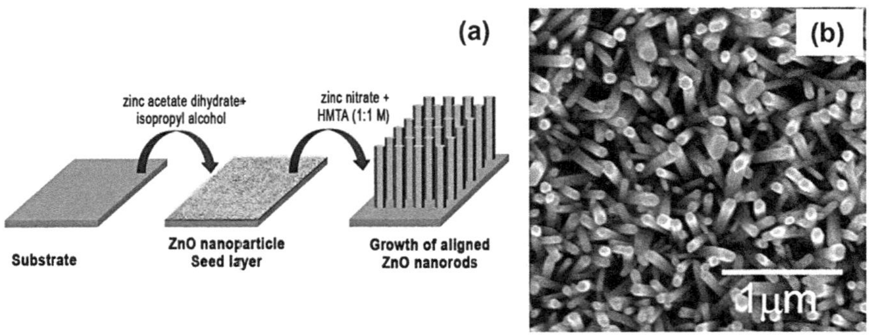

FIGURE 5.3 (a) Hydrothermal growth mechanism of ZnO nanorods and (b) typical SEM micrograph of hydrothermally grown ZnO nanorods on fluorine-doped tin oxide-coated glass substrates. (b: Reproduced with permission from reference 13. © 2016 American Chemical Society.)

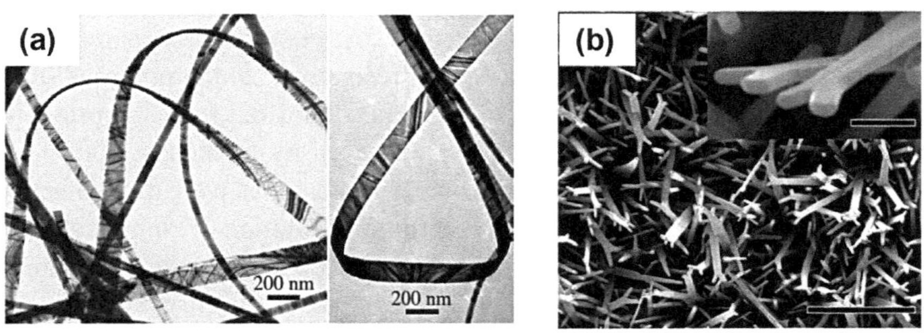

FIGURE 5.4 (a) Transmission electron microscopy micrographs of ZnO nanobelts[70] and (b) SEM micrograph of as-synthesized twinned ZnO nanobelts. (a: Reproduced with permission from reference 70. © 2011 Elsevier Ltd.)

Hierarchical ZnO complex nanostructure can be synthesized from the vapor phase of a mixture of ZnO and SnO_2 powders of equal weight ratio (1:1).[65] The detail growth mechanism of such nanostructures was reported by several researchers.[65,72,73] At sufficiently high temperature, SnO_2 decomposes into Sn and O_2 and the Sn catalyst particles are responsible for initiating and leading the growth of ZnO nanostructures. Such growth process is dominated by VLS mechanism. Figure 5.5a and b show the SEM micrographs of hexagonal and preferentially oriented three-dimensional ZnO tripods and tetrapods grown on p-Si (100) substrates by a simple vapor–solid technique without using any catalysts. The detailed process was reported by Mondal et al.[74]

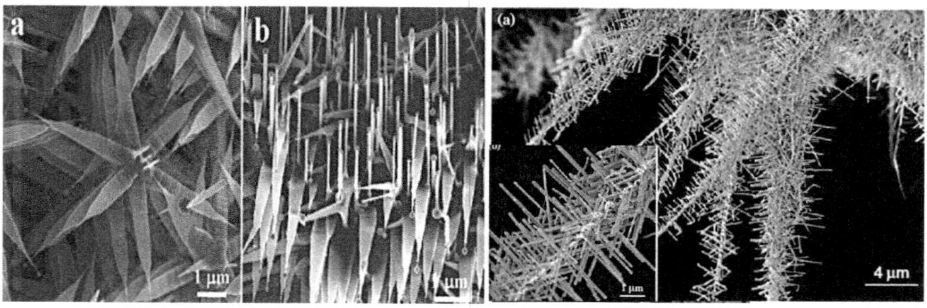

FIGURE 5.5 FE-SEM micrographs of ZnO (a) tripods and (b) tetrapods grown by the VS technique. (Reproduced with permission from reference 73. © 2009 AIP Publishing.)

"Comb-like" nanostructures structures of ZnO have been reported by several researchers.[77,78] The nanocombs shown in Figure 5.6a are random in nature. However, Figure 5.6b shows well-arranged equal spaced parallel comb-like structure made of a periodic array of rectangular ZnO nanobelts of width 400 nm and spacing of 700 nm.

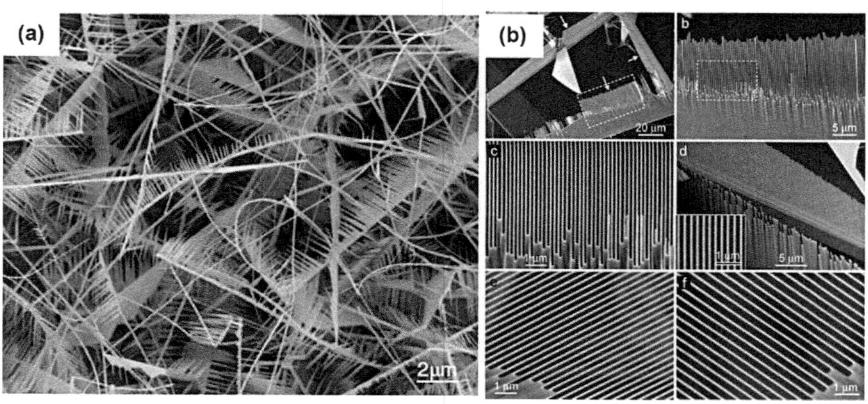

FIGURE 5.6 (a) SEM image of comb-like ZnO nanostructures formed due to the results of surface polarization-induced growth[75] and (b) SEM images of ZnO nanocombs synthesized by evaporating ZnO powder at 1400°C for 2 h. (a: Reproduced with permission from reference 75. © 2003 American Physical Society. b: Reproduced with permission from reference 76. © American Chemical Society.)

Figure 5.7a and b shows the SEM micrographs of ZnO nanosprings and nanorings grown by solid–vapor deposition process. For the synthesis of nanospring, pure ZnO powder loaded in an alumina boat was evaporated at high temperature (~1390°C) in an alumina tube.[79] The pressure was maintained at 250–300 mbar and the argon carrier gas flow rate was maintained 50 sccm. The ZnO nanosprings were deposited on a polycrystalline Al_2O_3 substrate in the temperature range of 500–600°C. Figure 5.7a shows SEM micrographs of ZnO nanosprings with a yield of > 50%, distributed densely on a substrate. The typical dimensions of an individual nanospring are 1–4 μm in diameter, 1–3 μm in pitch distance, and 10 μm long. The thickness and width of the coiled nanobelt are 10 nm and 0.2–1 μm, respectively. Figure 5.7b represents the different configurations of nanoloops or nanorings which are formed by rolling the nanobelts circularly. The formation of the nanoring structures can be explained by the polar nature of ZnO nanobelt surfaces. The polar nanobelt usually grows along $(10\bar{1}0)$ plane, with side surfaces ± $(1\bar{2}10)$ and the directions of top/bottom surfaces are along ± (0001). The nanobelt has two different polar charges on its top and bottom surfaces. Usually in a nanobelt, Zn-terminated (0001) planes (top surface) are positively charged and O-terminated (0001) planes (bottom surface) are negatively charged. The schematic representation of nanoloop or nanoring formation is presented in Figure 5.7b.

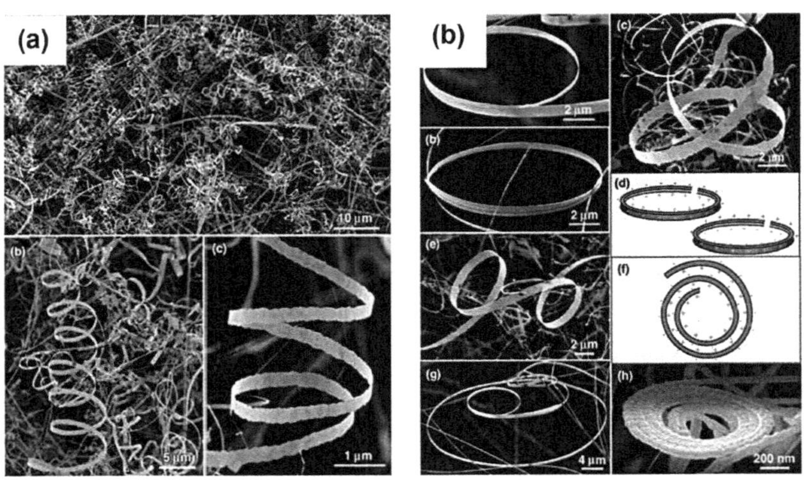

FIGURE 5.7 (a) SEM micrographs ZnO nanosprings and (b) SEM micrographs of ZnO nanorings formed at the ends of nanobelts. The schematic models of the nanoloops/nanorings formation are also presented. (Reproduced with permission from reference 79. © 2003 American Chemical Society.)

The growth of ZnO nanostructures of various shapes and sizes using solution-based approach has gained much attention compared to other techniques due to the low cost and easy synthesis process. The morphological and structural characteristics of the ZnO nanostructures can be monitored by adjusting the growth parameters, such as the chemical precursors, stoichiometry, temperature, and pH of the solutions. Various controllable ZnO nanoarchitectures, such as flower-like structures, nanocorals, nanopencils, nails or towers, and radialized bundled tubular structures can be synthesized.[80–83] Figure 5.8 shows the SEM images of flower-like ZnO nanostructures grown by chemical methods.

FIGURE 5.8 SEM micrographs of ZnO nanoflowers grown by chemical methods. (Reproduced with permission from reference 80. © 2010 Elsevier B.V.)

5.3 ZnO NANOSTRUCTURE–POLYMER COMPOSITE

Polymer nanocomposites are the composites consisting of a polymer matrix filled with nanosized metal or semiconductor particles.[51–56] Nanocomposites are especially interesting due to their high stability, enable control of particle size by controlling of their nucleation, surface passivated, and polymer matrix prevents both oxidation and coalescence. The influence of polymer matrix plays a major role in the optical and electrical properties, especially for semiconductor materials. In recent years, significant progress has been made in the synthesis of various types of ZnO nanostructure–polymer nanocomposites for the fabrication of electronic and optoelectronic devices.[51–56] Several investigations have been carried out for the synthesis of ZnO nanostructures embedded in conducting as well insulating polymer matrices. ZnO nanostructures embedded in conducting polymer matrices have been studied for LEDs, solar cells, photodetectors, memories, and sensing devices.[13,20–23,40,41,84,85] On the other hand, ZnO nanostructures grown or embedded in insulating matrices are particularly interesting for high dielectric materials, peizoelectronic devices, charge storage, and sensing applications.[42,52,53] Here, we focused the growth and characteristics of ZnO nanostructures encapsulated on insulating polymer matrices for several important applications.

5.3.1 ELECTRONIC DEVICE APPLICATIONS

ZnO NPs embedded in the insulating polymer matrix are attractive for resistive switching memory and floating gate memory applications. Pradhan et al.[84] studied electrical bistability in thin films of ZnO NPs embedded in an insulating polymer matrix. ZnO NPs, capped with PMMA, were grown using a chemical process from zinc acetate and PMMA solutions. In such reaction, insulating polymer restricts the growth of ZnO NPs and regulates the size of them.

Figure 5.9a represents the SEM micrographs of ZnO NPs encapsulated in PMMA matrix. For electrical measurements, ZnO NPs were dispersed in chloroform solution of PMMA and spin coated on strip indium tin oxide (ITO)-coated substrates with a spin speed of 1000 rpm. By keeping the concentration of the solution fixed (2 mg/ml), ZnO NPs concentration in the PMMA solution was varied up to 60 wt.%. Aluminum (Al) metal electrodes were vacuum evaporated on top of the films. The active area of the devices was 6 mm^2. Figure 5.9b shows the typical I–V curve of the device of sweeping voltage between ±2.0 V. The characteristics in the reverse bias show electrical bistability nature which is associated with a memory phenomenon. Figure 5.9c shows the switching behaviors of PMMA-, ZnO–PMMA- (25%), and ZnO–PMMA (50%)-based devices. Interestingly, bare PMMA device did not show any change in conductivity during the voltage sweep. On the other hand, the devices with ZnO NPs demonstrated electrical bistability with a reproducible change between a high and a low conducting state. It has been also observed that the on/off ratio at a particular voltage depends on ZnO concentration. The study demonstrated ZnO–PMMA nanocomposite films for random access memory applications under "write–read–erase–read" voltage sequence.

Kim et al.[85] demonstrated floating gate memory device structure using ZnO NPs grown in polyimide (PI) matrix. To grow the ZnO NPs, at first, 10-nm thick Zn was deposited on a p-type Si substrate and on SiO_2/Si with a SiO_2 layer thickness of 13.5 nm. Then, 50-nm thick polyamic acid was spin coated on the Zn film. To form ZnO NPs, the PI/Zn layer was cured at 300–400°C.

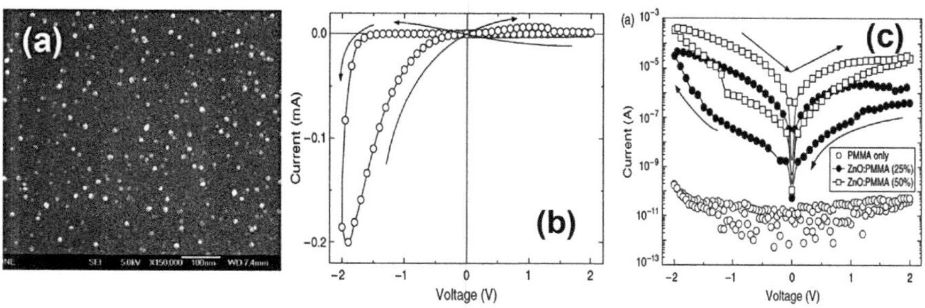

FIGURE 5.9 (a) SEM micrographs ZnO nanoparticles embedded in poly(methyl methacrylate) (PMMA) matrix, (b) current–voltage (I–V) characteristics of the sample made of thin films of PMMA: ZnO (30%), and (c) I–V characteristics of three devices: only PMMA, ZnO–PMMA (25%) and ZnO–PMMA (50%).[84]

Figure 5.10 shows the TEM image of ZnO NPs in the PI matrix with an average size of approximately 6 nm. Figure 5.10b shows the cross-sectional-view TEM image of ZnO NPs embedded in PI layer. Figure 5.10c represents the capacitance versus voltage characteristics (C–V) of the samples grown on SiO_2/p-Si substrates. Inset of the Figure 5.10c shows the typical device structure of floating gate memory device. The hysteresis in C–V characteristics indicates the memory behavior which arises due to the storage of charges in ZnO NPs embedded in PI matrix.

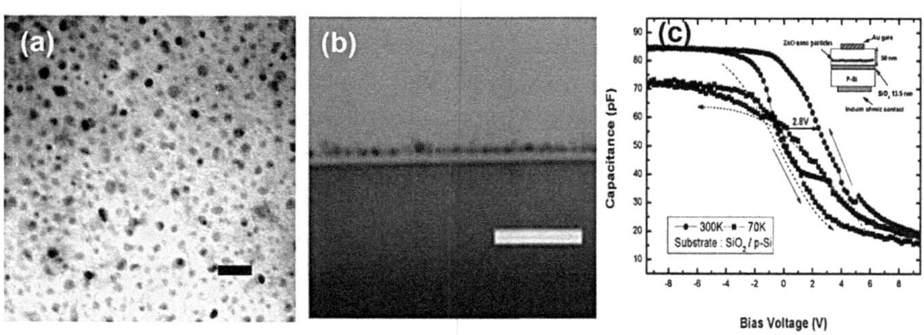

FIGURE 5.10 (a) Planar-view transmission electron microscopy (TEM) images of ZnO nanoparticles in polyimide matrix after curing at 400°C for 1 h. (b) Cross-sectional view of ZnO nanoparticles in polyimide matrix. (c) C–V curves of floating gate capacitors with ZnO nanoparticles measured at 70 and 300vK when Zn metal was deposited with SiO_2 substrate.[85]

Kathalingam et al.[53] studied bistable switching memory devices using ZnO NPs embedded in PMMA matrix. The memory devices were fabricated by spin-coating ZnO nanoparticle–polymer layers on conducting ITO-coated glasses substrates as shown in Figure 5.11a. Figure 5.11b showed the I–V measurements of the typical polymer device demonstrating electrical bistability nature. The ON to OFF current ratio of the bistable device was found to be ~ 10^3. They also studied retention behavior (current–time response) of the fabricated devices.

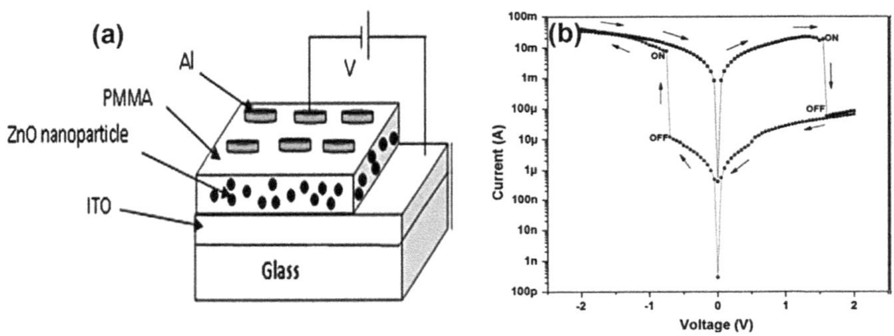

FIGURE 5.11 (a) Schematic diagram of the ZnO nanoparticles (NPs)-embedded polymer memory device. (b) Current–voltage curves of the Al/ZnO NP–PMMA/indium tin oxide (ITO) device.[53]

5.3.2 OPTOELECTRONIC DEVICE APPLICATIONS

ZnO nanostructures have excellent optoelectronic properties and have been studied for several optoelectronic device applications such as solar cell, LED, LASER, UV photodetector, and so forth. However, most of the application ZnO nanostructures such as nanorods or NPs were incorporated in conjugated polymer matrices.[21,85,86] Here, we report the optoelectronic properties of ZnO nanostructures embedded in insulating polymer matrices.

Hmar et al.[52] studied ZnO nanosheet–multiwalled carbon nanotube (ZnO–MWCNT) nanostructures for flexible, transparent, high dielectric, and photoconductive thin films for device applications. ZnO nanosheets–MWCNT nanocomposites were synthesized by a solgel chemical method. Figure 5.12 shows the typical SEM micrograph of ZnO nanosheet–MWCNT composites. The electron micrograph revealed the formation of ZnO nanosheets of thickness 20–30 nm with several micron lengths.

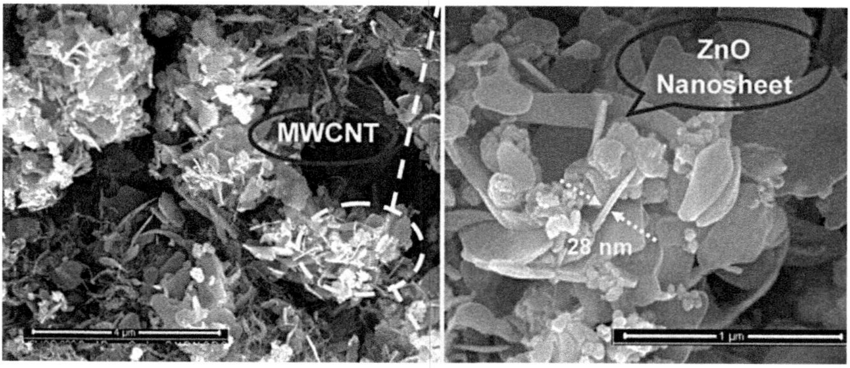

FIGURE 5.12 SEM images of ZnO nanosheets–multiwalled carbon nanotube (MWCNT) nanocomposites. (Reproduced with permission from reference 52. © 2015 Elsevier B.V.)

ZnO–MWCNT nanostructures were dispersed in PVA polymer solutions and spin coated on ITO-coated polyethylene terephthalate (PET) substrates to fabricate a flexible, transparent, high dielectric, and photoconductive thin film device. The schematic diagram of flexible ITO/ZnO–MWCNT–PVA/Al device structures is shown in Figure 5.13a. The flexibility and transparency of a typical device are presented in Figure 5.13c and b, respectively. Figure 5.13d shows the ac conductivity versus frequency plots of ITO/ZnO–MWCNT–PVA/Al devices with different volume fraction of ZnO–MWCNT nanostructures ($f_{ZnO-MWCNT}$ ~ 0.002–0.01). It has been observed, at equal volume fraction (0.2%), the low-frequency conductivity (at 40 Hz) for ZnO–PVA, ZnO–MWCNT–PVA, and pure PVA thin films were found to be 1.6×10^{-7}, 2.5×10^{-6}, and 1.1×10^{-7} Sm^{-1}, respectively. The higher DC conductivity value for ZnO–MWCNT–PVA nanocomposite samples is attributed to the formation of interconnected paths in PVA matrix.

Figure 5.14d also showed that as the volume fraction of ZnO–MWCNT increases, the low frequency conductivity also enhances. The photoconductive behaviors of the samples have been investigated under illumination of broadband light source of intensity 80 mW/cm^2. Figure 5.14a shows the transient photocurrents of MWCNT–PVA, ZnO–PVA, and ZnO MWCNT–PVA nanocomposites. Interestingly, although PVA is an insulating polymer, ZnO–PVA and ZnO–MWCNT–PVA show detectable photocurrent under illumination. The ON/OFF ratios (J_{light}/J_{dark}) for ZnO–PVA and ZnO–MWCNT–PVA nanocomposite samples were found to be 1.1 and 9.2, respectively. In MWCNT–ZnO–PVA composites, carbon nanotubes create favorable conducting paths between ZnO nanostructures in polymer matrix, which results superior photoconductive behaviors (Fig. 5.14b). Figure 5.14c shows transient photocurrents of ITO/ZnO–MWCNT–PVA/Al samples at various volume fractions of ZnO–MWCNT nanostructures. The variation of J_{light}/J_{dark} ratio with $f_{ZnO-MWCNT}$ (Fig. 5.14d) showed that the ratio increases up to $f_{ZnO-MWCNT}$ ~0.009 which is close to the percolation threshold and beyond that value, it decreases. Such photoconductive behavior was observed due to the high conductive nature of nanocomposites near percolation threshold.

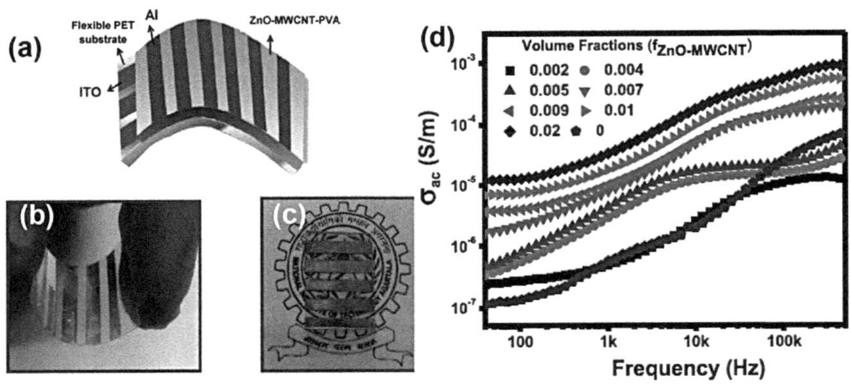

FIGURE 5.13 **(See color insert.)** (a) Schematic representation of ITO/ZnO-MWCNT-polyvinyl alcohol (PVA)/Al flexible device. Digital photograph of a typical device demonstrating (b) flexibility and (c) transparency (d) real part of ac conductivity (σ_{ac}) versus frequency plot at different $f_{ZnO-MWCNT}$. (Reproduced with permission from reference 52. © 2015 Elsevier B.V.)

5.3.3 PIEZOTRONIC DEVICE APPLICATIONS

Recently, ZnO nanostructures have been used in piezotronic applications, which open a new research area in science and engineering[42,51,53]. ZnO has a highly stable non-centrosymmetric hexagonal wurtzite structure. Such unique crystal structure creates relatively large piezoelectric coefficient, high modulus of elasticity, and high piezoelectric tensor.[42] Harvesting electrical energy from mechanical energy using piezoelectric nanomaterials provides a new opportunity for advancement of self-powered systems. Such technology can harvest energy from the sources around us such as ambient mechanical vibrations, noise, and human movement, and so forth and convert it to electric energy using the piezoelectric effect.

Several researchers have been working on piezoelectric NGs using ZnO nanostructures. Here, we reviewed some research work based on ZnO nanostructure/insulating polymer composites for piezotronic device applications.

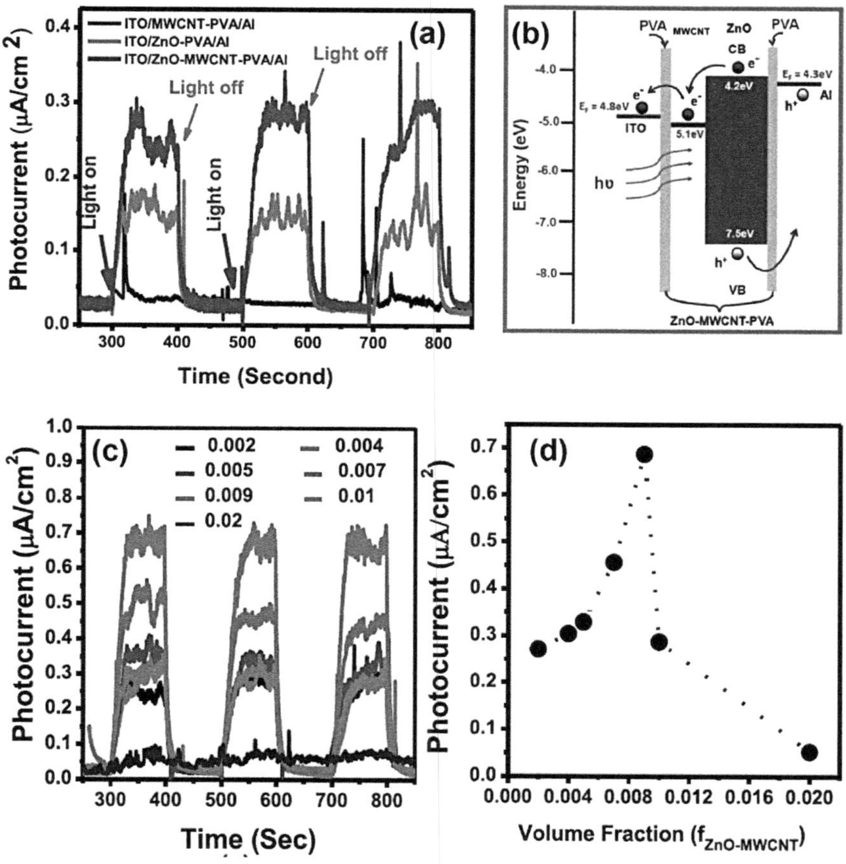

FIGURE 5.14 **(See color insert.)** (a) Transient photocurrents of ITO/MWCNT–PVA/Al, ITO/ZnO–PVA/Al, and ITO/ZnO–MWCNT-PVA/Al samples under illumination of broad band light of intensity 80 mW/cm^2. The volume fraction of ZnO–MWCNT, CNT, and ZnO in PVA are equal to 0.2%. (b) Carrier transport mechanism in ITO/ZnO–MWCNT–PVA/Al device. (c) Transient photocurrents of ITO/ZnO–MWCNT–PVA/Al devices at various $f_{ZnO-MWCNT}$ under illumination. (d) Change of photocurrent versus $f_{ZnO-MWCNT}$ plots. (Reproduced with permission from reference 52. © 2015 Elsevier B.V.)

Cauda et al.[88] studied flexible NGs using various ZnO nanostructures/PDMS composite films. Various ZnO microstructures such as microwires, multipods, and desert roses were synthesized through hydrothermal route. To study the piezotronic effect, these three different microparticles and round-shaped commercial NPs were dispersed into PDMS with filler concentration of 20, 30, 40, and 50 wt.%.

Figure 5.15 shows SEM micrographs of ZnO microstructure–PDMS composites and ZnO microstructures (inset). The device structure of ZnO–PDMS nanogenerator and photograph of the actual device are represented in Figure 5.16a. The piezoelectric voltage generation (peak-to-peak output voltage *Vpp*) of ZnO–PDMS composites of various weight percentages of ZnO microstructures is shown in Figure 5.16b. The highest performances was obtained with ZnO microwires of a 40 wt.% concentration in PDMS matrix and the maximum output voltage and power was found to be 10 V and 55 mW, respectively.

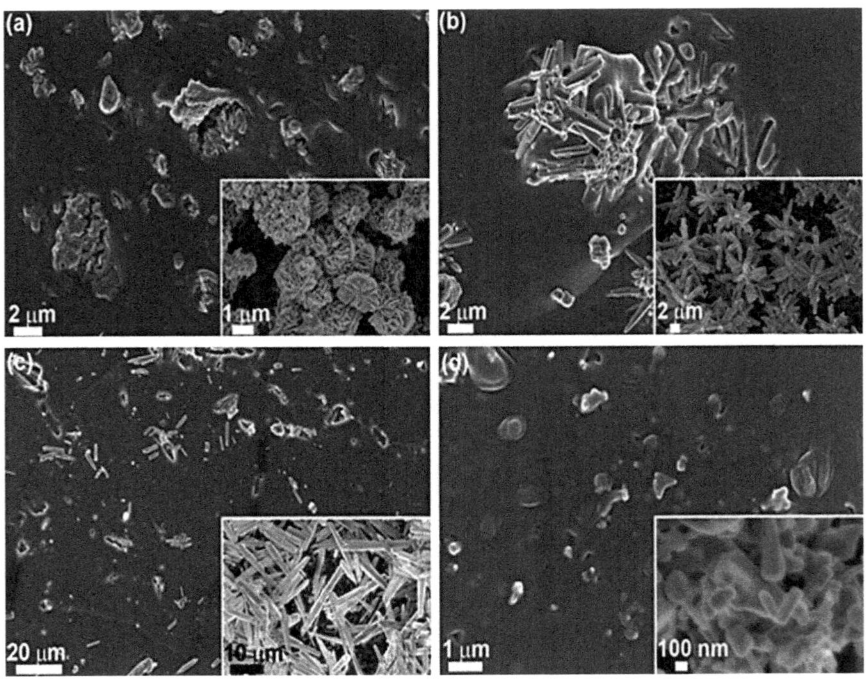

FIGURE 5.15 SEM micrographs of ZnO–polydimethylsiloxane (PDMS) composites with different ZnO morphologies of 40 wt.%: (a) desert roses; (b) multipods; (c) microwires; and (d) commercial powder. ZnO microparticles prior to the incorporation in the PDMS matrix are presented in the inset. (Reproduced with permission from reference 88. © 2015 Elsevier Ltd.)

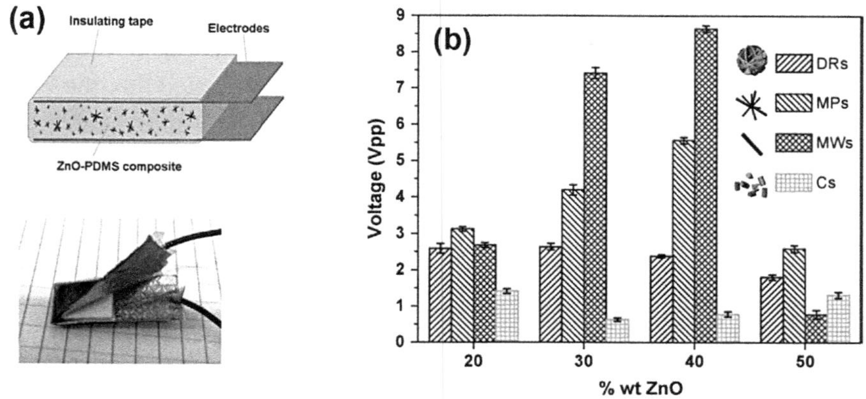

FIGURE 5.16 (a) Schematic representation of ZnO–PDMS nanogenerator (top) and digital photograph of the actual device (bottom) (b) Peak-to-peak output voltages of the ZnO–PDMS devices obtained after compression. The devices were fabricated with four different ZnO structures at different ZnO filler concentrations (20, 30, 40, and 50 wt.%). All measurements were performed at 75 Hz of oscillating frequency and 0.2 mm peak-to-peak displacement. (Reproduced with permission from reference 88. © 2015 Elsevier Ltd.)

Figure 5.17a shows the electrical output waveforms of ZnO microwire–PDMS composites (ZnO ~ 40 wt.%) at various oscillating frequencies. The waveforms were measured by a mechanical displacement of 0.2 mm to the moving stage of the shaker and increasing the oscillating frequency between 10 and 100 Hz. The NGs were also exploited to charge a home-made carbon-based supercapacitor. Figure 5.17b shows the charging voltage–time curves of ZnO MWs–PDMS composite NG, at 40 wt.% of filler content, under compressive force with an oscillating frequency of 75 Hz. They observed large capacitance with ZnO MWs-based NGs of capacitance value 90 mF and the supercapacitor was charged from 0 to 4 mV within 60 s.

From the previous discussion, it has been observed that ZnO has been demonstrated as one of the most promising materials for piezoelectric energy harvesting materials due to its environmentally safe nature, abundance, mechanical robustness, and low cost. ZnO has a hexagonal wurtzite crystal structure and inherently reveals piezoelectric property along the c-axis. However, such piezoelectric property can be enhanced by doping with elements such as lithium (Li) which can induce the ferroelectric phase transition in ZnO. Shin et al.[42] studied flexible piezoelectric NGs using composite solution prepared by mixing hydrothermally grown

Li-doped ZnO NWs and PDMS polymer. They have prepared a composite solution by mixing hydrothermally grown Li-doped ZnO NWs and PDMS polymer. The solution was spin coated on ITO-coated flexible PET substrates followed by curing and high voltage (0–105 kV) poling steps to fabricate the samples.

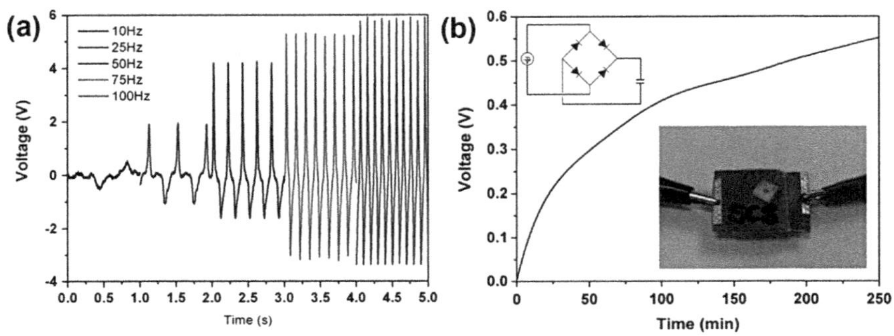

FIGURE 5.17 (a) Piezoelectric output voltage generations at various frequencies of the ZnO microwire–PDMS composites (ZnO ~ 40 wt.%). (b) Charging curve of the supercapacitor made by the ZnO–MWs composite nanogenerators (NG) with filler amount of 40 wt.%. The equivalent circuit diagram of NG coupled to the supercapacitor system and the typical photograph of the carbon-based supercapacitor device are shown at the upper and lower inset, respectively. (Reproduced with permission from reference 88. © 2015 Elsevier Ltd.)

Figure 5.18a shows the schematic diagram of the Li-doped ZnO NWs/PDMS polymer composite NG device fabrication process. The digital photograph of the actual device, SEM micrographs of the nanorods, and SEM cross-section view of the device are shown in Figure 5.18b, c, and d, respectively.

Figure 5.19a and b showed the piezoelectric output voltage generation for intrinsic ZnO NW-based NG and Li-doped ZnO NWs-based NGs. In the case of undoped ZnO NWs, the polarization in ZnO NWs inside the PDMS matrix is random before and after the poling process as schematically shown in Figure 5.19c. It is evident that in undoped ZnO NWs, the poling process has no impact on the piezoelectric output voltage. Interestingly, Li-doped ZnO NWs-based NGs demonstrated significantly enhanced output voltage after poling process (Fig. 5.19b). The superior piezoelectric output voltage generation after poling process in Li-doped samples is attributed to the alignment of randomly oriented electric dipoles and spontaneous polarization in ZnO NWs in the PDMS matrix.

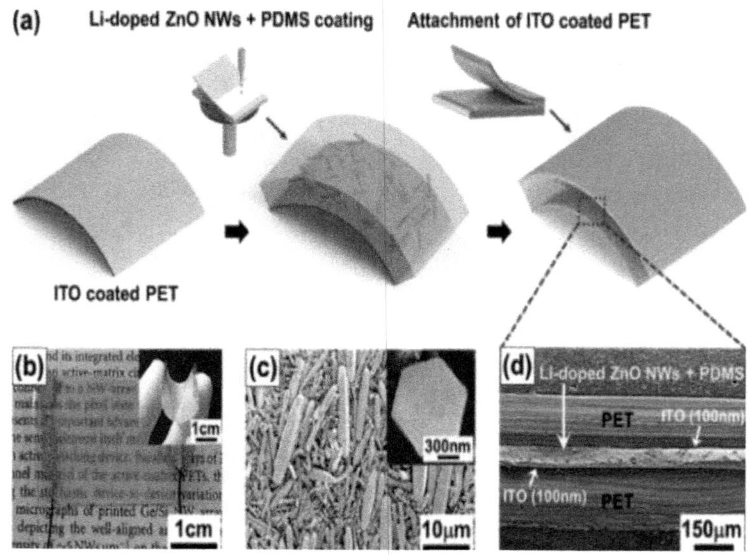

FIGURE 5.18 (a) Schematic representation of the NG fabrication process. (b) Digital photograph of the fabricated NG sample demonstrating semitransparency and flexibility (inset). (c) SEM image of Li-doped ZnO NWs. The magnified mage at the inset shows hexagonal-shaped NRs. (d) Cross-sectional SEM image of a typical NG device. (Reproduced with permission from reference 42. © 2014 American Chemical Society.)

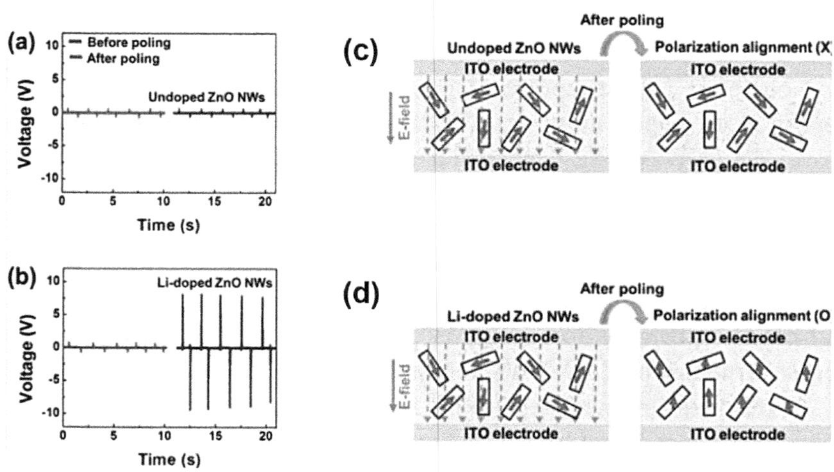

FIGURE 5.19 **(See color insert.)** Piezoelectric output voltage generations (a) undoped and (b) Li-doped NGs. The schematic diagram represents the polarization of NWs inside PDMS before/after poling process for (c) undoped and (d) Li-doped ZnO NWs. (Reproduced with permission from reference 42. © 2014 American Chemical Society.)

Such Li-doped ZnO NWs/PDMS polymer composite NG devices also demonstrated excellent durability and stability over a long period of observation. The piezoelectric output voltage of the devices was further enhanced by enlarging the size of the NGs. Figure 5.20a and b show the output voltage and output current of a NG of area 10 cm^2. The maximum output voltage and the current were found to be 180 V and 50 μA, respectively.

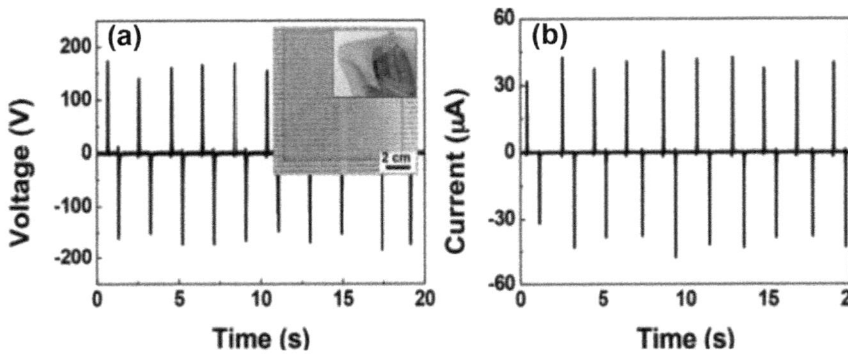

FIGURE 5.20 (a) Output voltage versus time plot and photograph of a fabricated large-scale NG with an area of 10 cm × 10 cm (inset). (b) Output current versus time plots of a large-scale NG device of are 10 cm × 10 cm. (Reproduced with permission from reference 42. © 2014 American Chemical Society.)

Nour et al.[54] have reported ZnO NWs/polymer composite-based NG deposited on paper substrates. To fabricate the device, first, they have grown ZnO NWs on Ag-deposited paper substrates by a chemical process. Afterward, ZnO NWs/PVDF polymer composite ink was pasted and sandwiched between two pieces of paper with ZnO NWs on one side of each piece of paper. They have also studied piezoelectric output voltage/current by utilizing a ZnO/PVDF ink with different ZnO nanostructures on paper and plastic substrates for comparison. Figure 5.21 shows the schematic device structure of ZnO NWs/polymer samples and digital photographs of various measurement steps of the samples. The maximum open circuit voltage for paper-based films was found to be 4.8 V. On the other hand, the open circuit output voltage from plastic-based NG has reached a maximum value of 0.04 V. The study demonstrated that the porous and soft nature of paper substrate improved the output voltage by a factor of 100 compared to the plastic.

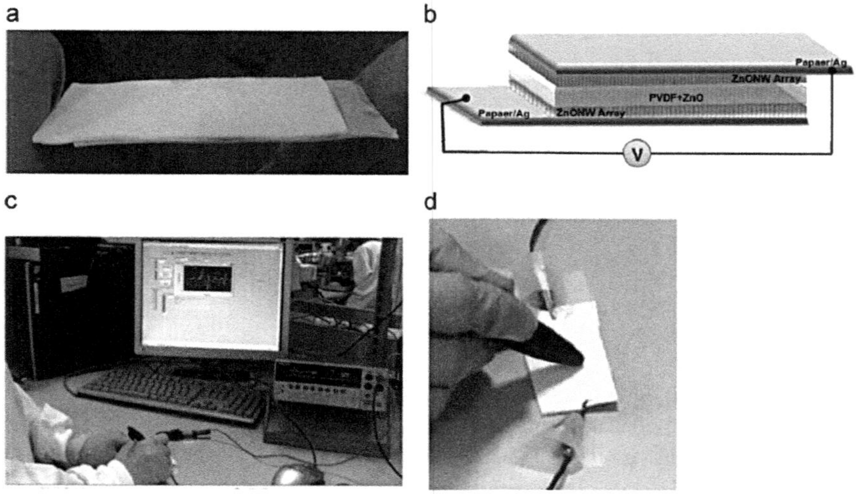

FIGURE 5.21 (a) Typical digital photograph the as fabricated NG sample and (b) schematic representation of the NG device. (c) and (d) Photographs of the electrical measurements of the samples. (Reproduced with permission from reference 54. © 2014 Elsevier Ltd.)

They have observed highest harvested electrical output power using a ZnO NWs/PVDF polymer ink pasted and sandwiched between two pieces of paper with ZnO NWs. Such configuration delivered piezoelectric output voltage as high as 4.8 V and a current 14.4 mA. The maximum output power density was obtained ~1.3 mW/mm^2 for fast handwriting mode (200–240 letters/min).

Li et al.[55] have studied ZnO NWs–PVDF hybrid nanogenerator by in situ synthesis of ZnO NWs in PVDF polymer matrix by a simple chemical process. Figure 5.22a shows the schematic representation of ZnO NWs growth in PVDF matrix and NG device fabrication. The corresponding SEM micrograph of the sample is shown in Figure 5.22b. For electrical characterization, two silver tapes were applied on the bottom of the PVDF surface and the top of the ZnO NWs as electrodes. In the optimized sample, the maximum piezoelectric output voltage of 3.2 V was observed. Figure 5.22c shows the circuit diagrams of the liquid crystal display (LCD) driven by the hybrid nanogenerator under application of forces. The study demonstrated that after touching the NG with finger pressure, ZnO NWs directly convert mechanical motion into electronic potential and LCD works.

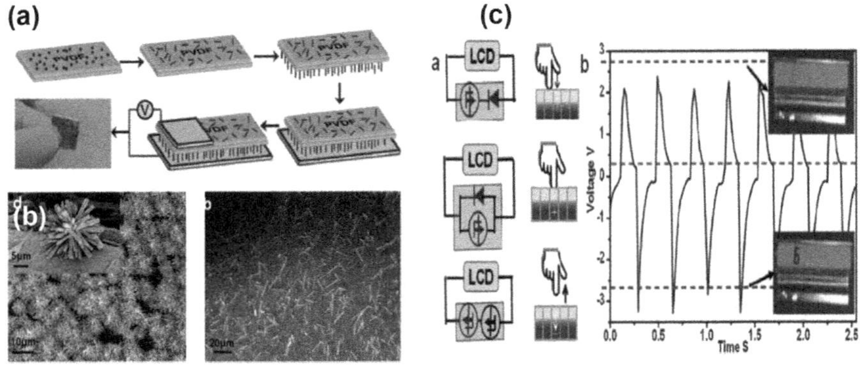

FIGURE 5.22 (a) Schematic diagram for the growth of ZnO NWs–polyvinylidene fluoride (PVDF) hybrid nanogenerator sample. (b) SEM micrographs of ZnO NWs-PVDF samples. (c) Demonstration of NG performance using a LCD display. (Reproduced with permission from reference 55. © 2014 Royal Society of Chemistry.)

5.3.4 DIELECTRIC MATERIALS FOR ELECTRONIC DEVICES

High dielectric constant polymer composites are attractive for several electronic device applications such as capacitors, electrical insulators, and others electrical equipment because of their flexibility, easy synthesis process, and low fabrication cost. ZnO nanostructures are attractive filler material to enhance the dielectric constant of the insulating polymer matrix. Wu et al.[89] reported high dielectric constant polymer composites by using three-dimensional zinc oxide superstructures as fillers.

Flower-like and walnut-like ZnO nanostructures (Fig. 5.23a and b) were prepared through a template-free solvothermal method. The dielectric properties of ZnO nanostructures/PVDF composites and commercial ZnO powder/PVDF composite were investigated by a broadband dielectric spectroscopy. Their results showed that the ZnO nanostructures/PVDF composites showed significantly high dielectric constant than ZnO commercial powder-based composites. Figure 5.24 shows the dielectric constant versus frequency plots of PVDF, PVDF/commercial ZnO, PVDF/flower-like ZnO, and PVDF/walnut-like ZnO samples.

Figure 5.24b shows the breakdown strength of various samples. From their study, the dielectric constants of the composite samples at 100 Hz for commercial ZnO, walnut-like ZnO, and flower-like ZnO nanostructures are 19.4, 104.9, and 221.1, respectively. The observed breakdown strengths of the above samples were found to be 45, 40, and 42 kV/mm, respectively.

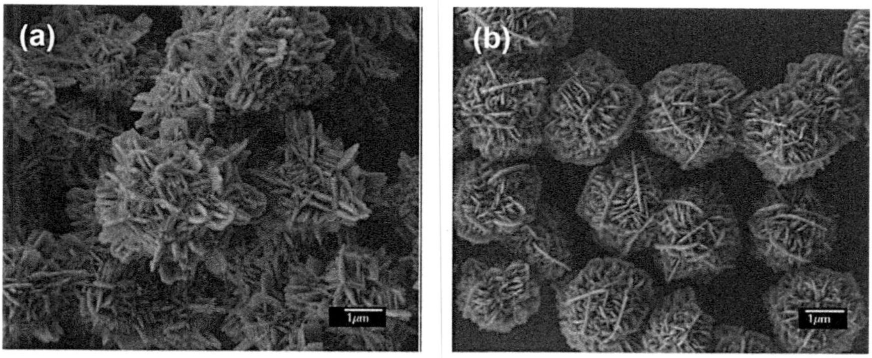

FIGURE 5.23 SEM images of (a) flower-like and (b) walnut-like ZnO structures grown by template-free solvothermal method. (Reproduced with permission from reference 89. © 2012 American Chemical Society.)

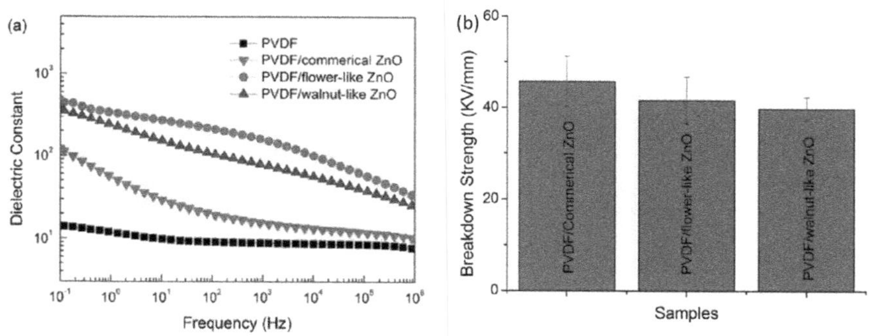

FIGURE 5.24 (See color insert.) (a) Frequency-dependent dielectric constants of flower-like and walnut-like ZnO structures and (b) breakdown strength of PVDF composites. (Reproduced with permission from reference 89. © 2012 American Chemical Society.)

Hmar et al.[52] studied the dielectric properties of ZnO–MWCNT nanostructure–PVA nanocomposites. The nanostructures were grown by hydrothermal process. Figures 5.12 and 5.13a show the ZnO–MWCNT nanostructures and the schematic representation of nanocomposite device, respectively. Figure 5.25a shows the dielectric permittivity (ε′) versus frequency dispersion spectra of ZnO–PVA, MWCNT–PVA, and ZnO–MWCNT–PVA nanocomposites. The dielectric dispersion curve of host PVA matrix is also plotted in Figure 5.25b. The volume fraction of ZnO nanostructures, MWCNT, and ZnO–MWCNT nanostructures was kept at constant value and equal to 0.2%. It was observed that ZnO–MWCNT–PVA composite demonstrated significant enhancement of dielectric constant over ZnO–PVA, MWCNT–PVA, and PVA thin films (Fig. 5.25a). The low-frequency dielectric constant (~40 Hz) of PVA polymer, MWCNT–PVA, ZnO–PVA, and ZnO–MWCNT–PVA nanocomposite films are 20, 29, 45, and 67, respectively. The superior dielectric behavior of ZnO–MWCNT–PVA nanocomposite was explained by the combined polarization effects of MWCNT and ZnO nanostructures in PVA matrix. The interfacial areas between ZnO nanosheets–MWCNT and ZnO nanosheets–PVA are also responsible for enhancement of dielectric constant of composite material. Figure 5.25b shows the variation of low frequency dielectric constant (at 40 Hz) of ZnO–MWCNT–PVA thin film devices with different volume fraction of ZnO–MWCNT nanostructures. It has been observed that the dielectric permittivity increases with increasing $f_{ZnO\text{–}MWCNT}$ and attained maximum value at $f_{ZnO\text{–}MWCNT}$ ~0.009 with dielectric permittivity ε′~266. Beyond $f_{ZnO\text{–}MWCNT}$ ~0.009, the dielectric permittivity decreases rapidly due to the increase of conductivity of the medium.[52] They have also studied the flexibility of the dielectric thin films.

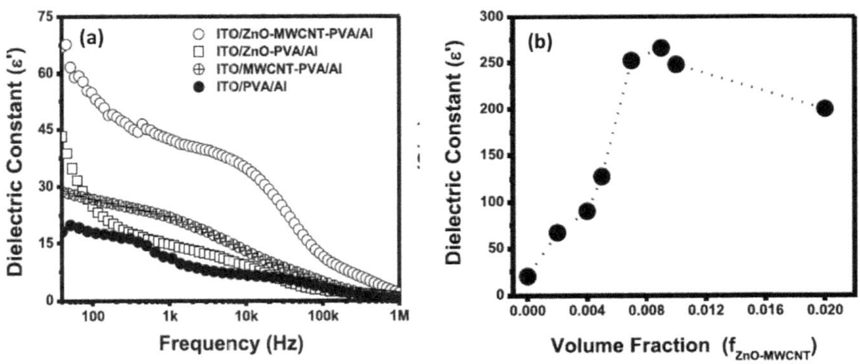

FIGURE 5.25 (a) Variation of real part of dielectric constant ε′ with frequency for ZnO–MWCNT–PVA, MWCNT–PVA, ZnO–PVA, and PVA thin film devices and (b) plot of ε′ versus volume fraction of ZnO–MWCNT at frequency ~40 Hz. (Reproduced with permission from reference 52. © 2015 Elsevier B.V.)

5.4 SUMMARY AND CONCLUSIONS

The major advantages of ZnO are direct wide bandgap (3.37 eV), large excitation binding energy (60 meV), noncentral symmetry which results piezoelectric properties, and overall bio-safe and biocompatibility. Here, we have presented some literature focused on the device applications of ZnO nanomaterials encapsulated in dielectric polymer matrices. ZnO nanorods/nanostructures embedded in PDMS and PVDF polymers demonstrated enhanced piezoelectric properties which are very attractive for harvesting electrical energy from mechanical forces. On the other hand, ZnO NPs grown in PMMA matrix showed charge storage behaviors which are essential for memory applications. ZnO flower-like nanostructures/PVDF films showed high dielectric properties and high electrical breakdown strengths. ZnO nanosheets–MWCNT composite nanostructures embedded in PVA polymer showed high dielectric properties as well as good photoconductive behaviors. The literature study demonstrated the importance of ZnO–dielectric polymer composite films for novel electronic and optoelectronic device applications.

KEYWORDS

- **ZnO nanostructure**
- **polymer nanocomposite**
- **poly(methyl methacrylate)**
- **polyvinylidene fluoride**
- **ZnO nanowires**

REFERENCES

1. Lin, B.; Fu, Z.; Jia, Y. Green Luminescent Center in Undoped Zinc Oxide Films Deposited on Silicon Substrates. *Appl. Phys. Lett.* **2001,** *79*, 943.
2. Srikant, V.; Clarke, D. R. On the Optical Band Gap of Zinc Oxide. *J. Appl. Phys.* **1998,** *83*, 5447.
3. Bagnall, D. M.; Chen, Y. F.; Zhu, Z.; Yao, T.; Shen, M. Y.; Goto, T. High Temperature Excitonic Stimulated Emission from ZnO Epitaxial Layers. *Appl. Phys. Lett.* **1998,** *73*, 1038.
4. Wu, J. J.; Liu, S. C. Low-Temperature Growth of Well-Aligned ZnO Nanorods by Chemical Vapor Deposition. *Adv. Mater.* **2002,** *14*, 215.
5. S. Desgreniers. High-density Phases of ZnO: Structural and Compressive Parameters. *Phys. Rev. B* **1998,** *58*, 14102.

6. Umar, A.; Hahn, Y. B. ZnO Nanosheet Networks and Hexagonal Nanodiscs Grown on Silicon Substrate: Growth Mechanism and Structural and Optical Properties. *Nanotechnology* **2006,** *17*, 9.
7. Guo, L.; Ji, Y. L.; Xu, H. Regularly Shaped, Single-Crystalline ZnO Nanorods with Wurtzite Structure. *J. Am. Chem. Soc.* **2002,** *124*(50), 14864.
8. Ko, S. H.; Lee, D.; Kang, H. W.; Nam, K. H.; Yeo, J. Y.; Hong, S. J.; Grigoropoulos, C. P.; Sung, H. J. Nanoforest of Hydrothermally Grown Hierarchical ZnO Nanowires for a High Efficiency Dye-Sensitized Solar Cell. *Nano Lett.* **2011,** *11*(2), 666.
9. Liao, S. H.; Jhuo, H. J.; Yeh, P. N.; Cheng, Y. S.; Li, Y. L.; Lee, Y. H.; Sharma S.; Chen, S. A. Single Junction Inverted Polymer Solar Cell Reaching Power Conversion Efficiency 10.31% by Employing Dual-Doped Zinc Oxide Nano-Film as Cathode Interlayer. *Sci. Rep.* **2014,** *4*, 6813.
10. Repins, I.; Contreras, M. A.; Egaas, B.; DeHart, C.; Scharf, J.; Perkins, C. L.; To, B.; Noufi, R. 19·9%-Efficient ZnO/CdS/$CuInGaSe_2$ Solar Cell with 81·2% Fill Factor. *Prog. Photovoltaics: Res. Appl.* **2008,** *16*, 235.
11. Majumder, T.; Debnath, K.; Dhar, S.; Hmar, J. J. L.; Mondal, S. P. Nitrogen-Doped Graphene Quantum Dot-Decorated ZnO Nanorods for Improved Electrochemical Solar Energy Conversion. *Energy Technol.* **2016,** *4*(8), 950.
12. Majumder, T.; Mondal, S. P. Advantages of Nitrogen-Doped Graphene Quantum Dots As A Green Sensitizer with ZnO Nanorods-Based Photoanodes for Solar Energy Conversion. *J. Electroanal. Chem.* **2016,** *769*, 48.
13. Dhar, S.; Majumder, T.; Mondal, S. P. Graphene Quantum Dot-Sensitized ZnO Nanorod/Polymer Schottky Junction UV Detector with Superior External Quantum Efficiency, Detectivity, and Responsivity. *ACS Appl. Mater. Interfaces* **2016,** *8*(46), 31, 822.
14. Hmar, J. J. L.; Majumder, T.; Dhar, S.; Mondal, S. P. Sulfur and Nitrogen Co-doped Graphene Quantum Dot Decorated ZnO Nanorod/Polymer Hybrid Flexible Device For Photosensing Applications. *Thin Solid Films* **2016,** *612*, 274.
15. Guo, D. Y.; Shan, C. X; Qu, S. N; Shen, D. Z. Highly Sensitive Ultraviolet Photodetectors Fabricated from ZnO Quantum Dots/Carbon Nanodots Hybrid Films. *Sci. Rep.* **2014,** *4*, 7469.
16. Dai, J.; Xu, C.; Xu, X.; Guo, J.; Li, J.; Zhu, G.; Lin, Y. Single ZnO Microrod Ultraviolet Photodetector with High Photocurrent Gain. *ACS Appl. Mater. Interfaces* **2013,** *5*, 9344.
17. Tang, Z. K.; Wong, G. K. L.; Yu, P.; Kawasaki, M.; Ohtomo A. Room-Temperature Ultraviolet Laser Emission from Self-Assembled ZnO Microcrystallite Thin Films. *Appl. Phys. Lett.* **1998,** *72*, 3270.
18. Ramamoorthy, K.; Arivanandhan, M.; Sankaranarayanan, K.; Sanjeeviraja, C. Highly Textured ZnO Thin Films: A Novel Economical Preparation and Approachment for Optical Devices, UV Lasers and Green LEDs. *Mater. Chem. Phys.* **2004,** *85*, 257.
19. Bagnall, D. M.; Chen, Y. F.; Zhu, Z.; Yao, T.; Koyama, S. Optically Pumped Lasing of ZnO at Room Temperature. *Appl. Phys. Lett.* **1997,** *70*, 2230.
20. Lim, J. H.; Kang, C. K.; Kim, K. K.; Park, I. K.; Hwang, D. K.; Park, S. J. UV Electroluminescence Emission from ZnO Light-Emitting Diodes Grown by High-Temperature Radio frequency Sputtering. *Adv. Mater.* **2006,** *18*, 2720.

21. Pan, C.; Dong, L.; Zhu, G.; Niu, S.; Yu, R.; Yang, Q.; Liu, Y.; Wang, Z. L. High-Resolution Electroluminescent Imaging of Pressure Distribution Using a Piezoelectric Nanowire LED Array. *Nat. Photonics* **2013,** *7*, 752.
22. Jiao, S. J.; Zhang, Z. Z.; Lu, Y. M.; Shen, D. Z.; Yao, B.; Zhang, J. Y.; Li, B. H.; Zhao, D. X.; Fan, X. W. Tang, Z. K. ZnO p–n Junction Light-Emitting Diodes Fabricated On Sapphire Substrates. *Appl. Phys. Lett.* **2006,** *88*, 31911.
23. Sun, X. W.; Huang, J. Z.; Wang, J. X.; Xu, Z. A ZnO Nanorod Inorganic/Organic Heterostructure Light-Emitting Diode Emitting at 342 nm. *Nano Lett.* **2008,** *8*(4), 1219.
24. Das, A.; Dost, R.; Richardson, T.; Grell, M.; Morrison, J. J.; Turner, M. L. A Nitrogen Dioxide Sensor Based on an Organic Transistor Constructed from Amorphous Semiconducting Polymers. *Adv. Mater.* **2007,** *19*, 4018.
25. Das, A.; Dost, R.; Richardson, T.; Grell, M.; Morrison, J. J.; Turner, M. L. A Room Temperature Ethanol Sensor Made from p-type Sb-doped SnO_2 Nanowires. *Nanotechnology* **2010,** *21*, 235501.
26. Kong, Y. C.; Yu, D. P.; Zhang, B.; Fang, W.; Feng, S. Q. Ultraviolet-Emitting ZnO Nanowires Synthesized by a Physical Vapor Deposition Approach. *Appl. Phys. Lett.* **2001,** *78*, 407.
27. Chen, C. Q.; Shi, Y.; Zhang, Y. S.; Zhu, J.; Yan, Y. J. Size Dependence of Young's Modulus in ZnO Nanowires. *Phys. Rev. Lett.* **2006,** *96*, 75505.
28. Chu, D.; Masuda, Y.; Ohji, T.; Kato, K. Formation and Photocatalytic Application of ZnO Nanotubes Using Aqueous Solution. *Langmuir* **2010,** *26*(4), 2811.
29. Martinson, A. B. F.; Góes, M. S.; Fabregat-Santiago, F.; Bisquert, J.; Pellin, M. J.; Hupp, J. T. Electron Transport in Dye-Sensitized Solar Cells Based on ZnO Nanotubes: Evidence for Highly Efficient Charge Collection and Exceptionally Rapid Dynamics. *J. Phys. Chem. A* **2009,** *113*(16), 4015–4021.
30. Yu, H.; Zhang, Z.; Han, M.; Hao, X.; Zhu, F. A. A General Low-Temperature Route for Large-Scale Fabrication of Highly Oriented ZnO Nanorod/Nanotube Arrays. *J. Am. Chem. Soc.* **2005,** *127*(8), 2378.
31. Xue, X. Y.; Chen, Z. H.; Xing, L. L.; Ma, C. H.; Chen, Y. J.; Wang, T. H. Enhanced Optical and Sensing Properties of One-Step Synthesized Pt–ZnO Nanoflowers. *J. Phys. Chem. C* **2010,** *114*(43), 18607.
32. Xing, L. L.; Ma, C. H.; Chen, Z. H.; Chen, Y. J.; Xue, X. Y. High Gas Sensing Performance of One-Step-Synthesized Pd–ZnO Nanoflowers due to Surface Reactions and Modifications. *Nanotechnology* **2011,** *22*, 215501.
33. Lao, C. S.; Park, M. C.; Kuang, Q.; Deng, Y.; Sood, A. K.; Polla, D. L.; Wang, Z. L. Giant Enhancement in UV Response of ZnO Nanobelts by Polymer Surface-Functionalization. *J. Am. Chem. Soc.* **2007,** *129*(40), 12096.
34. Ronning, C.; Gao, P. X.; Ding, Y.; Wang, Z. L.; Schwen, D. Manganese-Doped ZnO Nanobelts for Spintronics. *Appl. Phys. Lett.* **2004,** *84*, 783.
35. Hughes, W. L.; Wang, Z. L. Controlled Synthesis and Manipulation of ZnO Nanorings and Nanobows. *Appl. Phys. Lett.* **2005,** *86*, 43106.
36. Kong, X. Y.; Ding, Y.; Yang, R.; Wang, Z. L. Single-Crystal Nanorings Formed by Epitaxial Self-Coiling of Polar Nanobelts. *Science* **2004,** *303*(5662), 1348.
37. Zhang, J.; Yang, Y.; Xu, B.; Jiang, F.; Li, J. Shape-Controlled Synthesis of ZnO Nano- and Micro-Structures. *J. Cryst. Growth* **2005,** *280*, 509.

38. Sun, X.; Li, Q.; Jiangc, J.; Mao, Y. B. Morphology-Tunable Synthesis of ZnO Nanoforest and its Photoelectrochemical Performance. *Nanoscale* **2014,** *6*, 8769.
39. Saito, N.; Haneda, H.; Sekiguchi, T.; Ohashi, N.; Sakaguchi, I.; Koumoto, K. Low-Temperature Fabrication of Light-Emitting Zinc Oxide Micropatterns Using Self-Assembled Monolayers. *Adv. Mater.* **2002,** *14*(6), 418.
40. You, J.; Meng, L.; Song, T. B.; Guo, T. F.; Yang, Y.; Chang, W. H.; Hong, Z.; Chen, H.; Zhou, H.; Chen, Q.; Liu, Y.; Marco, N. D.; Yang, Y. Improved Air Stability of Perovskite Solar Cells Via Solution-Processed Metal Oxide Transport Layers. *Nat. Nanotechnol.* **2016,** *11*, 75.
41. Ambade, S. B.; Ambade, R. B.; Eom, S. H.; Baek, M. J.; Bagde, S. S.; Mane, R. S.; Lee, S. H. Co-functionalized Organic/Inorganic Hybrid ZnO Nanorods as Electron Transporting Layers for Inverted Organic Solar Cells. *Nanoscale* **2016,** *8*, 5024.
42. Shin, S. H.; Kim, Y. H.; Lee, M. H.; Jung, J. Y.; Seol, J. H.; Nah, J. Lithium-Doped Zinc Oxide Nanowires Polymer Composite for High Performance Flexible Piezoelectric Nanogenerator. *ACS Nano* **2014,** *8*(10), 10844.
43. Liao, J.; Li, Z.; Wang, G.; Chen, C.; Lv, S.; Li, M. ZnO Nanorod/Porous Silicon Nanowire Hybrid Structures as Highly-Sensitive NO_2 Gas Sensors at Room Temperature. *Phys. Chem. Chem. Phys.* **2016,** *18*, 4835.
44. Wang, J.; Li, X.; Xia, Y.; Komarneni, S.; Chen, H.; Xu, J.; Xiang, L.; Xie, D. Hierarchical ZnO Nanosheet-Nanorod Architectures for Fabrication of Poly(3-hexylthiophene)/ZnO Hybrid NO_2 Sensor. *ACS Appl. Mater. Interfaces* **2016,** *8*(13), 8600.
45. Maleka, M. F.; Sahdan, M. Z.; Mamat, M. H.; Musa, M. Z.; Khusaimi, Z.; Husairi, S. S.; Md Sin, N. D.; Rusop, M. A Novel Fabrication of MEH-PPV/Al: ZnO Nanorod Arrays-Based Ordered Bulk Heterojunction Hybrid Solar Cells. *Appl. Surf. Sci.* **2013,** *275*, 75.
46. Park, J.; Park, G.; Ko, H. J.; Ha, J. S. The Effect of ITO/Mo/MoO3 Anode Multilayer Film on Efficient Hole Extraction in MEH–PPV/ZnO NP Hybrid Solar Cells. *Ceram. Int.* **2014,** *40*(10), 16281.
47. Bilgaiyan, A.; Dixit, T.; Palani, I. A.; Singh, V. Performance Improvement of ZnO/P3HT Hybrid UV Photo-detector by Interfacial Au Nanolayer. *Physica E Low Dimens. Syst. Nanostruct.* **2017,** *86*, 136.
48. Shen, Q; Ogomi, Y.; Das, S. K.; Pandey, S. S.; Yoshino, K.; Katayama, K.; Momose, H.; Toyoda, T.; Hayase, S. Huge Suppression of Charge Recombination in P3HT–ZnO Organic–Inorganic Hybrid Solar Cells by Locating Dyes at the ZnO/P3HT Interfaces. *Phys. Chem. Chem. Phys.* **2013,** *15*, 14370.
49. Kumar, S.; Pradhan, S.; Dhar, A. Enhanced Performance with the Incorporation of Organo-Metal Trihalide Perovskite in Nanostructured ZnO Solar Cell. *Procedia Eng.* **2016,** *14*(1), 1.
50. Santhanakrishna, A. K.; Takshi, A. Photoelectric Memory Effect in an Organic Bulk Heterojunction Device. *J. Phys. Chem. C* **2015,** *119*(30), 17253.
51. Shin, S. H.; Kim, Y. H.; Lee, M. H.; Jung, J. Y.; Seol, J. H.; Nah, J. Lithium-Doped Zinc Oxide Nanowires–Polymer Composite for High Performance Flexible Piezoelectric Nanogenerator. *ACS Nano* **2014,** *8*(10), 10844.
52. Hmar, J. J, L.; Majumder, T.; Roy, J. N.; Mondal, S. P. Flexible, Transparent, High Dielectric and Photoconductive Thin Films Using ZnO Nanosheets-Multi-walled Carbon Nanotube-Polymer Nanocomposites. *J. Alloys Compd.* **2015,** *651*, 82.

53. Kathalingam, A.; Rhee, J. K. Fabrication of Bistable Switching Memory Devices Utilizing Polymer–ZnO Nanocomposites. *J. Electron. Mater.* **2012,** *41*, 82162.
54. Nour, E. S.; Sandberg, M. O.; Willander, M.; Nur, O. Handwriting-Enabled Harvested Piezoelectric Power Using ZnO Nanowires/Polymer Composite on Paper Substrate. *Nano Energy* **2014,** *9*, 221.
55. Li, Z.; Zhang, X.; Li, G. In situ ZnO Nanowire Growth to Promote the PVDF Piezo Phase and the ZnO–PVDF Hybrid Self-Rectified Nanogenerator as a Touch Sensor. *Phys. Chem. Chem. Phys.* **2014,** *16*, 5475.
56. Karthikeyan, B.; Pandiyarajan, T.; Mangalaraja, R. V. Enhanced Blue Light Emission in Transparent ZnO: PVA Nanocomposite Free Standing Polymer Films. *Spectrochim. Acta A Mol. Biomol. Spectrosc.* **2016,** *152*, 485.
57. Guo, F.; Yang, B.; Yuan, Y.; Xiao, Z.; Dong, Q.; Bi, Y.; Huang, J. A Nanocomposite Ultraviolet Photodetector Based on Interfacial Trap-Controlled Charge Injection. *Nat. Nanotechnol.* **2012,** *7*, 798.
58. Lee, Y. J.; Wang, J.; Cheng, S. R.; Hsu, J. W. P. Solution Processed ZnO Hybrid Nanocomposite with Tailored Work Function for Improved Electron Transport Layer in Organic Photovoltaic Devices. *ACS Appl. Mater. Interfaces* **2013,** *5*(18), 9128.
59. Shih, C. C.; Lee, W. Y.; Chiu, Y. C.; Hsu, H. W.; Chang, H. C.; Liu, C. L.; Chen, W. C. High Performance Transparent Transistor Memory Devices Using Nano-Floating Gate of Polymer/ZnO Nanocomposites. *Sci. Rep.* **2015,** *6*, 20129.
60. Yaralioglu, G. G.; Wygant, I. O.; Marentis, T. C.; Yakub, B. T. K. Ultrasonic Mixing in Microfluidic Channels Using Integrated Transducers. *Anal. Chem.* **2004,** *76*(13), 3694.
61. Zheng, Y.; Zheng, L.; Zhan, Y.; Lin, X.; Zheng, Q.; Wei, K. Ag/ZnO Heterostructure Nanocrystals: Synthesis, Characterization, and Photocatalysis. *Inorg. Chem.* **2007,** *46*(17), 6980.
62. Janotti, A.; Walle, C. G. V. D. Fundamentals of Zinc Oxide as a Semiconductor. *Rep. Prog. Phys.* **2009,** *72*, 126501.
63. Klingshirn, C. ZnO: Material, Physics and Applications. *Chem. Phys. Chem.* **2007,** *8*, 782.
64. Wang, Z. L. Zinc Oxide Nanostructures: Growth, Properties and Applications. *J. Phys.: Condens. Matter* **2004,** *16*, R829.
65. Yang, J.; Wang, D; Yang, L; Zhang, Y; Xing, G; Lang, J. Effects of Supply Time of Ar Gas Current on Structural Properties of Au-Catalyzed ZnO Nanowires on Silicon (100) Grown by Vapor–Liquid–Solid Process. *J. of Al. Com.* **2008,** *450*, 508.
66. Gao, P. X.; Wang, Z. L. Self-Assembled Nanowire–Nanoribbon Junction Arrays of ZnO. *J. Phys. Chem. B* **2002,** *106*, 12653.
67. Gao, P. X.; Ding, Y.; Wang, Z. L. Crystallographic Orientation-Aligned ZnO Nanorods Grown by a Tin Catalyst. *Nano Lett.* **2003,** *3*, 1315.
68. Pan, Z. W.; Dai, Z. R.; Wang, Z. L. Nanobelts of Semiconducting Oxides. *Science* **2001,** *291*, 1947.
69. Jie, J.; Wang, G.; Han, X.; Yu, Q.; Liao, Y.; Li, G.; Hou, J. G. Indium-Doped Zinc Oxide Nanobelts. *Chem. Phys. Lett.* **2004,** *387*, 466.
70. Zhang, Q.; Wei, B. Synthesis and Growth Mechanism of Macroscopic ZnO Nanocombs and Nanobelts. *Vacuum* **2011,** *86*(4), 398.
71. Shi, Y.; Bao, S.; Shi, R.; Huang, C.; Amini, A.; Wu, Z.; Zhang, L.; Wang, N.; Cheng, C. Y-shaped ZnO Nanobelts Driven from Twinned Dislocations. *Sci. Rep.* **2016,** *6*, 22494.

72. Lao, J. Y.; Wen, J. G.; Ren, Z. F. Hierarchical ZnO Nanostructures. *Nano Lett.* **2002,** *2*, 1287.
73. Mandal, S.; Dhar, A.; Ray, S. K. Growth and Photoluminescence Characteristics of ZnO Tripods. *J. Appl. Phys.* **2009,** *105*(3), 033513.
74. Zhu, L.; Hong, M.; Ho, G. W. Hierarchical Assembly of SnO_2/ZnO Nanostructures for Enhanced Photocatalytic Performance. *Sci. Rep.* **2015,** *5*, 11609.
75. Wang, Z. L.; Kong, X. Y.; Zuo, J. M. Induced Growth of Asymmetric Nanocantilever Arrays on Polar Surfaces. *Phys. Rev. Lett.* **2003,** *91*, 185502.
76. Pan, Z. W.; Mahurin, S. M.; Dai, S.; Lowndes, D. H. Nanowire Array Gratings with ZnO Combs. *Nano Lett.* **2005,** *5*, 4723.
77. Huanga, Y.; Zhanga, Y.; Hea, J.; Daia, Y.; Gua, Y.; Jia, Z.; Zhou, C. Fabrication and Characterization of ZnO Comb-like Nanostructures. *Ceram. Int.* **2006,** *32*, 561.
78. Xu, X.; Wu, M.; Asoro, M.; Ferreira, P. J.; Fan, D. L. One-Step Hydrothermal Synthesis of Comb-like ZnO Nanostructures. *Cryst. Growth Des.* **2012,** *12*(10), 4829.
79. Kong, X. Y.; Wang, Z. L. Spontaneous Polarization-Induced Nanohelixes, Nanosprings, and Nanorings of Piezoelectric Nanobelts. *Nano Lett.* **2003,** *3*, 1625.
80. Xia, C.; Ning, W.; Longc, W.; Lin, G. Synthesis of Nanochain-Assembled ZnO Flowers and Their Application to Dopamine Sensing. *Sens. Actuators B* **2010,** *147*, 629.
81. Xinga, Y. J.; Xia, Z. H.; Zhanga, X. D.; Songa, J. H.; Wangb, R. M.; Xub, J.; Xuea, Z. Q.; Yu, D. P. Nanotubular Structures of Zinc Oxide. *Solid State Commun.* **2004,** *129*, 671.
82. Mali, S. S.; Betty, C. A.; Bhosale, P. N.; Patil, P. S.; Hong, C. K. From Nanocorals to Nanorods to Nanoflowers Nanoarchitecture for Efficient Dye-Sensitized Solar Cells at Relatively Low Film Thickness: All Hydrothermal Process. *Sci. Rep.* **2014,** *4*, 5451.
83. Yao, I.-C.; Lin, P.; Tseng, T.-Y. Nanotip Fabrication of Zinc Oxide Nanorods and Their Enhanced Field Emission Properties. *Nanotechnology* **2009,** *20*(12), 125202.
84. Pradhan, B.; Majee, S. K.; Batabyal, S. K.; Pal, A. J. Electrical Bistability in Zinc Oxide Nanoparticle-Polymer Composites. *J. Nanosci. Nanotechnol.* **2007,** *7*, 4534.
85. Kim, E. K.; Kim, J. H.; Lee, D. U.; Kim, G. H.; Kim, Y. H. Characterization of Nano-Floating Gate Memory with ZnO Nanoparticles Embedded in Polymeric Matrix. *Jpn. J. Appl. Phys.* **2006,** *45*, 9A.
86. Guo, F.; Yang, B.; Yuan, Y.; Xiao, Z.; Dong, Q.; Bi, Y.; Huang, J. A. Nanocomposite Ultraviolet Photodetector Based on Interfacial Trap-Controlled Charge Injection. *Nat. Nanotechnol.* **2012,** *7*, 798.
87. Fonseca, R. D.; Correa, D. S.; Paris, E. C.; Tribuzi, V.; Dev, A.; Voss, T.; Aoki, P. H. B.; Constantino, C. J. L.; Mendonca, C. R. Fabrication of Zinc Oxide Nanowires/Polymer Composites by Two-Photon Polymerization. *J. Polym. Sci. B Polym. Phys.* **2014,** *52*, 333.
88. Cauda, V.; Stassia, S.; Lamberti, A.; Morello, M.; Pirri, C. F.; Canavese, G. Leveraging ZnO Morphologies in Piezoelectric Composites for Mechanical Energy Harvesting. *Nano Energy* **2015,** *18*, 212.
89. Wu, W.; Huang, X.; Li, S.; Jiang, P.; Toshikatsu, T. Novel Three-Dimensional Zinc Oxide Superstructures for High Dielectric Constant Polymer Composites Capable of Withstanding High Electric Field. *J. Phys. Chem. C* **2012,** *116*, 24887.

CHAPTER 6

COMPUTING MODELING OF FILLING PROCESSES OF NANOPORES INTO TEMPLATES ALUMINUM OXIDE BY ATOMS OF VARIOUS MATERIALS

A. V. VAKHRUSHEV[1,2,*], A. Y. FEDOTOV[1,2], A. V. SEVERYUKHIN[1], and R. G. VALEEV[3]

[1]*Institute of Mechanics, Ural Branch, Russian Academy of Sciences, Izhevsk, Russia*

[2]*Kalashnikov Izhevsk State Technical University, Izhevsk, Russia*

[3]*Physical-Technical Institute Ural Branch of Russian Academy of Science, Izhevsk, Russia*

Corresponding author. E-mail: vakhrushev-a@yandex.ru

CONTENTS

ABSTRACT

The aim of this work is to study the structure of the nanofilm coating and the substrate during epitaxial deposition. Often the nature of the structure of objects that determines their properties, changes the optical, electrical, and physical–mechanical parameters. Theoretical studies were carried out using molecular dynamics. The potential used was Lennard-Jones potential. The temperature and pressure in the nanosystems was maintained by a Nose–Hoover thermostat and barostat. Periodic boundary conditions are used. The velocity field at the initial time was selected according to the Maxwell distribution in the form of nanosized elements. Silting of substrate was carried out as uniform deposition of atoms on the normal to the substrate. Deposited atoms are added to the stage for overgrowth in the area above the substrate. Zinc and sulfur atoms are considered as deposited atoms. About 5% copper atoms were added in some cases. Its position on the substrate was determined by a uniform random distribution law. The amount of added atoms per unit time and the total number were manageable process parameters. The initial rate was constant for the deposited atoms. Speed settings are only changed in the interaction with the substrate-deposited atoms. To carry out theoretical research, the software package called large-scale atomic/molecular massively parallel simulator (LAMMPS) for parallel computing processes is used. The substrate of aluminum oxide (solid and porous) and the deposited nanofilms of zinc, sulfur, and copper are considered in the analysis of the structure. In all cases, the materials were amorphous in structure.

6.1 INTRODUCTION AND THE PROBLEM STATEMENT

Owing to its hexagonal-ordered arrangement of pores vertically aligned to the film surface, porous anodic aluminum oxide is quite often used as a template to synthesize different nanostructures: nanopoints, nanowires, nanorings, nanotubes, and so forth.[1,2] AAO can also be successfully used as a carrier of catalytically-active nanoparticles[3,4] as well as nanostructures of semiconductors.[5,6] This gives the possibility to form the ordered aggregates of nanostructures of semiconductor fluorescent material of the same size and shape that allows representing each nanostructure as a separate light emitter. The coherent addition of radiation from each light source results in significant light intensity increase.[7]

As it was pointed out before, the lighting properties of electroluminescent light sources (ELS) depend on the thickness of the fluorescent material layer and its structure. In the case of ELS formed as nanocomposites of the type "semiconductor/dielectric matrix," the template thickness also plays an important role since the precipitated material penetrates the matrix pores to the depth of up to 10 μm. That is why the mechanism of nanostructure growth in matrixes of different thickness can differ, and the distribution of alloying element and structure of the fluorescent material obtained by the method of thermal precipitation of powder mixture can differ as well.[8]

Despite of the wide application of nanofilms, the questions of detailed investigation of their composition, structure, and processes flowing in them still arise. The problems of studying electrochemical and magnetic effects[9,10] in porous AAO and the application of similar templates as optic sensors[11] are topical. Similar problems are actively studied and solved by other authors both experimentally[12] and to decrease the costs with the help of theoretical methods and mathematical modeling apparatus.[13,14] Monte Carlo algorithms,[15] condensation models, equations of liquid and gas motion, finite element analysis[16] as well as molecular dynamics (MD) with different types of potentials are used as theoretical investigation methods. The application of MD apparatus is often the most justified as it allows observing the structure evolution in materials in detail and registering the whole set of phenomena and processes in nanofilms. The understanding of mechanisms and multilayer examination of nanofilm formation and functioning, interaction of nanostructures and nanoparticles they contain as well as the development of engineering ideas, and approaches to manage and use these processes, will give the possibility to properly design nanocomposites and find perspective areas of their application.

The work objective is to develop the algorithms and methods for modeling the processes of precipitation of nanosized films onto the templates of porous aluminum oxide and to study the kinetics of the foregoing processes and structure of nanosized films with the help of mathematical modeling methods. In the course of the work, we also investigated the influence of dimensional parameters of the pores in aluminum oxide matrix on the siltation processes and the formation of nanofilm coatings. Gold, silver, chromium, copper, iron, gallium, germanium, titanium, platinum, and vanadium were used as precipitated materials while modeling.

6.2 PROCESS MECHANISM AND MATHEMATICAL MODELS

The problem of precipitating nanofilms onto porous templates of aluminum oxide was solved by MD method.[17] MD method has been widely used when modeling the behavior of nanosystems due to the simplicity of implementation, satisfactory accuracy, and low costs of computational resources. The solution of Newton's differential equation of motion for each particle forms the basis of this method.

One of the main shortcomings of MD method is the complexity in comparing the modeling results with the actual experimental data. For stability and convergence of solving the problem of nanosystem modeling by MD method, it is necessary to select an integration step, which would satisfy the smallest and most mobile atoms. In most cases, the step of 0.5–2.0 fs provides the adequate results. Thus, to compare the results with the experiment, it is necessary to model the behavior of nanosystems for 10^{-15} and more iteration steps that require high computational costs and is impossible at this stage of computer development. Combined approaches and hybrid methods help solving this problem. Nevertheless, the complexity of transition from some managing derivatives to others when combining several approaches makes it difficult to use the combined algorithms. Moreover, the loss in particularization of describing the system and some atomic–molecular effects due to the increased size of the objects themselves and integration time is observed, which does not take place when using MD method.

Depending on the potential type and external forces comprised by the system, the problem considered during the modeling process can have different accuracy and different thermodynamic parameters. The issue of obtaining and searching the potential parameters is complicated and labor consuming. The ab initio calculations or experimental data can serve as data sources.

The sets of quite fairly selected parameters for similar molecules are accumulated into special databases and libraries, which define the type of force fields influencing the atoms. The potential type and potential energy of nanostructured photoelectronic systems contribute decisively to the type, character, and value of the interactions of nanosystem objects. The potentials are distinguished as multiparticle and pair, axisymmetric and spatial. The potential type when solving the problem is mostly determined by the availability of parameters in the libraries and force fields of databases for modeling nanostructured systems.

The potential selection for mathematical modeling is a complicated and complex task. The majority of empirical potentials describe the volumetric properties of materials well, but, some are also successfully used to describe the surface properties.

The actual correlations between elastic constant metals and semiconductors can be obtained taking into account pair and multiparticle interactions. The following approaches consider the multiparticle interaction: Stillinger–Weber potential,[18] Abel-Tersoff potential,[19] embedded atom method (EAM),[20–22] and modified EAM (MEAM)[23] are most widely spread when modeling metal and semiconductor systems.

In the EAM, the whole energy of atom system is given as follows:[22]

$$E = \sum_i F_i \left[\sum_{j \neq i} \rho_j \left(R_{ij} \right) \right] + \frac{1}{2} \sum_{j \neq i} \varphi_{ij} \left(R_{ij} \right), \tag{6.1}$$

where $\sum_i F_i \left[\sum_{j \neq i} \rho_j \left(R_{ij} \right) \right]$—function of i is that of atom embedding, depending on the total electron density in the region of i atom embedding, $\varphi_{ij}(R_{ij})$—energy of pair interaction. Each atom of the system is considered as a particle embedded into the electron gas produced by other atoms of the system being modeled. The energy required for embedding depends on the electron density in the embedding point. The embedding function thus introduced allows determining the exchange and correlation energy of the system electron gas.

The sense of the embedding function can be defined as the energy necessary for embedding one atom into the homogeneous gas with density ρ. However, there are other transformations,[24] which allow changing the functions (Eq. 6.1) under the condition that the resultant energy and interatomic forces will not change.

The approximation that the function of electron density of one atom is a spherically symmetric function depending only on the distance between the atoms is used in EAM. This approximation significantly limits the field of EAM application and makes taking into account the systems, in which the directedness of bond covalent component can be neglected. Besides, the electron density in the region of i atom embedding is determined in EAM as linear superposition of electron densities of the system of other atoms. This approximation significantly simplifies the electron density computation.

Currently, EAM potentials are developed for the majority of metals and some binary systems. The potentials of ternary systems are also calculated.[25] However, such "ternary" potentials do not always qualitatively reproduce physical properties of materials.

Based on the methods described, the semiempirical approach combining the advantages of multiparticle potentials and EAM was proposed. The theory of MEAM was developed applying the electron density functional theory (DFT).[23] At present, the DFT method is considered as the most acknowledged approach to describing electron properties of solids. In the EAM method, full electron density is represented as linear superposition of spherically averaged atoms. This shortcoming is eliminated in the MEAM.

In MEAM, the full energy of the system is recorded as the total of energies of single atoms:[26]

$$E = \sum_i E_i = \sum_i \left(F_i(\overline{\rho}_i) + \frac{1}{2} \sum_{j \neq i} \varphi_{ij}(r_{ij}) \right), \tag{6.2}$$

where E_i—full energy of the system, E_i—energy of i atom used to calculate the interaction forces of the atoms in motion equations, F_i—embedding function for i atom located in the medium with background electron density $\overline{\rho}_i$, $\varphi_{ij}(r_{ij})$—pair potential between i and j atoms located at the distance R_{ij}.

In Equation 6.2, embedding function $F_i(\overline{\rho}_i)$ is defined as follows:

$$F_i(\overline{\rho}_i) = A_i E_i^0 (\overline{\rho}_i) \ln(\overline{\rho}_i), \tag{6.3}$$

where A_i—adjustable parameter, E_i^0—sublimation energy, $\overline{\rho}_i$—background electron density. These parameters depend on the atom elementary type and are referred to as the atom of i type.

Background electron density $\overline{\rho}_i$ comprises the contributions of partial electron densities: spherically symmetric electron density $\rho_i^{(0)}$ and angular (orbital) contributions $\rho_i^{(1)}, \rho_i^{(2)}, \rho_i^{(3)}$. The electron densities described, correspond to s, p, d, f atom orbitals. In the EAM model, only spherically symmetric electron density $\rho_i^{(0)}$ is considered.

To obtain the full background density, the combination or functional dependence on single electron densities is necessary. This dependence is described in different ways, but the following expression comprising angular dependencies is frequently used:

$$\bar{\rho}_i = \frac{\rho_i^{(0)}}{\bar{\rho}_i^0} G(\Gamma_i), \tag{6.4}$$

In the expression (Eq. 6.4), $\bar{\rho}_i^0$ is the background electron density of the initial structure. There are also many varieties of function $G(\Gamma_i)$. Function Γ_i is calculated based on the following formula, the addition is carried on

$$\Gamma_i = \sum_{h=1}^{3} t_i^{(h)} \left(\frac{\rho_i^{(h)}}{\rho_i^{(0)}} \right)^2, \tag{6.5}$$

where $h = 0 - 3$, corresponds to s, p, d, f of the symmetry, $t_i^{(h)}$—weight factors, $\rho_i^{(h)}$—values determining the deviation of the electron density distribution from $\rho_i^{(0)}$ with different symmetry types.

The background electron density of the initial structure is calculated based on the following formula:

$$\rho_i^0 = \rho_{i0} Z_{i0} G\left(\Gamma_i^{\text{ref}}\right), \tag{6.6}$$

where ρ_{i0}—factor for the density depending on the element type, Z_{i0}—number of the nearest neighbors of i atom in its reference crystalline structure, Γ_i^{ref} is determined as follows:

$$\Gamma_i^{\text{ref}} = \frac{1}{Z_{i0}^2} \sum_{h=1}^{3} t_i^{(k)} s_i^{(k)}, \tag{6.7}$$

where $s_i^{(k)}$—shape parameter depending on the type of i element. Shape parameters for different structures are described in ref 23.

The pair potential between i and j atoms is determined from the expression for the energy (Eq. 6.1), if the dependence of energy on the distance to the nearest neighbors in the crystalline grid is known. The universal equation of state is written down as follows:[27]

$$E_{ii}^u\left(R_{ij}\right) = -E_{ij}\left(1 + a_{ij}^*\left(r_{ij}\right)\right) e^{-a_{ij}^*\left(r_{ij}\right)},\ a_{ij}^* = \alpha_{ij}\left(\frac{r_{ij}}{r_{ij}^0} - 1\right), \tag{6.8}$$

where E_{ij}, α_{ij} and r_{ij}^0 —parameters depending on the type of the certain atom.

From Equations 9.1 and 9.8 and the fact that the embedding functions $F_i(\bar{\rho}_i)$ and $F_j(\bar{\rho}_j)$ are already known at this stage, it is possible to write down the expression for calculating the pair potential between the atoms:

$$\varphi_{ij}\left(r_{ij}\right)=\frac{1}{Z_{ij}}\left[2E_{ij}^{u}\left(r_{ij}\right)-F_{i}\left(\frac{Z_{ij}}{Z_{i}}\rho_{i}^{\alpha(0)}\left(r_{ij}\right)\right)-F_{j}\left(\frac{Z_{ij}}{Z_{j}}\rho_{j}^{\alpha(0)}\left(r_{ij}\right)\right)\right], \tag{6.9}$$

where Z_{ij}—number of the nearest neighbors, $\rho_i^{a(0)}$ atomic electron density corresponding to s symmetry.

Thus, in MEAM method the interaction between two atoms is determined by the position of the rest of the atoms in the system. The model author[22,23] developed the scheme of atom shielding, in which two-particle functions of the model become dependent on the screening function, taking into account the position of the rest of the atoms. More details of the MEAM model as well as the expressions for calculating $\rho_i^{(0)}, \rho_i^{(1)}, \rho_i^{(2)}, \rho_i^{(3)}, \rho_i^{a(h)}$ and functions of $G(\Gamma_i)$ type can be found in the works of the model author[22,23] or in the previous publications of the authors of this paper on the similar topic.[28–30]

Software package for parallel computational processes LAMMPS is applied to conduct theoretical research. LAMMPS has been developed by the group of Sandia National Laboratories and is free software for mathematical models of different levels, including classical MD.[31]

The problem of modeling the formation of nanofilm coatings was solved in several steps. At the first stage, the template from amorphous aluminum oxide was formed. Aluminum and oxygen atoms are put into the computational cell in the required proportion (2:3) with periodic boundary conditions on each side (Fig. 6.1a). The template stabilizes and comes to rest under the action of potential forces under normal thermodynamic conditions (Fig. 6.1b). The template stabilization is conditioned by potential forces, in particular, as it is formed due to the self-organization of aluminum and oxygen atoms. At the same time, heat fluctuations and diffusion are present in the range of the set temperature in the template formed, but there is no essential reconstruction of its structure, the atoms slightly oscillate near the positions they occupy.

The hole is cut in the template at the second stage—the pore with the required radius and depth (the cutting of the template with the pore is demonstrated in Fig. 6.1c). Later this pore will be silted with atoms of different types (Fig. 6.1d). The general pattern of the problem of forming heterogeneous electro-optical coatings is given in Figure 6.1.

The boundary conditions and appearance of the system being modeled are demonstrated in Figure 6.2. Owing to the periodic boundary conditions in the directions x and y, only one pore was considered in this work.

In horizontal directions, the periodic boundary conditions envisage the parallel transfer of the computational cell. The system being modeled was affected by rigid boundary conditions from the top and bottom. When the atoms were approaching the upper boundary of the system investigated, their bounce from the rigid wall was imitated. The positions of atoms in the thin layer near the boundary of the computational cell were rigidly fixed from the bottom. This type of boundary condition did not allow the nanosystem atoms to leave the computational area in case of deviation from the main precipitation trajectory.

The convergence of numerical solution of the problem set frequently depends on the selection of the corresponding integration step. The step needs to be small enough to correctly reflect the system behavior. When using the methods of MD, the value of the mass of substances being modeled influence the integration step value. It is selected in the range from 0.5 to 2 fs. In this work, the integration step by time is 1 fs. The total time when modeling the system for the stabilization stage (Fig. 6.1a) was about 0.5 ns, for the relaxation stage (Fig. 6.1b and c)—0.2 ns and precipitation stage (Fig. 6.1d)—0.2 ns.

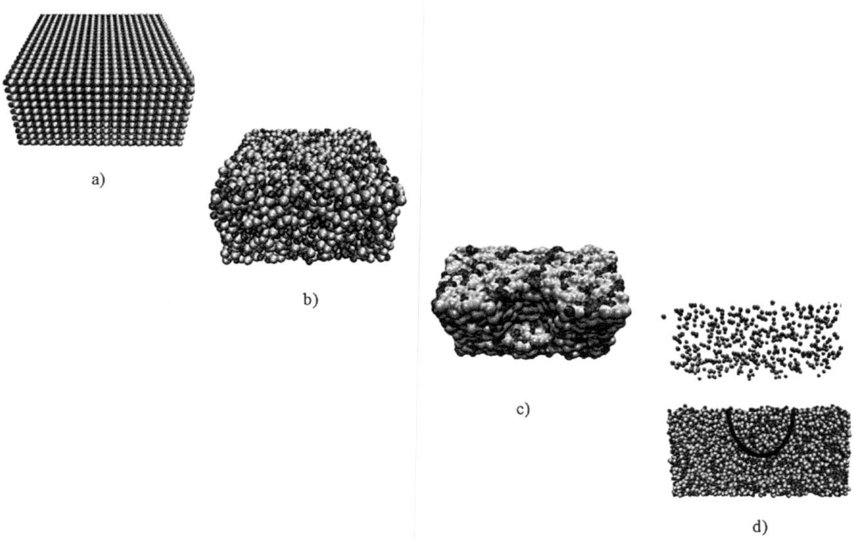

FIGURE 6.1 Steps of solving the problem of forming nanofilm coatings based on porous aluminum oxide: (a) initial state of the system, (b) template relaxation without the pore, (c) template relaxation with the pore, (d) precipitation.

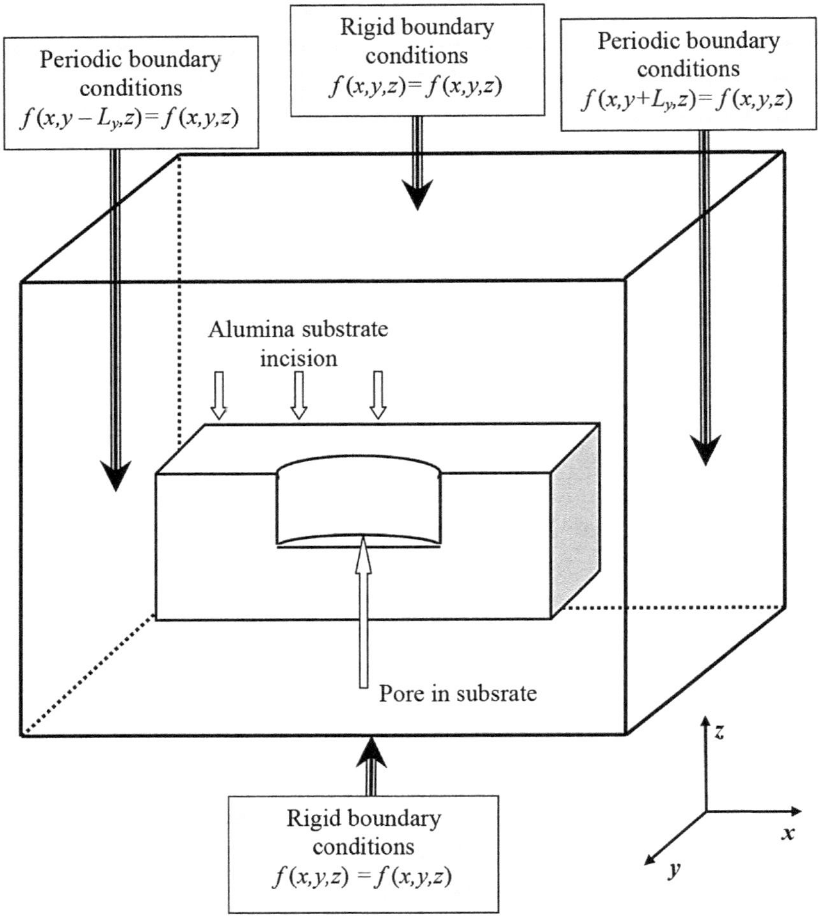

FIGURE 6.2 Boundary conditions and appearance of the system being modeled.

The porous template was silted by homogenous precipitation of atoms along the normal against the template. The atoms being precipitated were added in the region above the template during the siltation stage. Their position above the template was determined by the uniform random distribution law. The number of atoms added in a time unit and their total number were the process control parameters. The initial velocity of the precipitated atoms was constant. The velocity parameters were changed only under the interaction of the precipitated atoms with the template. To conduct the test calculations, a single nanostructure was considered in the air-free atmosphere and its dynamics during the relaxation self-organization of atoms.

The potential energy of the system of nanofilm coating formation on the templates from porous aluminum oxide has an electromagnetic nature and is set in the form of potentials (Eq. 6.1). It is often necessary to calculate the system kinetic energy. The expression for the system kinetic energy is as follows:[31]

$$E_k = \frac{\sum_{i=1}^{N} m_i (\mathbf{V}_i)^2}{2}, \tag{6.10}$$

where E_k—value of the kinetic energy at certain time moment, m_i—molecular mass of i atom, $\mathbf{V}_i = \mathbf{V}_i\ (t)$ —atom velocity vector, N—total number of particles or atoms.

The molecular system temperature has a close connection with the kinetic energy:

$$E_k = \frac{3Nk_B T}{2}, \tag{6.11}$$

where k_B—Boltzmann constant.

Equations 6.10 and 6.11 allow calculating the instant temperature value. Thus, the instant temperature of the molecular system is determined in nanosystems as the system average kinetic energy:

$$T = \frac{1}{3Nk_B} \sum_{i=1}^{N} m_i (\mathbf{V}_i)^2. \tag{6.12}$$

The influence of the initial temperature on the distribution of molecule velocities is an important factor in the process of nanosystem investigation, influencing the physicality of the proceeding processes. In the problems of forming nanosized elements, the velocity field at the initial time moment was selected in accordance with the Maxwell distribution. Maxwell distribution for the velocity vector of each atom $\mathbf{V} = (V_x, V_y, V_z)$ is the product of distributions in three directions:

$$f_V\left(V_x, V_y, V_z\right) = f_V\left(V_x\right) f_V\left(V_y\right) f_V\left(V_z\right), \tag{6.13}$$

where the distribution in one direction is defined by the following correlation:

$$f_V\left(V_j\right)=\sqrt{\frac{m}{2\pi k_B T_0}}\exp\left[\frac{-mV_j^2}{2k_B T_0}\right],\ j=\{x,y,z\}, \tag{6.14}$$

where T_0—temperature at the initial time moment. The mass and velocity components in (Eq. 6.14) will be individual for each atom with number *i*. Maxwell distribution has the form of normal distribution. As expected for the rest of the system, the average velocity equals zero in any direction.

The following expression is used to calculate the nanosystem pressure:

$$P=\frac{Nk_B T}{W}+\frac{1}{3W}\sum_{i=1}^{N}\mathbf{r}_i\cdot\mathbf{f}_i, \tag{6.15}$$

Where W—volume occupied by the nanosystems. Based on Equations 6.10 and 6.11, the first summand in Equation 6.15 can be calculated through the combination of masses and velocities of the system atoms. The second summand in Equation 6.15 represents the virial or the total of scalar products of vectors of force f_i, acting upon the atoms and their coordinates r_i.

The MD modeling was carried out at constant temperature. A constant temperature was maintained in the system with the help of thermostat algorithm. Thermostat is a means of energy extraction and cooling of fast atoms as well as a means of energy feeding when the nanosystem is not heated up enough. At present, thermostat algorithms are quite variable: collision thermostat, Berendsen thermostat, friction thermostat, and Nose–Hoover thermostat. Nose–Hoover thermostat was used in this work.

The temperature control with Nose–Hoover thermostat[32] consists in the introduction of effective friction forces proportional to the velocities of particles with dynamically changing coefficient ξ into the system:

$$\frac{d^2 r_i}{dt^2}=\frac{f_i}{m_i}-\xi\frac{dr_i}{dt},\ i=\overline{1,N}. \tag{6.16}$$

The equations for coefficient Q are solved by numerical integration by time together with the integration of motion equations:

$$\frac{d\xi}{dt}=\frac{1}{Q}\left(T-T_{ext}\right). \tag{6.17}$$

"Mass" coefficient Q determines the speed of reaching the required temperature. This parameter can be sorted out either intuitively or set through other values. The convergence to temperature T_{ext} is of oscillation type with period τ_T:

$$Q = \frac{\tau_T^2 T_{ext}}{4\pi^2}. \quad (6.18)$$

6.3 MODELING RESULTS AND THEIR ANALYSIS

The templates were silted with zinc and sulfur atoms in equal proportions—40,000 atoms of each type. The precipitation was uniform along the whole surface of the template and it was equally intensive by time. The atom velocity during the epitaxy was 0.05 nm/ps. Five percent of copper was added to the precipitated atoms in some computational experiments. Similar epitaxial compositions are conditioned by the fact that the investigation refers to certain technological processes used in practice when producing samples with unique optical properties.

The results of nanofilm epitaxial formation from zinc and sulfur atoms on the template of porous aluminum oxide are given in Figure 6.3. The height of the nanofilm obtained was 7.2 nm. The precipitated atoms formed the nonuniform relief and the inequalities were observed on the surface. The pore was not silted in the template, but zinc and sulfur atoms partially got inside it (Fig. 6.3b). The partially blocked hole is observed immediately above the pore in Figure 6.3. The precipitation picture was similar for monolithic templates of aluminum oxide.

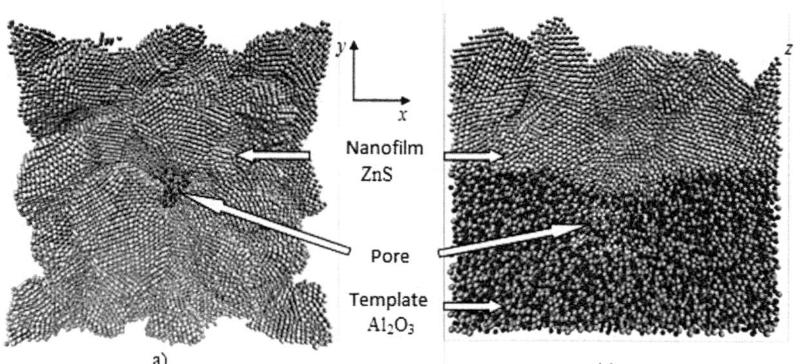

FIGURE 6.3 Results of Zn and S atoms precipitation onto the template of porous aluminum oxide, precipitation time—0.5 ns.

You can judge about the material structure (crystalline or amorphous) by the ideality parameter of the crystalline lattice calculated in Equation 6.8. This parameter is calculated for the whole group of atoms, then its average value is determined. In Figure 6.3, you can see the values of ideality parameter of the crystalline lattice for the template of aluminum oxide.

Four cases of the computational experiment were considered in the paper: monolithic template with the precipitation of Zn and S atoms, monolithic template with the precipitation of Zn, S and 5% Cu atoms, porous template with the precipitation of Zn and S atoms, and porous template with the precipitation of Zn, S, and 5% Cu atoms. Besides, the structure of the template and nanofilm formed were evaluated.

Figure 6.4 demonstrates the change in the template crystalline lattice structure at the stage of the structure reconstruction (increased region in Fig. 6.3). The analysis of the dependencies in Figures 6.3 and 6.4 shows that the ideality parameter of the crystalline lattice for all four cases is similar. The time of 250 ps corresponds to the moment when the atoms stop precipitating, the reconstruction of their coordinates to the more energy favorable position starts. The template practically does not change its structure at the time moments of 250–500 ps, only the insignificant temperature fluctuations of atoms near the crystalline lattice nodes are observed. The value of parameter C for the template is rather large that indicates the structure of the amorphous material. From Figure 6.4, it is seen that the structure parameter for the investigations carried out is slightly different.

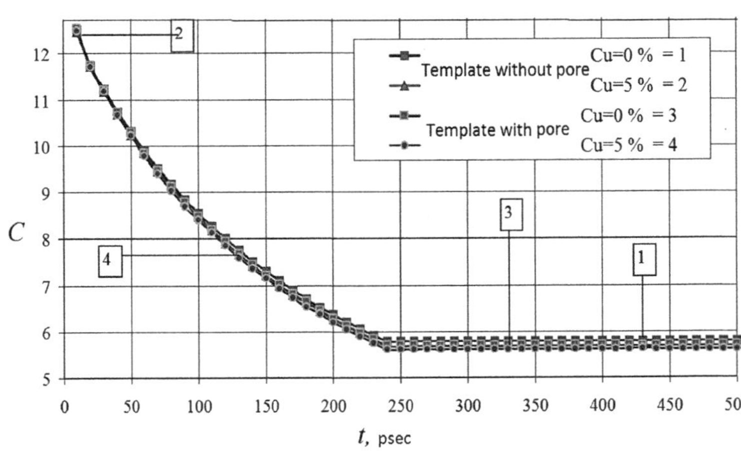

FIGURE 6.4 **(See color insert.)** Average parameter of the crystalline lattice structure for the template.

For all four options of computational experiments, the structure parameter change was also calculated for the precipitated atoms (Figs. 6.5 and 6.6). The distributions of these parameters are characterized by the active reconstruction of atom positions at the sputtering moments and further little changing structure. The increased region of the precipitation end (Fig. 6.7) allows considering the differences in the properties of the materials formed. The computational experiment without the pore with the addition of 5% of copper atoms has the least value of the crystalline structure parameter during the condensation process. The stabilized sputtered material has the least value of the parameter for the template without the pore and without the addition of copper atoms. The amorphous structure of the template and precipitated nanofilms is characteristic for all the considered cases.

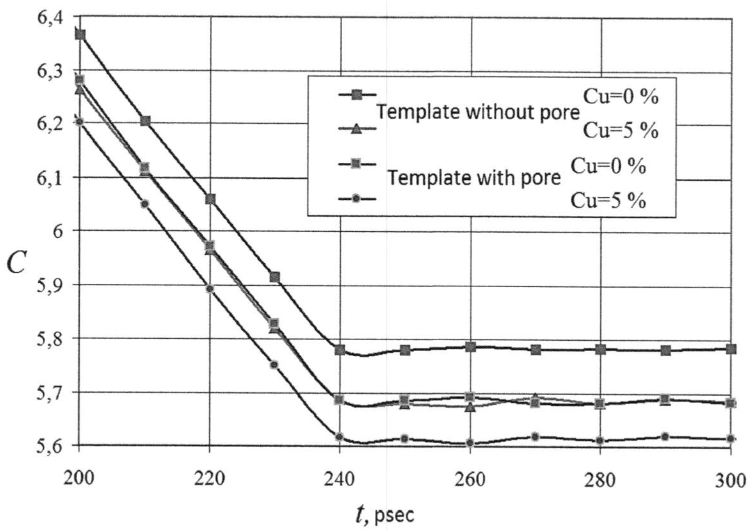

FIGURE 6.5 **(See color insert.)** Average parameter of the crystalline lattice structure for the template at the structure reconstruction stage.

The amorphous aluminum oxide templates with the following dimensions: length—12.4 nm, width—12.4 nm, height—6.2 nm were used in the modeling process. The total number of atoms in the template after the pore formation was about 60,500. The template was at rest before the precipitation process, at the beginning its temperature was 300 K and it was further maintained at the same level. The graph of the template temperature changes as well as the kinetic energy for the stabilization and relaxation stages is given in Figure 6.8.

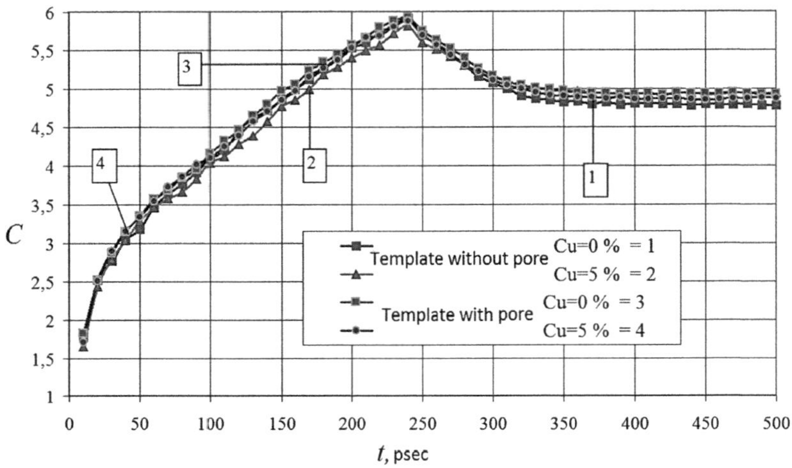

FIGURE 6.6 **(See color insert.)** Average parameter of the crystalline lattice structure for the precipitated atoms.

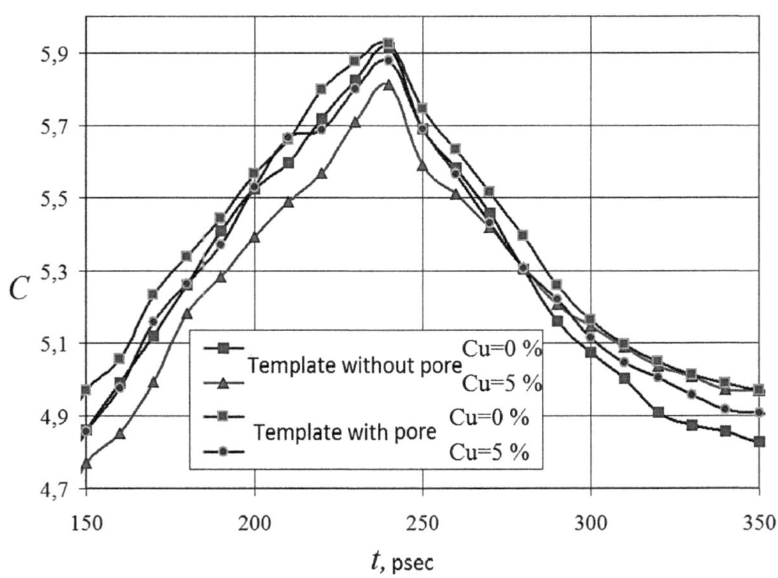

FIGURE 6.7 **(See color insert.)** Average parameter of the crystalline lattice structure for the precipitated atoms at the structure reconstruction stage.

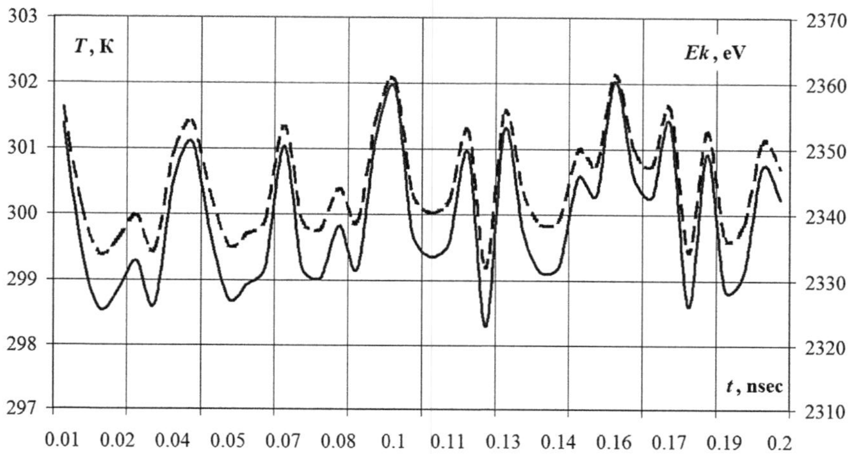

FIGURE 6.8 Dependence of temperature (left ordinate) and kinetic energy (right ordinate) of the porous aluminum template for the relaxation stage.

The pore with the radius of 2 nm and depth of 4 nm was cut in the aluminum oxide template. The pore siltation with aluminum oxide was not observed at rest without the atom precipitation. The lower template layer was fixed to avoid its vertical movement at the precipitation stage. The rest of the atoms were not fixed and could move freely in any direction.

Different types of atoms were precipitated onto the aluminum oxide template in this work. The number of precipitated atoms was 20,000. The precipitation was uniform along the whole template surface and with the same intensity in time. The atom velocity at epitaxy was 0.05 nm/ps. The physical characteristics of substances used when modeling the process of pore siltation and formation of film coatings are given in the Table 6.1.

The results of epitaxial nanofilm formation from silver atoms on the template of porous aluminum oxide are given in Figure 6.9. The atom precipitation was uniform, the formation of large agglomerates in the air environment was not observed. The film formed on the template was uniform with little sinking in the pore region. The pore was not completely silted with silver atoms, it was observed that part of the silver atoms got inside the pore near its upper part. The rest of the pore was hollow during the entire epitaxy stage. The template central layer 0.2 nm thick with silver nanofilm formed on it is shown in Figure 6.9 at the right. The figure analysis confirms the incomplete pore siltation.

TABLE 6.1 Physical Prɔperties of the Precipitated Elements.

Symbol	Name	Standard atomic weight (amu)	Crystal lattice structure	The lattice perimeter (nm)	Melting temperature (K)
Au	Gold	196.967	Face-centered cubic	*a*=0.4078	1337.33
Ag	Silver	107.868	Face-centered cubic	*a*=0.4086	1235
Cr	Chromium	51.996	Body-centered cubic	*a*=0.2885	2130
Cu	Copper	63.546	Face-centered cubic	*a*=0.3615	1356
Fe	Iron	55.847	Body-centered cubic	*a*=0.2866	1812
Ga	Gallium	69.723	Orthorhombic	*a*=0.4519 *b*=0.7658 *c*=0.4526	302,93
Ge	Germanium	72.630	Face-centered diamond-cubic	*a*=0.566	1210.6
Ti	Titanium	47.867	Hexagonal close-packed	*a*=0.2951 *c*=0.4697	1933±20
Pd	Palladium	106.42	Face-centered cubic	*a*=0.3890	1827
Pt	Platinum	195.084	Face-centered cubic	*a*=0.3920	2041.4
V	Vanadium	50.9415	Body-centered cubic	*a*=0.3024	2160

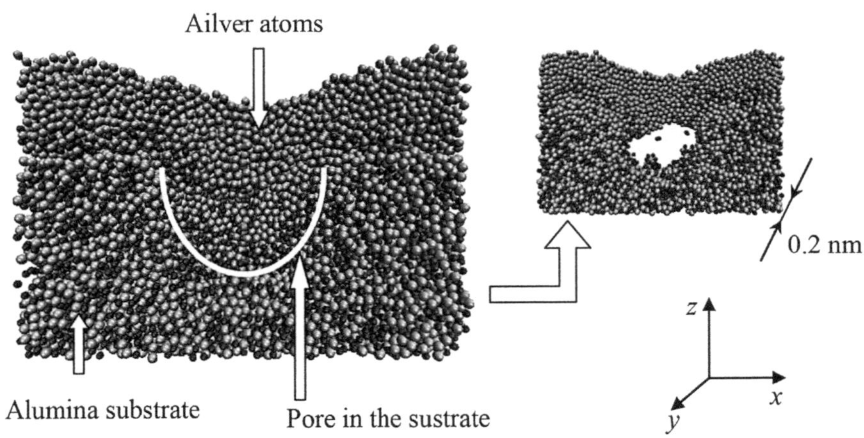

FIGURE 6.9 Results of silver atom precipitation onto the template of porous aluminum oxide, precipitation time—0.2 ns.

There are practically no silver atoms left in the air environment above the template by the time moment of 0.2 ns of precipitation stage.

The partial pulling of oxygen atoms out of the template and getting into the lower layers of the nanofilm on the template surface are observed. The slight migration of oxygen and aluminum atoms during the whole modeling stage takes place, by the temperature movements of the system atoms.

The horizontal section of nanofilm coating along the template surface after the precipitation of silver atoms for the precipitation time of 0.2 ns is demonstrated in Figure 6.10. The partial formation of crystalline structure formed by aluminum atoms is observed in some parts.

Silver atoms in the template center in Figure 6.10 are only on the surface, the pore is hollow inside. The picture asymmetry is explained by the pseudo stochastic behavior of the nanosystem produced by the temperature laws of initial distribution of atom velocities.

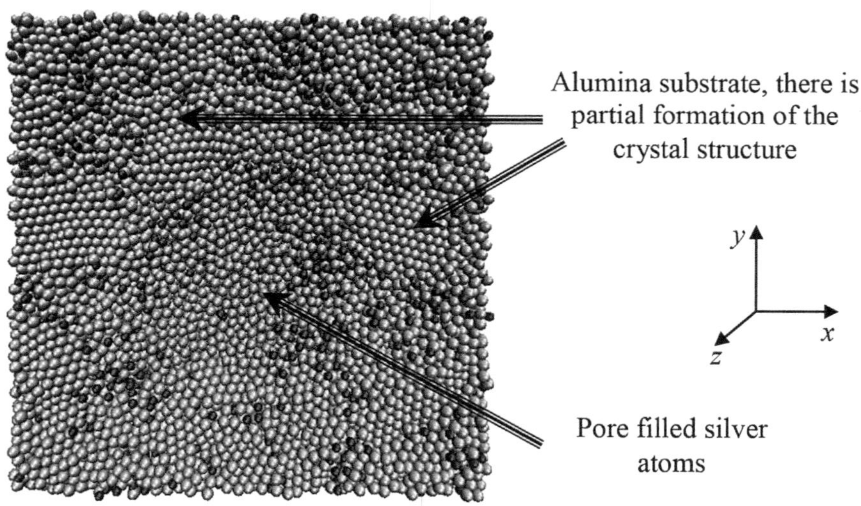

FIGURE 6.10 Horizontal section of nanofilm coating surface after the precipitation of silver atoms along the template surface, precipitation time—0.2 ns.

The results of gold and magnesium atom precipitation onto the template of porous aluminum oxide are similar to the previously described process of sliver atom epitaxy. In such cases, the template is coated quite uniformly by bending in the pore position. At the same time, the pore is not silted completely, the precipitated atoms produce something like a cork or cover near its surface.

In the process of template siltation, chromium, copper, and iron atoms start grouping into nanostructures in the air, before reaching the template surface. A significant coarsening of agglomerates is not observed, they continue moving to the template. The example of iron atom precipitation is shown in Figures 6.11 and 6.12.

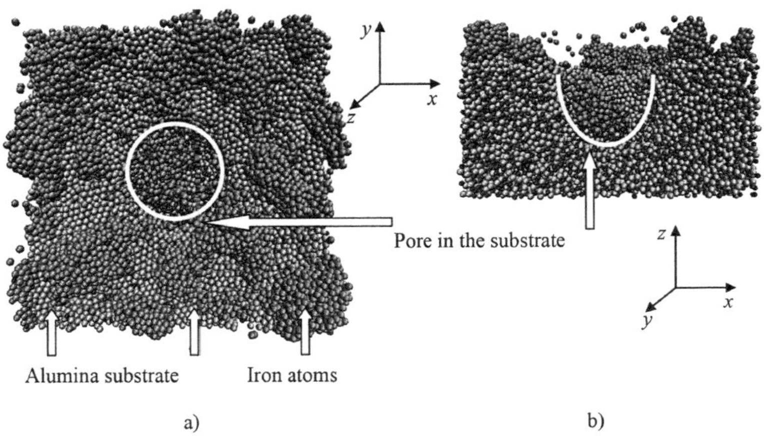

FIGURE 6.11 Siltation of the template of porous aluminum oxide with iron atoms: (a) top view, (b) vertical section along the pore center, precipitation time–0.05 ns.

The siltation of the porous aluminum oxide template with iron atoms for a precipitation time of 0.05 ns is demonstrated in Figure 6.11. As iron atoms already started grouping in the air environment above the template, the template was silted following the island principle. Small iron nanostructures gradually grew on the template and grouped into the bigger ones. The formation of one of the iron nanostructures was observed inside the pore, it is especially vividly seen on the vertical section along the pore center (Fig. 6.11b). Oxygen atoms from the template upper layers actively interact with iron atoms forming amorphous oxide structures. Figure 6.12 (axis z is inclined by 45° to the observer) illustrates that the surface of iron nanofilm formed on the template is uneven. In contrast to the epitaxy of silver and gold atoms, iron does not completely cover the template. Besides, rather significant height differences of several nanometers are sometimes observed in the nanofilm. The same results as for the iron atoms are characteristic for the coatings of porous templates with chromium atoms, except that the nanofilm is even and chromium atom agglomeration in the air environment is less intensive.

The precipitation of gallium and germanium atoms is similar to each other in the physical process. The result for the time of 0.2 ns is given in Figure 6.13. The pore in these cases is also not silted completely. The nanofilm is formed by regions on the template surface, the non-silted template is also seen as large regions in the Figure 6.12. Small gallium nanoparticles are seen on the template surface. Oxygen atoms are more intensely pulled out of the upper layers of the template than in the previously considered processes.

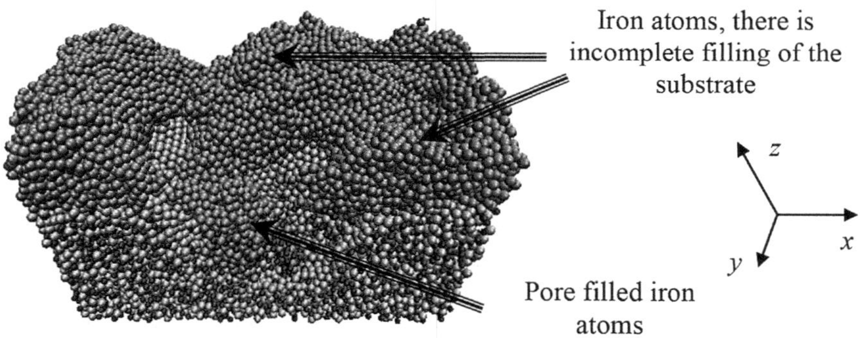

FIGURE 6.12 Siltation of the template of porous aluminum oxide with iron atoms: vertical section along the pore center, precipitation time—0.2 ns.

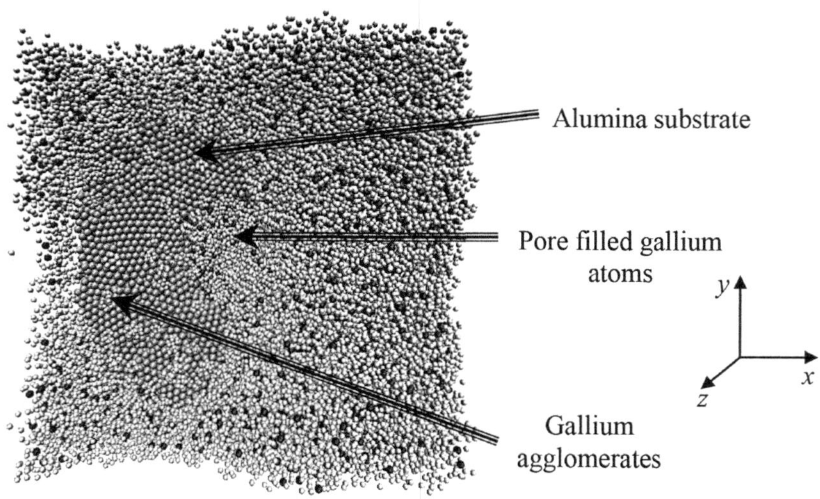

FIGURE 6.13 Siltation of the template of porous aluminum oxide with gallium atoms, precipitation time—0.2 ns.

When precipitated, germanium atoms penetrated more intensely into the pore region, but further in the process of nanofilm formation they were pulled out onto the surface. The surface profile of the obtained nanofilm from germanium atoms was not uniform with height differences and large non-silted regions.

An interesting effect was observed during the epitaxy of palladium and platinum atoms onto the template of porous aluminum oxide. The precipitation result for palladium atoms in two projections is given in Figure 6.14. In this case, the uniform nanofilm was formed with slight sinking in the pore region. However, the non-silted opening was observed immediately above the pore during the whole condensation stage. Palladium atoms got inside the pore only insignificantly, as it is seen from the vertical section along the pore center demonstrated in Figure 6.14b.

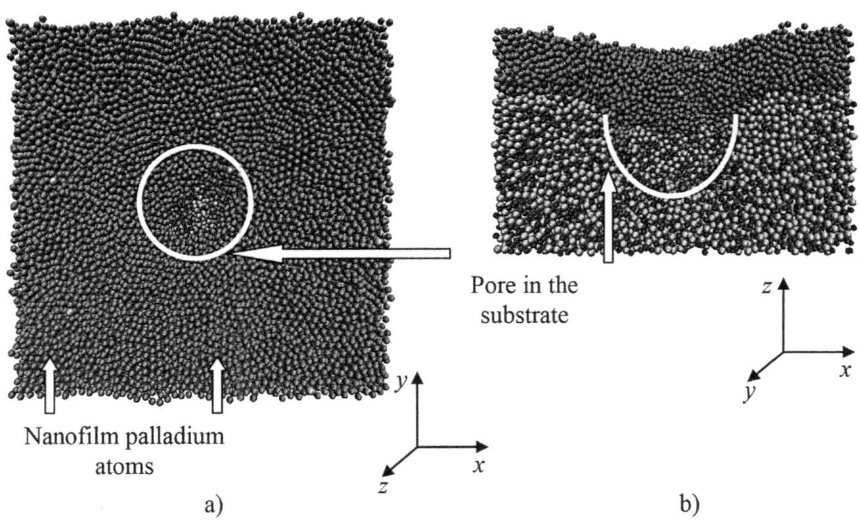

FIGURE 6.14 Siltation of the template of porous aluminum oxide with palladium atoms: (a) top view, (b) vertical section along the pore center, precipitation time—0.2 ns.

The siltation of the template of porous aluminum oxide with titanium atoms for the precipitation time of 0.2 ns is demonstrated in Figure 6.15. For this type of atoms, the nanofilm formed looked incoherent and coarse. The atoms got inside the pore only insignificantly and stayed on the template surface during the whole condensation stage.

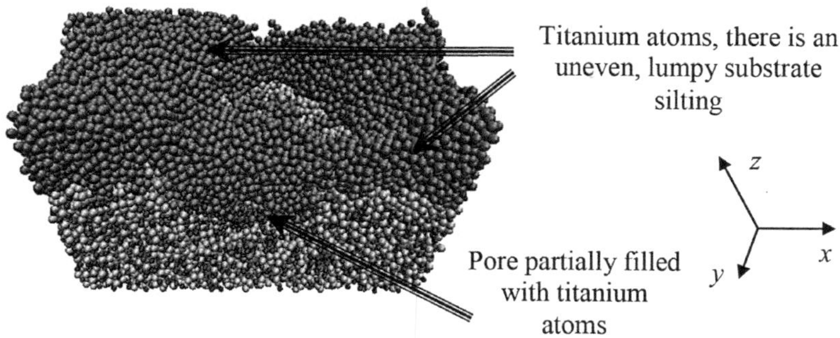

FIGURE 6.15 Siltation of the template of porous aluminum oxide with titanium atoms: vertical section along the pore center, precipitation time—0.2 ns.

The siltation of the template was observed for its epitaxy with vanadium atoms. A large number of vanadium atoms got inside the pore, but non-silted regions remained on the template surface itself. The translation results of the periodic computational cell relative to the perpendicular of the precipitation plane during the epitaxy of iron atoms are shown in Figure 6.16.

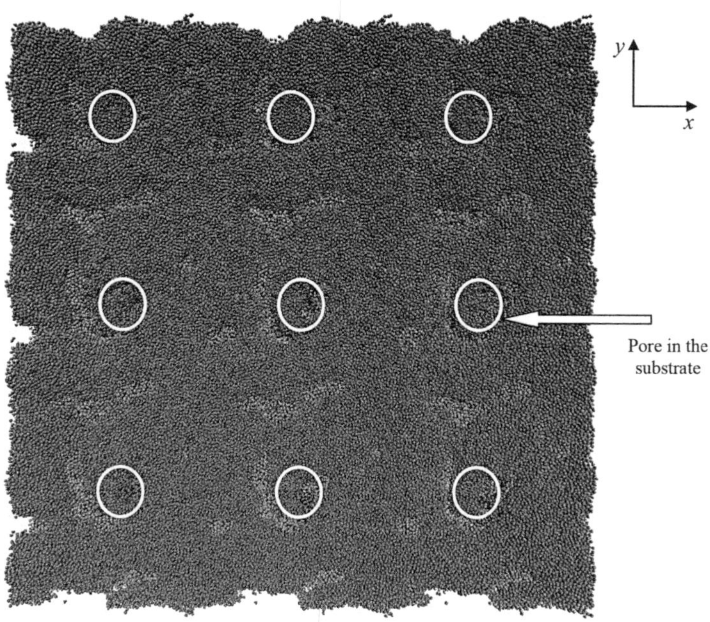

FIGURE 6.16 Siltation of the template of porous aluminum oxide with iron atoms: top view, precipitation time—0.05 ns.

The graphs of penetration of considered substance atoms into the pore on the aluminum oxide template are shown in Figure 6.17. The most penetration was demonstrated by gallium atoms, the least—by palladium ones. Gold demonstrated the best penetration among noble metals. After about 120 ps, the intensity of all atom penetration goes down and further it changes insignificantly that is seen in Figure 6.17. Iron and gallium atoms demonstrated the highest intensity at the initial modeling stage (approximately up to 90 ps).

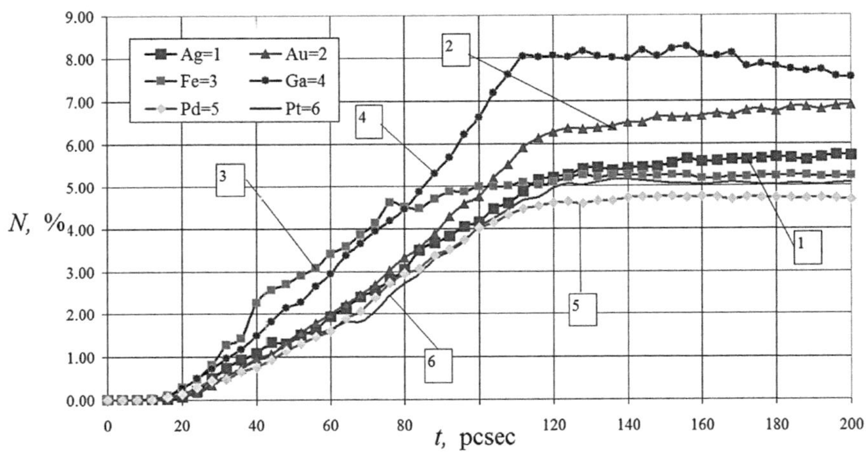

FIGURE 6.17 **(See color insert.)** Percentage of precipitated atoms that got into the pore to the total number of precipitated atoms.

Since the initial distribution of atoms has a stochastic character, it would be interesting to see which considered elements will penetrate the pore to the maximum depth. From the graph in Figure 6.18 it is seen that the considered atom types completely penetrate the pore at the initial stage (to the pore height). The penetration intensity at the initial stage (up to 20 ps) is similar.

The dynamics of the pore filling with precipitated atoms is given in Figure 6.19, where you can see the graphs of the depth of the mass center of atoms, which penetrated the pore. The depth of mass center is calculated only relatively to the atoms which filled the pore. Therefore, at the initial time moments (up to 20 ps), the greater shifting of the mass center is observed. Further, the number of atoms penetrating the pore grows, the dependencies in Figure 6.19 shift in the direction of the middle of the pore depth. The shift of the mass center reaches the most significant depth when iron and gallium atoms are precipitated.

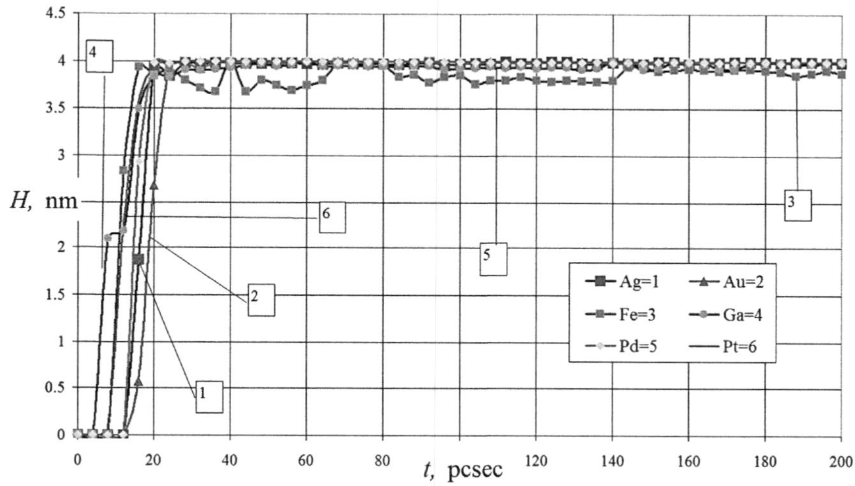

FIGURE 6.18 **(See color insert.)** Penetration depth into the pore of precipitated atoms.

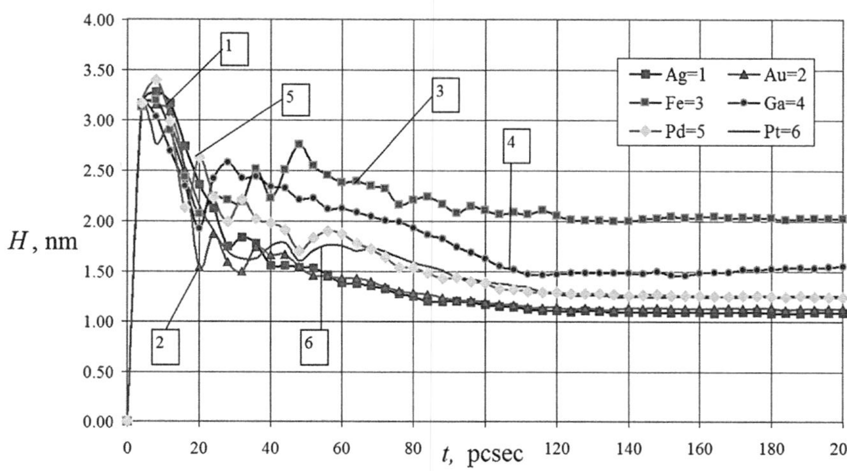

FIGURE 6.19 **(See color insert.)** Depth of the mass center of atoms, which penetrated the pore.

The pore radius in the template varied for the following series of computational experiments, the depth being the same (4 nm). Gallium atoms were used for precipitation as one of the most suitable to form nano-structured objects on the template. The percentage of Ga atoms, which got into the pore in relation to the total number of precipitated atoms are demonstrated in Figure 6.20. Analyzing the graphs in Figure 6.21, you can see that the number of atoms actively grows in the time interval of 20–120 ps. The pore siltation after 120 ps of condensation is followed by the reconstruction of the atomic structure that corresponds to the stabilization of dependencies and the slight decrease in the percentage of atoms, which penetrated the pore.

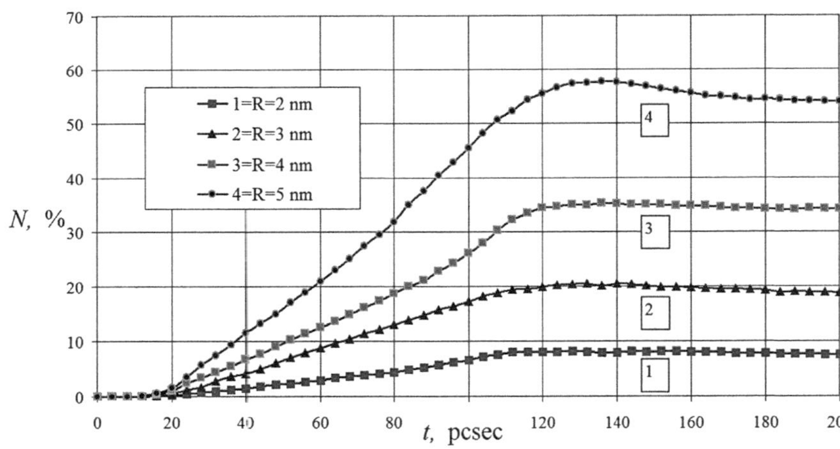

FIGURE 6.20 Percentage of Ga atoms, which got into the pore in relation to the total number of precipitated atoms.

Besides, during the siltation of pores of different radii with gallium atoms, the mass center of precipitated atoms stabilizes at different depths of the pore. For the pores with the radii of 2 and 3 nm, the mass center is formed above the middle of the pore depth. With the size growth, the mass center starts forming in one place—near the middle of the pore depth. This fact allows saying that the further radius growth (over 5 nm) will not significantly influence the mass center, the pore is already rather compactly filled with the precipitated atoms.

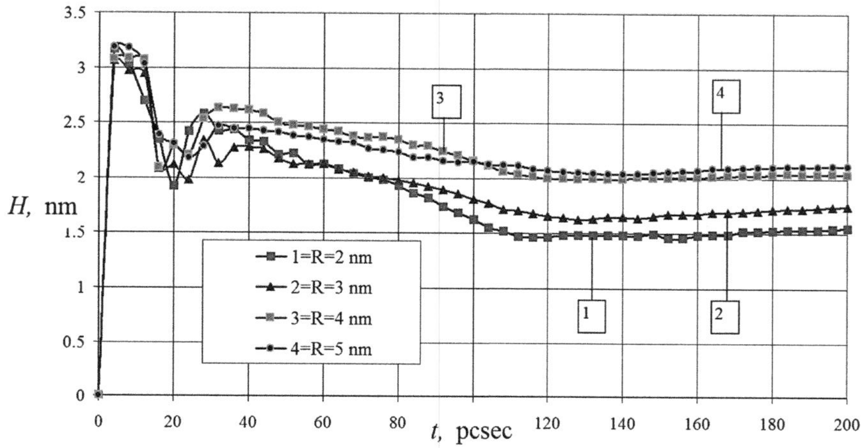

FIGURE 6.21 Percentage of Ga atoms, which got into the pores with different radii in relation to the total number of precipitated atoms.

6.5 CONCLUSIONS

The equations of MD with the use of multiparticle interaction potential—MEAM are given. The difference of MEAM potential, embedded atom potential, and the theory of electron density functional are considered.

The setting of the problem for studying the precipitation processes of nanosized films onto the templates of porous aluminum oxide is presented. The precipitation methods of nanosized films are used for certain technological processes and applied to forecast and design nanofilm materials.

The variants of epitaxial siltation of porous templates based on aluminum oxide with various types of atoms are illustrated. Different processes of nanostructure interaction and siltation mechanisms of templates and pores are registered for different types of precipitated atoms:

1. For the siltation with silver, gold, and magnesium atoms the uniform pore covering with nanofilm without the penetration of these atoms inside was observed. The nanofilm was slightly sinking in the pore region.
2. Chromium and iron atoms demonstrated the formation of nanostructures in the air environment above the template already,

the template was silted following the island principle. Small iron nanostructures were gradually growing on the template and grouping into the bigger ones. The formation of iron nanostructure was observed inside the pore.

3. When precipitating the gallium and germanium atoms, the pore was also not silted completely, the nanofilm was formed by regions on the template surface. Small gallium nanoparticles were seen on the template surface. When precipitated, germanium atoms penetrated the pore region more intensely, but they were pulled out onto the surface further in the process of nanofilm formation.
4. During the epitaxy of palladium, a uniform nanofilm was formed with slight sinking in the pore region. But the non-silted opening was observed immediately above the pore during the whole condensation stage.
5. Siltation of the template of porous aluminum oxide with titanium atoms was characterized by incoherent and coarse appearance of the nanofilm formed. The atoms got inside the pore only insignificantly and stayed on the template surface.
6. Siltation of the template was observed for its epitaxy with vanadium atoms. A large number of vanadium atoms got inside the pore, but non-silted regions remained on the template surface itself.

Single atoms, which reached the pore bottom, were observed for all types of precipitated atoms. The most complete and dense pore siltation was registered in the process of gallium epitaxy. The pore filled with atoms can be considered as a quantum point and used to obtain optic and electric effects.

When studying the siltation with gallium atoms of the coatings with pores of different sizes, it was found out that the number of atoms actively grows in the pore in the time interval of 20–120 ps. The pore siltation after 120 ps of condensation time is accompanied by the reconstruction of atomic structure that contributes to the stabilization of dependencies and a slight decrease in the percentage of gallium atoms, which penetrate the pore. Besides, the mass center of precipitated atoms is stabilized at different depths of the pore. For the pores with the radii of 2 and 3 nm, the mass center is formed above the middle of the pore depth. With the pore growth, the mass center starts to be formed in one place—near the middle of the pore depth. The further growth of the pore radius (over 5 nm) does not significantly influence the mass center, the pore is already quite compactly packed with the precipitated atoms.

The modeling results can be used when developing and optimizing technological processes of optic coating formation and the analysis of physical properties of nanofilms and nanostructures formed in pores and on surfaces of aluminum oxide templates.

ACKNOWLEDGMENT

The investigation was carried out in the frameworks of the state task to Kalashnikov ISTU (project No 1239) and financial support of the Russian Science Foundation (project No 15–19–10,002). In the frameworks of the state task, Kalashnikov ISTU corrected and extended the mathematical model, carried out computational experiments, and obtained main research results.

KEYWORDS

- **epitaxy**
- **nanofilms**
- **molecular dynamics modeling**
- **crystalline structure**

REFERENCES

1. Ying, J. Y. Nanoporous Systems and Templates the Unique Self-Assembly and Synthesis of Nanostructures. *Sci. Spectra.* **1999,** *18,* 56–63.
2. Li, A. P.; Muller, F.; Birner, A.; Nielsch, K.; Gosele, U. Hexagonal Pore Arrays with a 50–420 nm Interpore Distance Formed by Self-Organization in Anodic Alumina. *J. Appl. Phys.* **1998,** *84*(11), 6023–6026. DOI: 10.1063/1.368911.
3. Doroshenko, M. N.; Gerasimchuk, A. I.; Mazurenko, E. A. Kataliticheskoe vliyanie poverkhnosti na formirovanie nanotrubok germaniya PE MOCVD-metodom [The Catalytic Effect of the Surface on the Formation of Germanium Nanotubes PE CVD-method]. *Khimiya, fizika i tekhnologiya poverkhnosti* (*Chem. Phys. Surf. Technol.*) **2013,** *4*(4), 366–372.

4. Mu, C.; Yu, Y.; Liao, W.; Zhao, X.; Xu, D.; Chen, X.; Yu, D. Controlling Growth and Field Emission Properties of Silicon Nanotube Arrays by Multistep Template Replication and Chemical Vapour Deposition. *Appl. Phys. Lett.* **2005,** *87*(11), 1–3. DOI: 10.1063/1.2042545.
5. Melnik, Yu. V.; Nikolaev, A. E.; Stepanov, S. I.; Zubrilov, A. S.; Nikitina, I. P.; Vassilevski, K. V.; Tsvetkov, D. V.; Babanin, A. I.; Musikhin, Yu. G.; Tretyakov, V. V.; Dmitriev, V. A. AlN/GaN and AlGaN/GaN Heterostructures Grown by HVPE on SiC Substrates. *Mat. Res. Soc. Symp. Proc.* **1998,** *482,* 245–249.
6. Nikolaev, A. E.; Melnik, Y. V.; Kuznetsov, N. I.; Strelchuk, A. M.; Kovarsky, A. P.; Vassilevski, K. V.; Dmitriev, V. A. GaN pn-Structures Grown by Hydride Vapor Phase Epitaxy. *Mat. Res. Soc. Symp. Proc.* **1998,** *482,* 251–256.
7. Xu, H. J.; Li, X. J. Structure and Photoluminescent Properties of a ZnS/Si Nanoheterostructure Based on a Silicon Nanoporous Pillar Array. *Semicond. Sci. Technol.* **2009,** *24*(7), 75008.
8. Masuda, H. Highly Ordered Nanohole Arrays in Anodic Porous Alumina. In *Ordered Porous Nanostructures and Applications. Part of the Series Nanostructure Science and Technology;* Springer: US, 2005; 37–55.
9. Nakagawa, N.; Yoshioka, H.; Kuroda, C.; Ishida, M. Electrode Performance of a Thin-Film YSZ Cell Set on a Porous Ceramic Substrate by rf Sputtering Technique. *Solid State Ionics* **1989,** *35*(3–4), 249–255.
10. Lysenkov, D.; Engstler, J.; Dangwal, A.; Popp, A.; Müller, G.; Schneider, J. J.; Janardhanan, V. M.; Deutschmann, O; Strauch, P; Ebert, V.; Wolfrum, J. Nonaligned Carbon Nanotubes Anchored on Porous Alumina: Formation, Process Modeling, Gas-Phase Analysis, and Field-Emission Properties. *Small* **2007,** *3*(6), 974–985.
11. Toccafondi, C.; Zaccaria, R. P.; Dante, S.; Salerno, M. Fabrication of Gold-Coated Ultra-Thin Anodic Porous Alumina Substrates for Augmented SERS. *Materials* **2016,** *9(*6), 403.1–403.12.
12. Zaghdoudi, W.; Gaidi, M.; Chtourou, R. Microstructural and Optical Properties of Porous Alumina Elaborated on Glass Substrate. *J. Mater. Eng. Perform.* **2013,** *22*(3), 869–874.
13. Wadley, H. N. G.; Zhou, X.; Johnson, R. A.; Neurock, M. Mechanisms, Models and Methods of Vapor Deposition. *Prog. Mater. Sci.* **2001,** *46*(3–4), 329–377.
14. Song, H.; Ilegbusi, O. J.; Trakhtenberg, L. I. Modeling Vapor Deposition of Metal/Semiconductor-Polymer Nanocomposite. *Thin Solid Films* **2005,** *476*(1), 190–195.
15. Bruschi, L.; Mistura, G.; Nguyen, P. T. M.; Do, D. D.; Nicholson, D.; Parkd, S. J.; Lee, W. Adsorption in Alumina Pores Open at One and at Both Ends. *Nanoscale* **2015,** *7,* 2587–2596.
16. Amsellem, O.; Borit, F.; Jeulin, D.; Guipont, V.; Jeandin, M.; Boller, E.; Pauchet, F. Three-Dimensional Simulation of Porosity in Plasma-Sprayed Alumina Using Microtomography and Electrochemical Impedance Spectrometry for Finite Element Modeling of Properties. *J. Therm. Spray Technol.* **2012,** *21*(2), 193–201.
17. Lennard-Jones, J. E. On the Determination of Molecular Fields. II. From the Equation of State of a Gas. *Proc. Roy. Soc. A* **1924,** *106,* 463–477.
18. Stillinger, F. H.; Weber, T. A. Computer Simulation of Local Order in Condensed Phases of Silicon. *Phys. Rev. B* **1985,** *31,* 5262–5271.

19. Tersoff, J. New Empirical Approach for the Structure and Energy of Covalent Systems. *Phys. Rev. B* **1988,** *37*(12), 6991–7000.
20. Daw, M. S.; Baskes, M. I. Semiempirical, Quantum Mechanical Calculations of Hydrogen Embrittlement in Metals. *Phys. Rev. Lett.* **1983,** *50*(17), 1285–1288.
21. Daw, M. S. Model of Metallic Cohesion: The Embedded-Atom Method. *Phys. Rev. B* **1989,** *39*(11), 7441–7452.
22. Daw, M. S.; Baskes, M. I. Embedded-Atom Method: Derivation and Application to Impurities, Surfaces, and Other Defects in Metals. *Phys. Rev. B* **1984,** *29*(12), 6443–6453.
23. Baskes, M. I. Modified Embedded-Atom Potentials for Cubic Materials and Impurities. *Phys. Rev. B* **1992,** *46*(5), 2727–2742.
24. Ruda, M.; Farkas, D.; Abriata, J. Interatomic Potentials for Carbon Interstitials in Metals and Intermetallics. *Scr. Mater.* **2002,** *46*, 349–355.
25. Tomar, V.; Zhou, M. Classical Molecular-Dynamics Potential for the Mechanical Strength of Nanocrystalline Composite fcc Al+α-Fe_2O_3. *Phys. Rev. B* **2006,** *73*(17), 174116.1–174116.16.
26. Jelinek, B.; Houze, J.; Kim, S.; Horstemeyer, M. F.; Baskes, M. I.; Kim, S. G. Modified Embedded-Atom Method Interatomic Potentials for the Mg-Al Alloy System. *Phys. Rev. B* **2007,** *75*(5), 54106.
27. Kim, Y.-M.; Lee, B.-J.; Baskes, M. I. Modified Embedded-Atom Method Interatomic Potentials for Ti and Zr. *Phys. Rev. B* **2006,** *74*(1), 14101.
28. Vakhrushev, A. V.; Fedotov, A. Yu.; Severyukhin, A. V.; Suvorov, S. V. Modelling of Processes of Special Nanostructured Layers of Epitaxial Structures for Sophisticated Photovoltaic Cells. *Chem. Phys. Mesoscopics* **2014,** *16*(3), 364–380.
29. Vakhrushev, A. V.; Severyukhin, A. V.; Fedotov, A. Yu.; Valeev, R. G. Issledovanie protsessov osazhdeniya nanoplenok na podlozhku iz poristogo oksida alyuminiya metodami matematicheskogo modelirovaniya [Research Nanofilms Deposition Processes on a Substrate of Porous Aluminum Oxide by Means of Mathematical Simulation]. *Vychislitelnaya mehanika sploshnyh sred* (*Comput. Continuum Mech.)* **2016,** *9*(1), 59–72.
30. Vakhrushev, A. V.; Fedotov, A. Yu.; Severyukhin, A. V.; Valeev, R. G. Modelirovanie protsessov osazhdeniya nanoplenok na podlozhku poristogo oksida alyuminiya [Simulation of Deposition Films on Nano Porous Alumina Substrate]. *Himicheskaya fizika i mezoskopiya* (*Chem. Phys. Mesoscopics)* **2015,** *17*(4), 511–522.
31. LAMMPS Molecular Dynamics Simulator http://lammps.sandia.gov (accessed May 25, 2016).
32. Hoover, W. Canonical Dynamics: Equilibrium Phase-Space Distributions. *Phys. Rev. A* **1985,** *31*(3), 1695–1697.

CHAPTER 7

CALCULATION OF THERMAL CONDUCTIVITY COEFFICIENT OF HOMOGENEOUS NANOSYSTEMS

A. V. SEVERYUKHIN[1], O. YU. SEVERYUKHINA[1,2], and A. V. VAKHRUSHEV[1,2,*]

[1]*Institute of Mechanics, Ural Branch, Russian Academy of Sciences, Izhevsk, Russia*

[2]*Kalashnikov Izhevsk State Technical University, Izhevsk, Russia*

[*]*Corresponding author. E-mail: vakhrushev-a@yandex.ru*

CONTENTS

ABSTRACT

This chapter provides a physical basis and numerical methods of calculation of thermal conductivity of homogeneous nanosystems. Equations describing many-body potentials modified embedded atom method (MEAM), environment-dependent interatomic potential (EDIP), and so forth are considered. The features and differences of the interaction potentials used in molecular dynamics calculations are shown in the chapter. The temperature dependences of the heat conductivity coefficient for various types of materials are determined. Calculations of thermophysical characteristics of homogeneous nanosystems based on silicon and gold are performed. In this work, the formalism of Green–Kubo, which connects an autocorrelation function of a heat flux with a thermal conductivity, is used. The simulation was performed using the software package LAMMPS. The curves of the temperature dependence of the thermal conductivity coefficient for systems of various dimensions are presented. Comparison of the data, obtained with use of capacities of MEAM and EDIP, with experimental data is carried out. It is revealed that the shape of the curves and the values obtained in the simulation are in good agreement with experimental data. The nature of the curve of the temperature dependence of coefficient of thermal conductivity of gold corresponds to the submitted theoretical calculations for metals. This indicates the possibility of using the presented modeling techniques for predicting the thermophysical characteristics of various substances.

7.1 INTRODUCTION

At the end of the 20th century, the new cross-disciplinary field of science and technology which has received the name "Nanotechnologies," associated with studying and control of groups of atoms and molecules which characteristic sizes (or the size of any one or two directions) do not exceed 100 nm (1 nm = 10^{-9} m) was formed. In this case, as experimental studies show, the mechanical, electromagnetic, chemical, and other properties of nanoelements are several times different from the properties of materials in which atoms and molecules are organized by an ordinary way, known long ago from classical physics. It demands deep understanding of the physical and chemical processes[1] proceeding in nanoscale. However, the small scale of processes complicates their research by experimental methods and leads to the need for use of mathematical modeling.

Mathematical modeling is a powerful tool in the design of various types of nanosystems and analysis of the processes occurring in them.[2–6] It is especially important to use methods of mathematical modeling in new areas of science and technology, in which experience of exploitation and experimental data have not been accumulated yet. The main tasks of mathematical modeling in nanosystems are formation of nanoelements (a form, structure, and properties), interaction of individual elements of a nanosystem (formation and loading), definition of structure of the isolated nanosystem in quasistatic and dynamic states, calculation of parameters of a nanosystem at interaction with the environment and calculation of macroparameters of a nanosystem. The last task in the list of tasks, namely calculation of macroparameters of nanosystems depending on their size, structure, and the chemical composition, is the most actual at the moment. It is caused by the ongoing rapid transition from the study of the fundamental properties of nanosystems to the solution of problems of their production and application, in which the macroparameters of nanosystems are of decisive importance. In the submitted chapter, in development of the previous works of authors,[7,8] the problem of calculation of coefficient of thermal conductivity of homogeneous nanosystems is considered.

7.2 THEORETICAL BASIS

The issue of heat transfer in solids has been widely studied since the 18th century. In 1807, the French physicist and mathematician Jean Baptiste Joseph Fourier presented his work "Théorie de la propagation de la Chaleur dans les solides."[9] According to the Fourier law, the heat flux can be determined from the formula:

$$\vec{q} = -k\nabla\vec{T} \tag{7.1}$$

In this equation, the symbol $\vec{q}$ denotes the heat flux vector characterizing the heat flux passing through unit area per unit time, k is the thermal conductivity coefficient, and $\nabla\vec{T}$ is understood as temperature gradient. Even though more than 200 years have passed since its discovery, the Fourier law still provides a basis for modeling the processes of heat transfer in solids.

The physics of heat transfer has undergone considerable evolution within last two centuries. At the beginning of the 19th century, it was believed that an object called "calorie" transfers heat from a hot body to a cold one.[10] Classical thermodynamics was developed mainly based on this understanding. However, in 1914, the Dutch physicist Peter Josef Wilhelm Debye proposed a model in which heat transfer occurs due to the propagation of vibrational energy. It was the first correlation of heat transfer with the kinematics of fundamental particles. In the Debye model, the lattice vibrations are regarded as a gas of phonons—quasiparticles theoretically grounded by the Soviet physicist Igor Yevgenyevich Tamm in 1932. This type of heat transfer process is commonly called a lattice or phonon thermal conductivity.

Heat transfer can be carried out by two mechanisms: by free electrons and due to thermal vibrations of atoms (phonon thermal conductivity). Over the last century, an experiment has shown that heat is transmitted through the stochastic motion of elementary particles, such as electrons, atoms, and molecules.

In solids, the internal energy of the crystal lattice of the material increases with increase in temperature, and as a result, the amplitude of atoms vibrations also increases. Since the bond between the lattice atoms is sufficiently large, in the case of thermal excitation in one place of the lattice, it will be transmitted in the form of an elastic wave between them. The elastic wave reaching the surface of the body is reflected from it. Then there is a formation of standing wave as a result of the superposition of two waves (direct and reflected). Such a standing wave is called a normal mode and has a certain frequency ω. If we consider a system containing N atoms, then in such a system, it is possible to excite $3N$ normal modes. Accordingly, the frequency of the resulting oscillations can be denoted ω_i, where $i \in 1\ldots3N$. On a surface, there should be nodal points in which the vibration amplitude of standing wave equals to zero. The distance between nodes should be such that it contains an integer number of half-wavelengths (see Fig. 7.1).

Then the maximum oscillation frequency will be calculated according to the formula:

$$\omega_{\max} = 2\pi\upsilon_s / l_{\min} \cong \pi\upsilon_s / d,$$

where υ_s—speed of sound, d—distance between the atoms.

The minimum frequency of oscillation will be determined on the basis of the linear crystal size—L:

$$\omega_{min} = 2\pi \upsilon_s / l_{max} \cong \pi \upsilon_s / L.$$

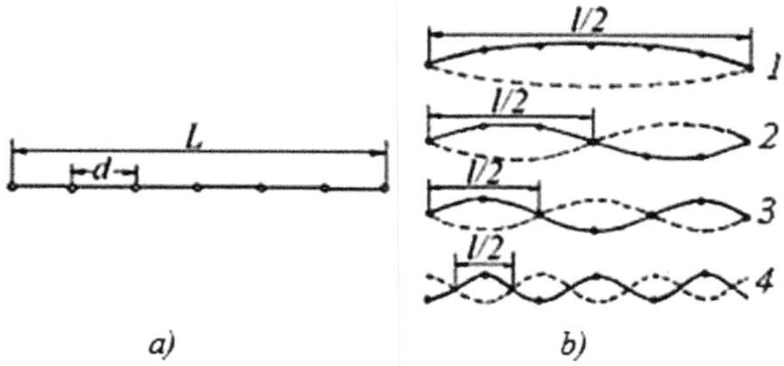

FIGURE 7.1 A one-dimensional model of solids (a) is a chain of atoms of length L and (b) are standing waves with a wavelength l: 1—$l = l_{max} = 2L$, 2—$l = L$, 3— $l = 2L/3$, 4— $l = l_{min} = 2d$.

On the basis of quantization of energy of normal modes and proportionality of the displacements of atoms by the forces of interatomic interaction (harmonic approximation), and as a consequence no interference of normal vibrations, their energy can be written in the form:

$$E_i = \hbar\omega_i\left(n + \frac{1}{2}\right), n = 0, 1, 2, ...,$$

where $\hbar = \dfrac{h}{2\pi}$—reduced Planck constant, ω_i—frequency of i-th normal mode.

Then the phonon will be understood as the minimum energy quantum of the normal mode of atoms.

7.2.1 THERMAL CONDUCTIVITY OF DIELECTRICS (LATTICE THERMAL CONDUCTIVITY)

It is possible to explain the temperature dependence and the finite value of the thermal conductivity of the lattice only by refusing to represent the phonon gas as ideal.

If we consider an ideal phonon gas, then even with the smallest change in temperature, a flow of phonons transporting thermal energy should arise. Let us consider the equation of Fourier's law for the one-dimensional case—heat distribution along a certain axis. It can be written as follows:

$$\vec{q} = -k\frac{dT}{dx}\vec{i}, \tag{7.2}$$

where $\frac{dT}{dx}$ is temperature gradient along the selected axis.

It can be seen from Formula (7.2) that in order for the heat flux to be finite for an infinitesimal change in temperature ($dT \to 0$), it is necessary that the value of the thermal conductivity coefficient be infinitely large. However, the coefficient of thermal conductivity in real bodies has a finite value. The reason for this is the interaction between phonons (phonon–phonon scattering), as a result of which part of the phonon energy is transferred to the lattice, which ensures the establishment of thermal equilibrium and limits the value of the thermal conductivity of the lattice.

If we apply the kinetic theory of gases to a phonon gas, then the thermal conductivity of the phonon gas will be written as:

$$k \approx \frac{1}{3}\rho C_V \langle l \rangle \langle \upsilon \rangle,$$

where ρ—gas density, C_V—volumetric heat capacity, $\langle l \rangle$ —average mean free path of phonons, and $\langle \upsilon \rangle$ —average speed of phonons.

In accordance with the Dulong–Petit law, the specific heat of solids does not depend on the temperature, but the change in the phonon concentration will be proportional to the temperature in the temperature range $T \gg \Theta_D$ (where $\Theta_D = \frac{\hbar\omega_{max}}{k_B}$ is the characteristic Debye temperature and k_B is the Boltzmann constant). Thus, the coefficient of thermal conductivity at a temperature $T \gg \Theta_D$ will depend only on the mean free path of the phonons and will be inversely proportional to the temperature and phonon concentration, that is, $k \sim \frac{1}{T}$. This agrees with the experimental data. Figure 7.2 shows the temperature dependence of the thermal conductivity of a single crystal of artificial sapphire (Al_2O_3).

In the temperature range $T \gg \Theta_D/20$, the mean free path of phonons will not depend on temperature, but will be determined directly by the dimensions of the considered system. Thus, the temperature dependence of the thermal conductivity coefficient is reduced to $k \sim T^3$. This qualitatively agrees with the experimental data.

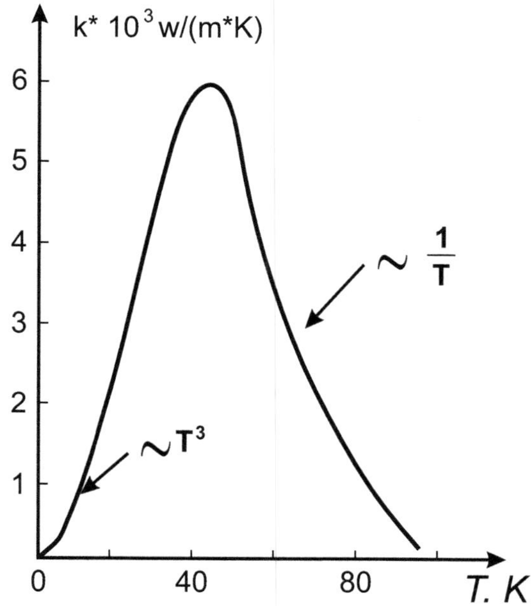

FIGURE 7.2 Temperature dependence of the thermal conductivity of a single crystal of artificial sapphire (Al_2O_3).

7.2.2 THERMAL CONDUCTIVITY OF METALS

In contrast to dielectrics, both heat transfer mechanisms work in metals. Thus, the coefficient of thermal conductivity for a given class of substances can be written as:

$$k = k_e + k_{ph}.$$

In pure metals, the process of heat transfer is carried out mainly by free electrons, whose concentration per unit volume in metals is very high. To determine the qualitative dependence of the electronic component of the thermal conductivity on temperature, it is possible to represent the entire set of metal electrons in the form of an electron gas, in analogy with the phonon gas described above. Then, using the kinetic theory of gases, we obtain:

$$k_e = \frac{1}{3} C_e \upsilon_F l_e$$

where C_e is heat capacity of electronic gas, equal to $C_e = \frac{\pi^2 T n k_B}{2E_F}$, where k_B is the Boltzmann constant, n is the concentration of free electrons; υ_F is the velocity of the electron gas corresponding to the Fermi energy E_F, and defined as $\upsilon_F = \sqrt{2E_F / m}$; l_e is the mean free path of the electrons. Thus, we get:

$$k_e = \frac{\pi^2 n k_B T l_e}{3m\upsilon_F}.$$

At a temperature $T \gg \Theta_D$, the mean free path of the electrons will be determined from the scattering of electrons by phonons. The concentration of phonons will be directly proportional to temperature. Then the mean free path of electrons will be inversely proportional to temperature $l_e \sim 1/T$, and the thermal conductivity will not depend on temperature. This situation is in good agreement with the experimental data presented in Figure 7.3.

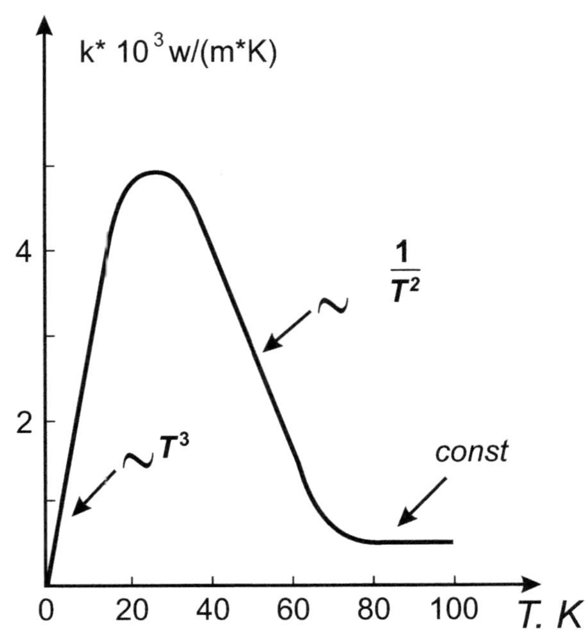

FIGURE 7.3 Temperature dependence of the thermal conductivity of cuprum (Cu).

As for the low-temperature range, there is a different picture according to the experimental data $l_e \sim 1/T^3$. Proceeding from this, we get $k_e \sim 1/T^2$. At temperature which is close to absolute zero T → 0, the phonon concentration is small, so that the thermal conductivity coefficient is directly proportional to the temperature $k_e \sim T$.

It should be noted that in pure metals under normal conditions, the thermal conductivity of the electron gas is much greater than the lattice thermal conductivity.

7.3 MODELING TECHNIQUE

The molecular dynamics was used as the modeling method. The method of molecular dynamics is widely used in modeling the behavior of nanosystems. In molecular dynamics calculations, the value of the thermal conductivity coefficient can be calculated in various ways.[11]

In this work, we use the Green–Kubo formalism, which connects the autocorrelation function of the heat flux with the coefficient of thermal conductivity.

The thermal conductivity coefficient in the Green–Kubo model is calculated by the following formula[12,13]:

$$k = \lim_{\tau\to\infty} \lim_{L\to\infty} \frac{1}{k_B T^2 L^d} \int_0^{\tau} \langle J(t) J(0) \rangle dt, \tag{7.3}$$

where *k*—coefficient of thermal conductivity of a *d*—dimensional system with a linear dimension *L*, *T*—temperature, k_B—Boltzmann constant, and *J*—component of the heat flux.

The time constant τ is the minimum time necessary for the autocorrelation function of the heat flux to decay to zero. This method is more suitable for anisotropic systems. The main disadvantage of this method is that the autocorrelation function of the heat flux takes a long time to attenuate to zero, and the resulting values of thermal conductivity depend on the size of the system.

The autocorrelation functions in Formula (7.3) are estimated in equilibrium, without a temperature gradient. Autocorrelation is a statistical relationship between random variables from one series, but taken with a shift, for example, for a random process—with a time shift. The autocorrelation function can be defined as:

$$\psi(\tau)=\int f(t)f(t-\tau)dt.$$

The total heat flux in the system is calculated as:

$$J(\mathrm{t})=\int j(x,t)dx\ ,$$

where $j(x,t)$ is the heat flux density.

The order of limits in Formula (7.3) is of great importance. At the correct orders of limits, it is possible to calculate correlation functions with any boundary conditions and to apply Formula (7.3). There are various forms of writing of Equation (7.3).[12–18]

There are situations when Formula (7.3) is not applicable. First, for small systems that are studied in mesoscopic physics, the thermodynamic limit does not make sense. Second, in many low-dimensional systems, heat transfer is anomalous and thermal conductivity deviates significantly from experimental values.[19,20] In such cases, it is impossible to take the limits in Formula (7.3). Therefore, the thermal conductivity is considered as a function of the length L. In the literature devoted to this topic,[19,20] usually in Formula (7.3), the upper limit of integration of t_c is changed to L. Another way to use the Green–Kubo formula for finite systems is to introduce infinite tanks, as done in refs 21–22.

In this chapter, the Green–Kubo method was chosen to calculate the thermal conductivity of silicon nanostructures. To calculate the thermal conductivity, we used the following Formula (7.3):[23]

$$k=\frac{1}{Vk_BT^2}\int_0^\infty \langle J_x(0)J_x(t)\rangle dt=\frac{1}{3Vk_BT^2}\int_0^\infty \langle J(0)\cdot J(t)\rangle dt, \tag{7.4}$$

where V is system volume.

The equations describing the considered atomic structure constitute a system of differential equations that determine the motion of all atoms of the system[24]:

$$m_i\frac{d^2\vec{r}_i}{dt^2}=\vec{F}_i(t,\vec{r}(t)), i=1,2,...,N, \tag{7.5}$$

$$t_0=0, \vec{r}_i(t_0)=\vec{r}_{i0}, \frac{d\vec{r}_i(t_0)}{dt}=\vec{\upsilon}_i(t_0)=\vec{\upsilon}_{i0}, i=1,2,..,N, \tag{7.6}$$

where m_i is the mass of i-th atom, N is the number of atoms in the system, $\vec{r}_{i0}$ is the initial radius vector of i-th atom, $\vec{r}_i$ is the current radius vector of i-th atom, and $\vec{F}_i(t,\vec{r}(t))$ is the total force acting on the i-th atom. Expression (7.6) defines the initial conditions for the system under consideration, where $\vec{\upsilon}_i$ and $\vec{\upsilon}_{i0}$ are initial and current speed of the i-th atom, respectively.

The forces $\vec{F}_i(t,\vec{r}(t))$ in Equation (7.5) are determined from the relation:

$$\vec{F}_i(t,\vec{r}(t)) = -\frac{\partial U(\vec{r}(t))}{\partial \vec{r}_i(t)},\ i = 1,2,...,N, \tag{7.7}$$

where $\vec{r}(t) = \{\vec{r}_1, \vec{r}_2, ..., \vec{r}_N\}$ and $U(\vec{r}(t))$ is some potential function describing the interactions of all atoms of the system. The interaction potential of the system can be written in the form:

$$U(\vec{r}(t)) = E.$$

The choice of the interaction potential plays a very important role in molecular dynamics. This problem is especially acute for metals and semiconductors. Silicon refers to materials with covalent bond types. Depending on the temperature, it has a different structure and properties. Therefore, under normal conditions, it is characterized by a diamond structure. With increasing pressure, new crystallographic structures can be formed in it, a simple cubic, face-centered cubic structure. In this case, there is a qualitative transition from one structure to another; there is an increase in the coordination number. The presence of the described features makes the problem of choosing an interatomic potential extremely difficult.

A lot of potentials for silicon are available in the LAMMPS package: Lennard-Jones, MEAM_1NN, MEAM_2NN, SW, Tersoff, BOP, MEAM_SPLINE_1, MEAM_SPLINE_2, MEAM_SW, and EDIP.

Taking into account only the pairwise interatomic interaction in mathematical modeling of metallic and/or semiconductor systems, a number of problems arise. In ref 25, it was shown that using only the pairwise interaction potential in metal and/or semiconductor, nonphysical relation for the coefficients is performed, Cauchy (C12=C44).

Pair potentials cannot provide realistic values of physical characteristics of material.[26] For the correct description of properties of solid bodies, it is necessary to use many-body potentials. It is known that any of the existing potentials is not capable to reproduce a full set of characteristics of solid substances. Thus, the choice of potential for mathematical modeling is a complex challenge. Many empirical potentials well describe volume properties of materials, but, nevertheless, some with success are used for the description and surface properties.

The following approaches, considering many-body interaction, have become most widespread in the modeling of metallic and semiconductor systems:

- The Stillinger–Weber potential;[27]
- The Abel–Tersoff potential;[28]
- The embedded atom method (EAM);[25,29]
- The modified EAM (MEAM);[30]
- The environment-dependent interatomic potential (EDIP).

In the EAM, the bond energy of the atomic system is represented as:

$$E = \sum_i F_i\left[\sum_{j\neq i} \rho_j(R_{ij})\right] + \frac{1}{2}\sum_{j\neq i} \varphi_{ij}(R_{ij}), \tag{7.8}$$

where $\sum F_i\left[\sum_{j\neq i} \rho_j(R_{ij})\right]$ is an embedding function, depending on the contribution to the electron charge density from atom j at the location of atom i; and $\varphi_{ij}(R_{ij})$ is a pairwise potential function. The derivation of this equation using the density functional theory (DFT) can be found in ref 29.

Each atom of the system is considered as the particle embedded in the electronic gas created by other atoms of the modeling system. The energy necessary for embedding depends on electronic density in an embedding point. The function of embedding entered thus allows to define an exchange and correlation energy of electronic gas of the system.

The sense of embedding function can be defined as the energy necessary for embedding of one atom in homogeneous electronic gas with density ρ. However, there are also other transformations[31] allowing to change Function (7.8) on condition that the resulting energy and interatomic forces would not change.

In EAM, the following approximations are used:

1. Function of electronic density of one atom is spherically symmetric function depending only on distance between atoms. This approach significantly limits a scope of EAM and allows to

consider systems in which the orientation of a covalent component of bond can be neglected.

2. Electronic density in the location of atom i is defined as linear superposition of electronic density of other atoms of system $\sum_{j\neq i}\rho_j(R_{ij})$. This approach significantly simplifies calculation of electronic density.
3. Value of $\sum_{j\neq i}\rho_j(R_{ij})$ in metal systems in the location of atom i changes poorly in comparison with the electronic density of the atom ρ_i. Thus, $\sum_{j\neq i}\rho_j(R_{ij})$ in the location of atom i is replaced with a constant $\bar{\rho}$.[29] Energy of electronic gas is approximated by the function depending only on the size of average value of electronic density in the field of embedding, but not difficult functionalities, as in the DFT method.

At present, EAM potentials are derived for most metals and some binary systems. Potentials for triple systems have also been calculated.[32] However, such "triple" potentials do not qualitatively reproduce the physical properties of materials.

Based on the described techniques, the semiempirical approach uniting advantages of many-body potentials and the embedded-atom method has been offered. The theory of the MEAM is output with the application of the DFT.[33] The DFT method is considered the most recognized approach to the description of electronic properties of solids now. In the EAM method, full electronic density is presented in the form of linear superposition spherically averaged functions. This defect is eliminated in the MEAM.

In the MEAM method, the total energy of the system is written in the following form:

$$E=\sum_i\left(Fi\left(\frac{\bar{\rho}}{Z_i}\right)+\frac{1}{2}\sum_{j\neq i}\varphi_{ij}(R_{ij})\right)$$

where F_i is the embedded function of atom i; Z_i is the number of nearest neighbors of atom i in its reference crystal structure; and φ_{ij} is pair potential between atoms i and j, located at a distance R_{ij}.

In MEAM, an embedding function $F(\rho)$ is defined as:

$$F(\rho)\,AE_c\,\rho\ln\rho$$

where A is an adjustable parameter and E_c is the cohesive energy.

The pair potential between atoms i and j is determined by the formula:

$$\varphi_{ij}(R) = \frac{2}{Z_i}\left\{E_i^u(R) - F_i\left(\frac{\overline{\rho}_i^0(R)}{Z_i}\right)\right\}.$$

The total electron density includes the angular dependences and is written in the form:

$$\overline{\rho} = \rho^{(0)}G(\Gamma).$$

There are many versions of function G(Γ). However, the most widespread is:

$$G(\Gamma) = \sqrt{1+\Gamma}.$$

The function Γ is calculated by the formula:

$$\Gamma = \sum_{h=1}^{3} t^{(h)}\left(\frac{\rho^{(h)}}{\rho^{(0)}}\right)^2,$$

where $h = 0 - 3$, correspond to *s, p, d, f* symmetry; $t^{(h)}$ is weighting factors; and $\rho^{(h)}$ are the quantities that determine the deviation of the distribution of the electron density from the distribution in an ideal cubic crystal $\rho^{(0)}$:

$$s(h=0):\rho^{(0)} = \sum_i \rho^{a(0)}\left(r^i\right),$$

$$p(h=1):\left(\rho^{(1)}\right)^2 = \sum_{\alpha}\left[\sum_i \rho^{a(1)}\left(r^i\right)\frac{r_\alpha^i}{r^i}\right]^2,$$

$$d(h=2):\left(\rho^{(2)}\right)^2 = \sum_{\alpha,\beta}\left[\sum_i \rho^{a(2)}\left(r^i\right)\frac{r_\alpha^i r_\beta^i}{r^{2i}}\right]^2 - \frac{1}{3}\sum_i\left[\sum_i \rho^{a(2)}\left(r^i\right)\right]^2,$$

$$f(h=3):\left(\rho^{(3)}\right)^2 = \sum_{\alpha,\beta,\gamma}\left[\sum_i \rho^{a(3)}\left(r^i\right)\frac{r_\alpha^i r_\beta^i r_\gamma^i}{r^{3i}}\right]^2.$$

Here, $\rho^{a(h)}$ are radial functions that represent a decrease in the contribution of distances r^i, the superscript i indicates the nearest atoms, and α, β, and

γ are summation indices for each of the three possible directions. Finally, the individual contribution is calculated from the formula:

$$\rho^{a(h)}(r) = \rho_0 e^{-\beta^{(h)}\left(\frac{r}{r_e}-1\right)}.$$

EDIP was first proposed in ref 34. In fact, it is a unification of formalisms of Stillinger–Weber and Tersoff and includes two-body and three-body interactions:

$$E = \frac{1}{2}\sum_i \left(\sum_{j\neq i} V_2\left(r_{ij}, Z_i\right) + \sum_{j\neq i}\ \sum_{k\neq i, k>j} V_3\left(r_{ij}, r_{ik}, Z_i\right)\right),$$

$$V_2\left(r, Z\right) = A\left[\left(\frac{B}{r}\right)^p - e^{-\beta Z^2}\right]\exp\left(\frac{\sigma}{r-a}\right),$$

$$V_3\left(r_{ij}, r_{ik}, Z_i\right) = g\left(r_{ij}\right) g\left(r_{ik}\right) h\left(\cos\theta_{ijk}, Z_i\right),$$

$$Z_i = \sum_{m\neq i} f\left(R_{im}\right),$$

$$f(r) = \begin{cases} 1, r < c \\ \exp\left(\dfrac{\alpha}{1-x^{-3}}\right), c < r < a, \\ 0, r > a \end{cases}$$

$$h(l, Z) = \lambda\left[\left(1 - e^{-Q(Z)(l+\tau(Z))^2}\right) + \eta Q(Z)(l+\tau(Z))^2\right],$$

$$Q(Z) = Q_0 e^{-\mu Z},$$

$$\tau(Z) = u_1 + u_2\left(u_3 e^{-u_4 Z} - e^{-2u_4 Z}\right),$$

where E is the total energy of the system per atom, θ_{ijk} is the angle between the faces ij and ik, r_{ij} is the unit vector directed from atom i to atom j, a is the cutoff radius, where Z_i is the coordination number, V_2 (r, Z) is the two-body interaction function, V_3 (r_{ij}, r_{ik}, Z_i) is the three-body interaction function, $f(r)$, $g(r)$, $h(l,Z)$ are the cutoff, radial, and angular functions, respectively.

As shown in ref 35, this type of potential describes amorphous silicon better than the Stillinger–Weber potential.

The potential of EDIP also has advantages. It reproduces the elastic characteristics of silicon well and predicts a melting point close to the experimental data.

7.4 SIMULATION RESULTS

Systems of various dimensions from $4 \times 4 \times 4$ to $4 \times 4 \times 144$ (in unit cells) were considered during the simulation. Calculations were carried out for semiconductor systems based on silicon single crystals and metallic systems based on pure gold. During the simulation, such types of interaction potentials as MEAM and EDIP were used.

The simulation was carried out using the LAMMPS software package. The integration step was 0.1 fs. The NVT ensemble was used in the calculations. The temperature was maintained using the Nose–Hoover thermostat.

Thermalization by this method consists of introduction of the effective frictional forces proportional to the velocities of particles with a dynamically changing coefficient ξ.

$$\frac{d^2 r_i}{dt^2} = \frac{F_i}{m_i} - \xi \frac{dr_i}{dt}.$$

Equations for the coefficient ξ are solved by numerical integration with respect to time, along with the integration of the equations of motion:

$$\frac{d\xi}{dt} = \frac{1}{Q}(T - T_0).$$

The first-order Euler scheme was used to solve the last differential equation with respect to ξ. Figure 7.4 shows the behavior of temperature when using Nose–Hoover thermostat.

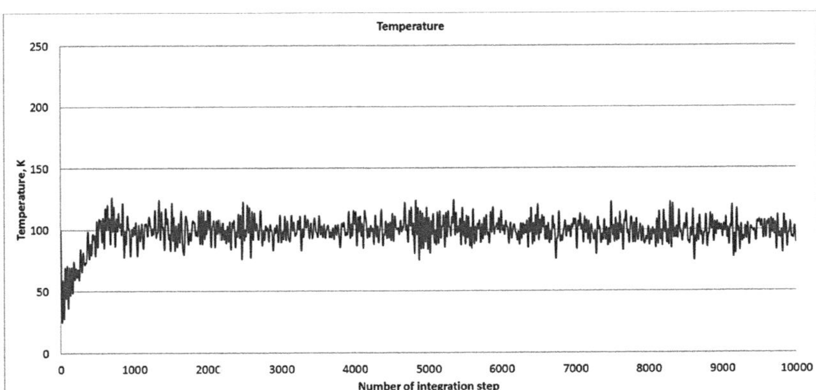

FIGURE 7.4 The behavior of temperature in the system when the Nose–Hoover thermostat is used.

The "mass" coefficient Q determines the rate at which the desired temperature is set. This parameter can be selected intuitively or set through other quantities. The convergence to a temperature of T_0 has the form of oscillations with period τ_T.

$$Q = \frac{\tau_T^2 T_0}{4\pi^2}.$$

The boundary conditions and the appearance of the simulated system are shown in Figure 7.5. Owing to periodic boundary conditions in all directions, only one nanowire of silicon was considered. Since the periodic boundary conditions provide for the mirror reflection of the calculated cell, then, in essence, the volume system was considered.

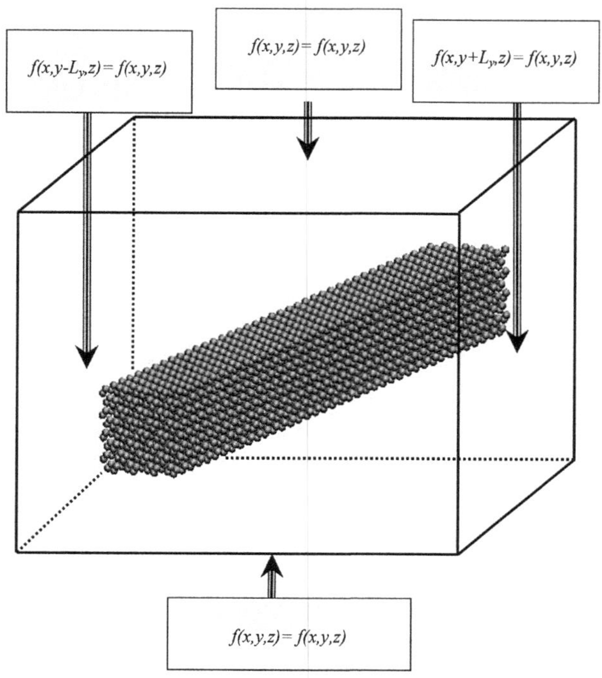

FIGURE 7.5 The boundary conditions and the appearance of the simulated system.

Despite the fact that the thermophysical characteristics of silicon systems are the subject of interest of many scientists, the data obtained by different groups of researchers are sometimes contradictory.

To verify the correctness of the proposed model, plots of the autocorrelation function of the heat flux were constructed, and the values of the function in the Green–Kubo formula were verified for convergence. Because of these operations, a conclusion was made about the correctness of the proposed model and the possibility of its use for calculating the thermophysical characteristics of nanosystems.

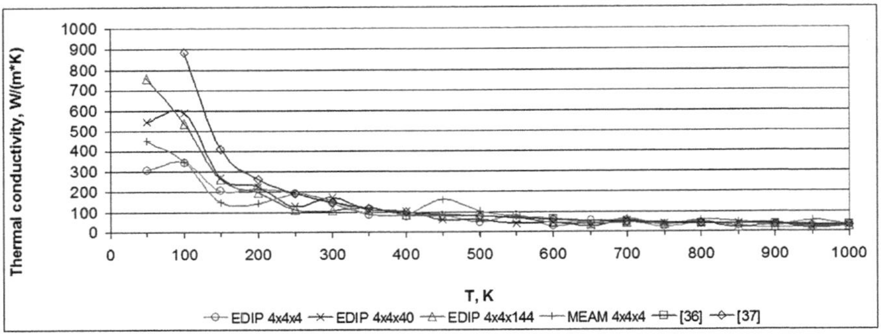

FIGURE 7.6 **(See color insert.)** Temperature dependence of the thermal conductivity of silicon, obtained from simulations and experiments [36,37].

Figure 7.6 shows the temperature dependence of the thermal conductivity of silicon for systems of different dimensions. As can be seen from the figure, the nature of the curves obtained during the simulation corresponds to the experimental data. The clearest correspondence is obtained at a temperature in the Debye temperature region and above it. This allows us to talk about the possibility of using the proposed model for calculating the thermophysical characteristics of nanosystems, as well as predicting the values of the calculated parameters for macrosystems.

It should be noted that at low temperatures, the size of the system affects the value of the thermal conductivity, which may be due to the use of the Green–Kubo method, but with increasing temperature, the effect of the size factor is neutralized.

The use of different types of potentials also affects the value of the thermal conductivity coefficient. This is confirmed by the graphs in Figure 7.7.

Based on the presented data, it can be concluded that at temperatures above the Debye temperature (645 ± 5 K for silicon), different potentials give similar values for the thermal conductivity. However, in the low-temperature region, the choice of the potential has a significant effect on the obtained values of the considered thermophysical characteristic.

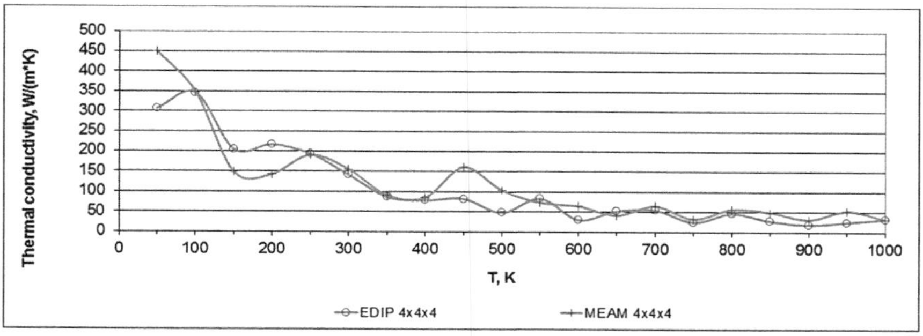

FIGURE 7.7 (See color insert.) Comparison of the values of the thermal conductivity of the $4\times4\times4$ system obtained using the modified embedded atom method (MEAM) and environment-dependent interatomic potential (EDIP) potentials.

Based on the results of modeling the process of thermal conduction in metal systems, particularly in gold crystals, similar dependencies were constructed. The curve for the temperature dependence of gold is shown in Figure 7.8.

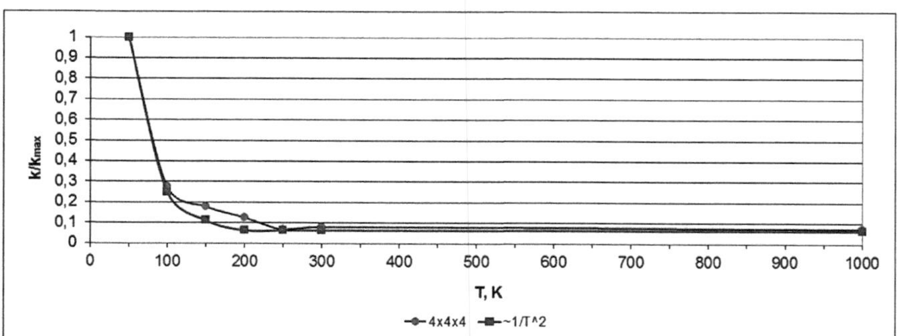

FIGURE 7.8 (See color insert.) Temperature dependence of the thermal conductivity of gold.

The behavior of the curve corresponds to the above-described laws for metals, which indicates good consistency with experimental data and theoretical calculations.

7.5 CONCLUSIONS

In this chapter, methods for calculating the thermophysical characteristics of a substance, in particular, the coefficient of thermal conductivity, for various substances are given. The theoretical temperature dependences of the coefficient of thermal conductivity of metals and dielectrics are described. A mathematical model of the process of heat conduction in nanosystems is presented. The methods for calculating the characteristics of this process are described. Moreover, the features and differences in the interaction potentials used in molecular dynamics calculations are shown in the chapter.

During the simulation, the thermal conductivity coefficients of the systems were calculated using materials such as silicon and gold. The obtained temperature dependences of the coefficients of thermal conductivity of silicon showed good agreement with the experimental data and the work of other researchers. The nature of the curve of the dependence of the thermal conductivity of gold on temperature corresponds to the theoretical calculations presented for metals, which in turn demonstrates the possibility of using the presented modeling techniques for predicting the thermophysical characteristics of various substances.

ACKNOWLEDGMENT

This work was carried out with financial support from the Research Program of the Ural Branch of the Russian Academy of Sciences: the Project 15–10–1–23.

KEYWORDS

- **Nanomaterials**
- **thermal conductivity**
- **simulation**
- **molecular dynamics**

REFERENCES

1. Suzdalev, I. P. *Nanotechnology: Physical Chemistry of Nanoclusters, Nanostructures and Nanomaterials;* KomKniga: Moscow, 2006; p 592.
2. Steinhauser, M. O. *Computational Multiscale Modeling of Fluids and Solids: Theory and Application;* Springer-Verlag: Berlin/Heidelberg, 2008; p 427.
3. Vakhrushev, A. V.; Lipanov, A. M.; Suetin, M. V. *Modeling of Processes of Accumulation of Hydrogen and Hydrocarbons by Nanostructures: Regular and Chaotic Dynamics;* Institute for Computer Research: Moscow-Izhevsk, 2008; p 118.
4. Alikin, V. N.; Vakhrushev, A. V.; Golubchikov, V. B.; Lipanov, A. M.; Serebrennikov, S. Y. *Development and Research of Aerosol Nanotechnologies;* Moscow: Mashinostroenie, 2010; p 196.
5. Suyetin, M. V.; Vakhrushev, A. V. Nanocapsule for Safe and Effective Methane Storage. *Nanoscale Res. Lett.* **2009,** *4*(11), 1267–1270.
6. Vakhrushev, A. V.; Fedotov, A. Y. Modeling the Formation of Composite Nanoparticles from the Gas Phase. *Int. Sci. J. Altern. Energy Ecol.* **2007,** *10*, 22–26.
7. Vakhrushev, A. V.; Severyukhin, A. V.; Severyukhina, O. Y.; Fedotov, A. Y. Investigation of the Thermophysical Properties of Silicon-Based Nanomaterials by the Green-Kubo Method Using the EDIP Potential. *Chem. Phys. Mesoscopy* **2016,** *18*(2), 187–198.
8. Severyukhin, A. V.; Severyukhina, O. Y.; Vakhrushev, A. V.; Fedotov, A. Y. *Study of Thermophysical Properties of Silicon Nanomaterials by the Green-Kubo method. Problems of Mechanics and Materials Science*, In Proceedings of the Institute of Mechanics, UrB RAS, Izhevsk, 2016; pp. 210–223.
9. Fourier, J. Théorie De La Propagation De La Chaleur Dans Les Solides. Institut de France, 1807.
10. Chen, G. *Nanoscale Energy Transport and Conversion: A Parallel Treatment of Electrons, Molecules, Phonons, and Photons;* Oxford University Press: USA, 2005.
11. http://lammps.sandia.gov/doc/Section_howto.html#calculating-thermal-conductivity (accessed April 6, 2016).
12. Green, M. S. Markoff Random Processes and the Statistical Mechanics of Time-Dependent Phenomena. II. Irreversible Processes in Fluids. *J. Chem. Phys.* **1954,** *22*, 398–413.
13. Kubo, R.; Yokota, M.; Nakajima, S. Statistical-Mechanical Theory of Irreversible Processes. II. Response to Thermal Disturbance. *J. Phys. Soc. Jpn.* **1957,** *12*, 1203–1211.
14. Mori, H. Statistical-Mechanical Theory of Transport in Fluids. *Phys. Rev.* **1958,** *112*, 1829–1842.
15. Green, M. S. Comment on a Paper of Mori on Time-Correlation Expressions for Transport Properties. *Phys. Rev.* **1960,** *119*, 829–830.
16. Kadanoff, L. P.; Martin, P. C. Hydrodynamic Equations and Correlation Functions. *Ann. Phys.* **1963,** *24*, 419–469.
17. Luttinger, J. M. Theory of Thermal Transport Coefficients. *Phys. Rev.* **1964,** *135*, A1505–A1514.
18. Visscher, W. M. Transport Processes in Solids and Linear-Response Theory. *Phys. Rev. A* **1974,** *10*, 2461–2472.

19. Lepri, S.; Livi, R.; Politi, A. Thermal Conduction in Classical Low-Dimensional Lattices. *Phys. Rep.* **2003,** *377*, 1–80.
20. Kundu, A.; Dhar, A.; Narayan, O. The Green-Kubo Formula for Heat Conduction in Open Systems. *J. Stat. Mech.* **2009,** *3*, L03001.
21. Allen, K. R.; Ford, J. Lattice Thermal Conductivity for a One-Dimensional, Harmonic, Isotopically Disordered Crystal. *Phys. Rev*. **1968,** *176*, 1046–1055.
22. Fisher, D. S.; Lee, P. A. Relation Between Conductivity and Transmission Matrix. *Phys. Rev. B* **1981,** *23*, 6851–6854.
23. http://lammps.sandia.gov/doc/compute_heat_flux.html (accessed April 6, 2016).
24. Vakhrushev, A. V.; Severyukhina, O. Y.; Severyukhin, A. V.; et al. Simulation of the Processes of Formation of Quantum Dots on the Basis of the Transition Metals. *Nanomech. Sci. Tech. Int. J.* **2012,** *3*, 51–75.
25. Daw, M. S.; Baskes, M. I. Embedded-Atom Method: Derivation and Application to Impurities, Surfaces, and Other Defects in Metals. *Phys. Rev. B* **1984,** *29*(12), 6443–6453.
26. http://www.fisica.uniud.it/~ercolessi/ (accessed April 6, 2016).
27. Stillinger, F. H.; Weber, T. A. Computer Simulation of Local Order in Condensed Phases of Silicon. *Phys. Rev. B* **1985,** *31*, 5262–5271.
28. Tersoff, J. New Empirical Approach for the Structure and Energy of Covalent Systems. *Phys. Rev. B* **1988,** *37*(12), 6991–7000.
29. Daw, M. S. Model of Metallic Cohesion: The Embedded-Atom Method. *Phys. Rev. B* **1989,** *39*(11), 7441–7452.
30. Baskes, M. I. Modified Embedded-Atom Potentials for Cubic Materials and Impurities. *Phys. Rev. B* **1992,** *46*, 2727–2742.
31. Ruda, M.; Farkas, D.; Abriata, J. Interatomic Potentials for Carbon Interstitials in Metals and Intermetallics. *Scripta Mater*. **2002,** *46*(5), 349–355.
32. Tomar, V.; Zhou, M. Classical Molecular-Dynamics Potential for the Mechanical Strength of Nanocrystalline Composite fcc Al+α-Fe_2O_3. *Phys. Rev. B* **2006,** *73*(17), 174, 116.
33. Hohenberg, P.; Kohn, W. Inhomogeneous Electron Gas. *Phys. Rev. B* **1964,** *136*(3), 864–871.
34. Justo, J. F.; Bazant, M. Z. Interatomic Potential for Silicon Defects and Disordered Phases. *Phys. Rev. B* **1998,** *58*(5), 2539–2550.
35. Clark, L. A.; Xianglong, Y.; Martin, Z. B.; Linn, W. H. Elastic Constants of Defected and Amorphous Silicon with the Environment-Dependent Interatomic Potential. *Phys. Rev. B* **2004,** *70*, 134–113.
36. Howell, P. C. Comparison of Molecular Dynamics Methods and Interatomic Potentials for Calculating the Thermal Conductivity of Silicon. *J. Chem. Phys.* **2012,** *137*, 224, 111.
37. Sheludyak, Y. E.; Kashporov, L. Y.; Malinin, L. A.; Tsalkov, V. N. Thermophysical Properties of Components of Combustible Systems, Handbook; Silin, N. A., Ed.; NGO "Inform TEI": Moscow, 1992; p 184.

CHAPTER 8

NANOMATERIALS, MOLECULAR ION MAGNETS, AND ULTRASTRONG AND SPIN-ORBIT COUPLINGS IN QUANTUM MATERIALS

FRANCISCO TORRENS[1,*] and GLORIA CASTELLANO[2]

[1]*Institut Universitari de Ciència Molecular, Universitat de València, Edifici d'Instituts de Paterna, P. O. Box 22085, E-46071 València, Spain*

[2]*Departamento de Ciencias Experimentales y Matemáticas, Facultad de Veterinaria y Ciencias Experimentales, Universidad Católica de Valencia San Vicente Mártir, Guillem de Castro-94, E-46001 València, Spain*

**Corresponding author. E-mail: torrens@uv.es*

CONTENTS

ABSTRACT

An ultrastrong coupling between photons and *qubits*, 10 times larger than ever seen before, opens the door to a domain of physics and applications deemed unattainable until recently. Nuclear magnetic resonance experiments confirmed the presence of higher order interactions and lattice distortions, in the quantum ferromagnetic phase transition, in the double perovskite Ba_2NaOsO_6.ç

8.1 INTRODUCTION

In earlier publications, fractal hybrid-orbital analysis,[1,2] resonance,[3] molecular diversity,[4] periodic table of the elements,[5,6] law, property, information entropy, molecular classification, simulators,[7–12] labor risk prevention, and preventive health care at work with nanomaterials (NMs)[13–15] were reviewed. In the present report, the aim is to understand NMs in plastic industry, coherent manipulation of three-quantum-binary-digit (*bit*) (*qubit*) states in a molecular single-ion magnet for quantum simulation, molecular nanoscience, ultrastrong coupling (USC) of a single artificial atom to an electromagnetic continuum in the nonperturbative regime, magnetism and local symmetry breaking in a Mott insulator with strong spin–orbit interactions in quantum materials, and C nanotube (C-NT) transistors opening up to innovative scales.

8.2 NANODESK-I STAKEHOLDER'S DAY: NANOMATERIALS IN PLASTIC INDUSTRY

Instituto Valenciano de Seguridad y Salud en el Trabajo organized NanoDESK-I Stakeholder's Day: advanced web-based tools to promote the application of nanotechnology and safe use of NMs in the plastic industry.[16] Fito proposed questions (Qs)/answers (As)/fact (F) on nano-specific challenges to the industry and regulatory issues/current knowledge on toxicological profile/exposure potential.[17]

Q1. Do we have a definition?

Q2. Do you know what engineered NMs (ENMs) mean?

A2. A size of 1–100 nm in at least one dimension.

Q3. An emerging risk: what do we know?

A3. Toxicity is studied but neither occupational exposure nor risk management (RM) strategies.

F1. Important ENMs for plastic nanocomposites (NCs) are: Ag, ZnO, SiO_2, $CaCO_3$, and nanoclay.

Espiña proposed F on ENMs applications in plastic industry and road-mapping exercise.[18]

F2. Important ENMs for plastic NCs are: Ag, metal oxide (Al_2O_3, Sb_2O_5/SnO_2, TiO_2, ZnO), SiO_2, $CaCO_3$, nanoclay (montmorillonite), and carbon nanotubes (C-NTs).

Fito proposed F on NanoDESK Project: applied research to promote safe/responsible nanotechnology uptake/use in plastic industry.

F3. Important ENMs for plastic NCs are: Ag, Al_2O_3, $CaCO_3$, nanoclay, Sb_2O_5, SiO_2, etc.

Domat proposed Q/A on recommended tools for risk assessment (RA) RM.[19]

Q4. What are we measuring?

Q5. Why do we measure it?

A5. Because most access is through inhalation route.

Q6. Why do we measure them?

A6. Their larger surface areas cause larger reactivity.

Q7. RA, does its general protocol work for nanoparticles (NPs)?

Ricarte raised the following question on multi-criteria decision support tools.[20]

Q8. What is the most appropriate nanofiller?

Gozalbes raised Q on quantitative structure–activity relationship models optimization/development.[21]

Q9. In Registration, Evaluation, Authorization and Restriction of Chemicals regulation, use NPs?

A9. Do it safely!

Aceti raised questions on occupational and environmental exposure estimation models.[22]

Q10. Why exposure models?

Q11. Why models?

Q12. What is the dustiness of the substance?

Q13. What kind of local exhaust ventilation is present?

Santamaría Coria proposed Q/A on observatory on safety issues of polymer-based NCs.[23]

Q14. What information can be obtained in such observatory?

Q15. Who does it supply information by sectors to such observatory?

A15. Enterprises and associations.

Additional questions were raised.

Q16. Is there an imminent risk or not in working with ENMs?

Q17. How to persuade industry with NPs risks and expensive prices?

Q18. What properties in one's day in day out do you think that are important for ENMs?

8.3 *QUBIT* STATES COHERENT MANIPULATION IN A MOLECULAR ION MAGNET

Coronado group enhanced coherence in molecular spin *qubits* through atomic clock transitions.[24] They studied the quantum spin dynamics of nearly isotropic Gd^{3+} entrapped in polyoxometalate molecules and diluted in crystals of a diamagnetic Y^{3+} derivative.[25] The full energy-level spectrum and orientations of the magnetic anisotropy axes were determined through continuous-wave electron paramagnetic resonance, through X-band (9–10 GHz) cavities and on-chip superconducting waveguides (WGs), and 1.5 GHz resonators. The results showed that seven allowed transitions between $2S+1$ spin states can be separately addressed. Spin coherence T_2 and spin–lattice relaxation T_1 rates were measured for every transition in properly oriented single crystals. Quantum spin coherence is limited by residual dipolar interactions with neighbor electronic spins. Coherent Rabi oscillations were observed for all transitions. Rabi frequencies rose with microwave power and agreed quantitatively with predictions based on the spin Hamiltonian of the molecular spin. They argued that the spin states of every Gd^{3+} could be mapped onto the states of three addressable *qubits* (or of a $d=8$-level *qudit*), for which seven allowed transitions formed a universal set of operations. Within the scheme, one of the coherent oscillations observed experimentally provides an implementation of a controlled–controlled-NOT (Toffoli) three-*qubit* gate.

8.4 MOLECULAR NANOSCIENCE: FROM FUNCTIONAL MOLECULES TO DEVICES

The molecular region of nanoscience was scarcely explored, maybe because the larger structural and electronic complexity of molecules, compared to that found in simpler atom-based nano-objects and nano-structures, makes them more difficult to study and manipulate at the nanoscale with available instrumental techniques. Albeit, it is in the molecular region where molecular chemists, biologists, physicists, and engineers working in nanosciences may find the best opportunities to interact and converge. Coronado group used the magnetic molecular systems to illustrate how molecular nanoscience can be useful to design functional molecules, materials, and devices, which can be of interest in emerging areas [e.g., quantum computing and molecular spin electronics (spintronics)].[26,27]

8.5 ULTRASTRONG COUPLING ACHIEVED BETWEEN LIGHT AND MATTER

The study of light–matter interaction led to important advances in quantum optics and enabled numerous technologies. Progress was made in increasing the strength of the interaction at the single-photon level. A major achievement was the demonstration of the *strong coupling regime*, a key advancement enabling progress in quantum information science. Lupascu group showed light–matter interaction over an order of magnitude stronger than previously reported, reaching the nonperturbative USC regime[28]. They achieved it through a superconducting artificial atom tunably coupled to the electromagnetic continuum of a one-dimensional (1D) WG. For the largest coupling, the spontaneous emission rate of the atom exceeds its transition frequency. In USC regime, the description of atom and light as distinct entities breaks down, and a new description in terms of hybrid states is required. Beyond light–matter interaction itself, the tunability of their system makes it a promising tool to study a number of important physical systems (e.g., spin-boson and Kondo models).

8.6 NUCLEAR MAGNETIC RESONANCE CONFIRMS SPIN–ORBIT COUPLING DRIVES MAGNETIC QUANTUM PHASE TRANSITION

Nuclear magnetic resonance experiments confirmed the presence of higher order interactions and lattice distortions, in the quantum ferromagnetic (FM) phase transition, in the double perovskite Ba_2NaOsO_6. The study of the combined effects of strong electronic correlations with spin–orbit coupling (SOC) represents a central issue in quantum materials research. Predicting emergent properties represents a huge theoretical problem since SOC presence implies that the spin is not a good quantum number. Existing theories propose the emergence of a multitude of exotic quantum phases, distinguishable by local point symmetry breaking or local spin expectation values, even in materials with simple cubic crystal structure, for example, Ba_2NaOsO_6. Experimental tests of the theories by local probes are highly sought for. Mitrović group local measurements, designed to concurrently probe spin and orbital/lattice degrees of freedom of Ba_2NaOsO_6, provided such tests.[29] They showed that a canted FM phase, which is preceded by local point symmetry breaking, was stabilized at low temperatures, as predicted by quantum theories involving multipolar spin interactions.

8.7 CARBON NANOTUBE TRANSISTORS OPEN UP TO INNOVATIVE SCALES

Giordani investigated the rising use of C-NTs to produce next-generation semiconductor technologies.[30]

ACKNOWLEDGMENT

Francisco Torrens belongs to the Institut Universitari de Ciència Molecular, Universitat de València. Gloria Castellanobelongs to the Departamento de Ciencias Experimentales y Matemáticas, Facultad de Veterinaria y Ciencias Experimentales, Universidad Católica de Valencia *San Vicente Mártir*. The authors thank for the support from Generalitat Valenciana (Project No. PROMETEO/2016/094) and Universidad Católica de Valencia San Vicente Mártir (Projects No. UCV.PRO.17-18.AIV-03).

KEYWORDS

- **quantum information**
- **quantum optics**
- **qubit**
- **superconducting device**

REFERENCES

1. Torrens, F. Fractals for Hybrid Orbitals in Protein Models. *Complex. Int.* **2001,** *8*, 1–13.
2. Torrens, F. Fractal Hybrid-Orbital Analysis of the Protein Tertiary Structure. *Complex. Int.* (in press).
3. Torrens, F.; Castellano, G. Resonance in Interacting Induced-Dipole Polarizing Force Fields: Application to Force-Field Derivatives. *Algorithms* **2009,** *2*, 437–447.
4. Torrens, F.; Castellano, G. Molecular Diversity Classification via Information Theory: A review. *ICST Tran. Compl. Syst.* **2012,** *12*(10–12), e4–e1–8.
5. Torrens, F.; Castellano, G. Reflections on the Nature of the Periodic Table of the Elements: Implications in Chemical Education. In *Synthetic Organic Chemistry*; Seijas, J. A., Vázquez Tato, M. P., Lin, S. K., Eds.; MDPI: Basel, Switzerland, 2015; vol. 18, pp 1–15.
6. Putz, M. V., Ed. *The Explicative Dictionary of Nanochemistry;* Apple Academic Press–CRC: Waretown, NJ, in press.
7. Torrens, F.; Castellano, G. Reflections on the Cultural History of Nanominiaturization and Quantum Simulators (Computers). In *Sensors and Molecular Recognition*; Laguarda Miró, N., Masot Peris, R., Brun Sánchez, E., Eds.; Universidad Politécnica de Valencia: València, Spain, 2015; vol. 9; pp 1–7.
8. Torrens, F.; Castellano, G. Ideas in the History of Nano/Miniaturization and (Quantum) Simulators: Feynman, Education and Research Reorientation in Translational Science. In *Synthetic Organic Chemistry*; Seijas, J. A., Vázquez Tato, M. P., Lin, S. K., Eds.; MDPI: Basel, Switzerland, 2016; vol. 19, pp 1–16.
9. Torrens, F.; Castellano, G. Nanominiaturization and Quantum Computing. In *Sensors and Molecular Recognition*; Costero Nieto, A. M., Parra Álvarez, M., Gaviña Costero, P., Gil Grau, S., Eds.; Universitat de València: València: Spain, 2016; vol .10, pp 1–5.
10. Torrens, F.; Castellano, G. Nanominiaturization, Classical/Quantum Computers/ Simulators, Superconductivity and Universe. In *Methodologies and Applications for Analytical and Physical Chemistry*; Haghi, A. K., Thomas, S., Palit, S., Main, P., Eds.; Apple Academic Press–CRC: Waretown, NJ, in press.
11. Torrens, F.; Castellano, G. Superconductors, Superconductivity, BCS Theory and Entangled Photons for Quantum Computing. In *Physical Chemistry for Engineering*

and Applied Sciences: Theoretical and Methodological Implication; Haghi, A. K., Aguilar, C. N., Thomas, S., Praveen, K. M., Eds.; Apple Academic Press–CRC: Waretown, NJ; in press.
12. Torrens, F.; Castellano, G. EPR Paradox, Quantum Decoherence, Qubits, Goals and Opportunities in Quantum Simulation. In *Innovations in Physical Chemistry*; Haghi, A. K., Ed.; Apple Academic Press–CRC: Waretown, NJ; vol. 5, in press.
13. Torrens, F.; Castellano, G. Book of Abstracts, Certamen Integral de la Prevención y el Bienestar Laboral, València, Spain, September 28–29, 2016; Generalitat Valenciana–INVASSAT: València, Spain, 2016; P–P3.
14. Torrens, F.; Castellano, G. Nanoscience: From a Two-Dimensional to a Three-Dimensional Periodic Table of the Elements. In *Methodologies and Applications for Analytical and Physical Chemistry*; Haghi, A. K., Thomas, S., Palit, S., Main, P., Eds.; Apple Academic Press–CRC: Waretown, NJ, in press.
15. Torrens, F.; Castellano, G. Book of Abstracts, Congreso Internacional de Tecnología, Ciencia y Sociedad, València, Spain, October 19–20, 2017; Universidad Cardenal Herrera CEU: València, Spain, 2017; P–P1.
16. Book of Abstracts, NanoDESK-I Stakeholder's Day, Burjassot, València, Spain, March 28, 2017, INVASSAT, Burjassot, València, Spain.
17. Fito, C. Book of Abstracts, NanoDESK-I Stakeholder's Day, Burjassot, València, Spain, March 28, 2017, INVASSAT,. Burjassot, València, Spain; O1.
18. Espiña, B. Book of Abstracts, NanoDESK-I Stakeholder's Day, Burjassot, València, Spain, March 28, 2017, INVASSAT,. Burjassot, València, Spain; O2.
19. Domat, M. Book of Abstracts, NanoDESK-I Stakeholder's Day, Burjassot, València, Spain, March 28, 2017, INVASSAT,. Burjassot, València, Spain; O4.
20. Ricarte, S. Book of Abstracts, NanoDESK-I Stakeholder's Day, Burjassot, València, Spain, March 28, 2017, INVASSAT,. Burjassot, València, Spain; O5.
21. Gozalbes, R. Book of Abstracts, NanoDESK-I Stakeholder's Day, Burjassot, València, Spain, March 28, 2017, INVASSAT,. Burjassot, València, Spain; O7.
22. Aceti, F. Book of Abstracts, NanoDESK-I Stakeholder's Day, Burjassot, València, Spain, March 28, 2017, INVASSAT,. Burjassot, València, Spain; O8.
23. Santamaría, C. E. Book of Abstracts, NanoDESK-I Stakeholder's Day, INVASSAT, Burjassot (València), March 28, 2017; O9.
24. Shiddiq, M.; Komijani, D.; Duan, Y.; Gaita-Ariño, A.; Coronado, E.; Hill, S. Enhancing Coherence in Molecular Spin Qubits Via Atomic Clock Transitions. *Nature* **2016,** *531*, 348–351.
25. Jenkins, M. D.; Duan, Y.; Diosdado, B.; García-Ripoll, J. J.; Gaita-Ariño, A.; Giménez-Saiz, C.; Alonso, P. J.; Coronado, E.; Luis, F. Coherent Manipulation of Three-Qubit States in a Molecular Single-Ion Magnet. *Phys. Rev. B* **2017,** *95*, 064423–064418.
26. Dugay, J.; Aarts, M.; Giménez-Marqués, M.; Kozlova, T.; Zandberger, H. W.; Coronado, E.; van der Zant, H. S. J. Phase Transitions in Spin-Crossover Thin Films Probed by Graphene Transport Measurements. *Nano Lett.* **2017,** *17*, 186–193.
27. Prieto-Ruiz, J. P.; Miralles, S. G.; Großmann, N.; Aeschlimann, M.; Cinchetti, M.; Prima-García, H.; Coronado, E. Design of Molecular Spintronics Devices Containing Molybdenum Oxide As Hole Injection Layer. *Adv. Electron. Mater.* **2017**, *3*, 1600366–160031-6.

28. Forn-Díaz, P.; García-Ripoll, J. J.; Peropadre, B.; Orgiazzi, J. L.; Yurtalan, M. A.; Belyansky, R.; Wilson, C. M.; Lupascu, A. Ultrastrong Coupling of a Single Artificial Atom to An Electromagnetic Continuum in the Nonperturbative Regime. *Nat. Phys.* **2017,** *13*, 39–43.
29. Lu, L.; Song, M.; Liu, W.; Reyes, A. P.; Kuhns, P.; Lee, H. O.; Fisher, I. R.; Mitrović, V. F. Magnetism and Local Symmetry Breaking in a Mott Insulator with Strong Spin Orbit Interactions. *Nat. Commun.* **2017,** *8*, 14407–14418.
30. Giordani, A. Carbon nanotube transistors open up to innovative scales. *Sci. Comput. World* **2017,** *2017*(153), 10–14.

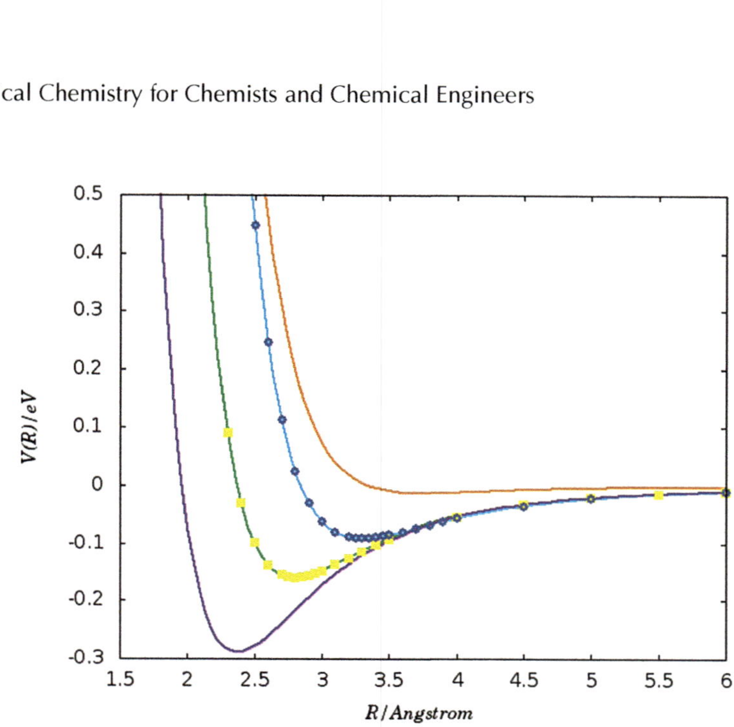

FIGURE 3.1 Potential energy curves for Li^+–Ar (magenta), Na^+–Ar (green), K^+–Ar (blue), and Ar–Ar (orange) interactions. The Ar–Ar curve is from ref 48, while the ion Ar ones correspond to fits of Equation 3.1 to ab initio points (also represented for Na^+–Ar and K^+–Ar).

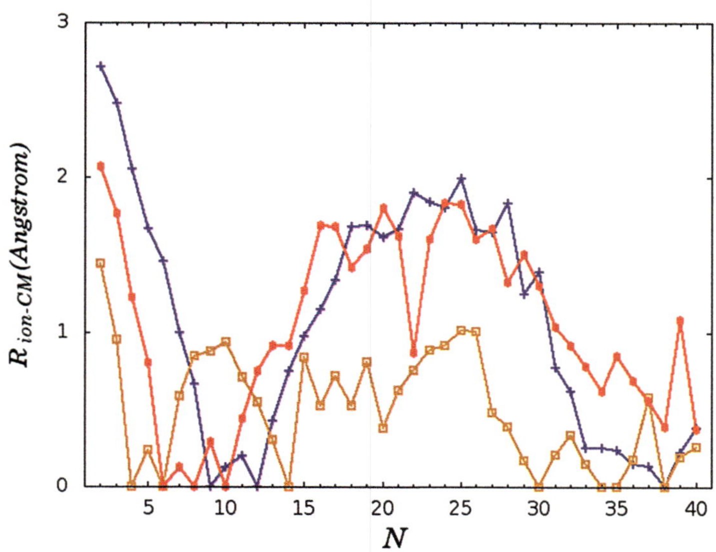

FIGURE 3.4 Distance separating the ion from the center of mass of the cluster: orange line and open squares, $Li^+(Ar)_N$; red line and solid squares, $Na^+(Ar)_N$; magenta line and crosses, $K^+(Ar)_N$.

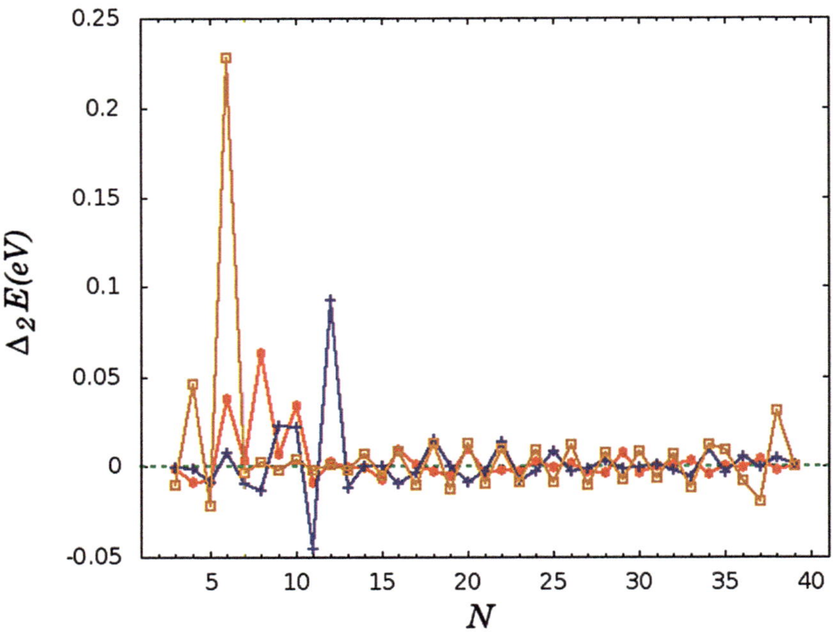

FIGURE 3.5 Second energy difference of the global minimum structures: orange line and open squares, $Li^+(Ar)_N$; red line and solid squares, $Na^+(Ar)_N$; magenta line and crosses, $K^+(Ar)_N$.

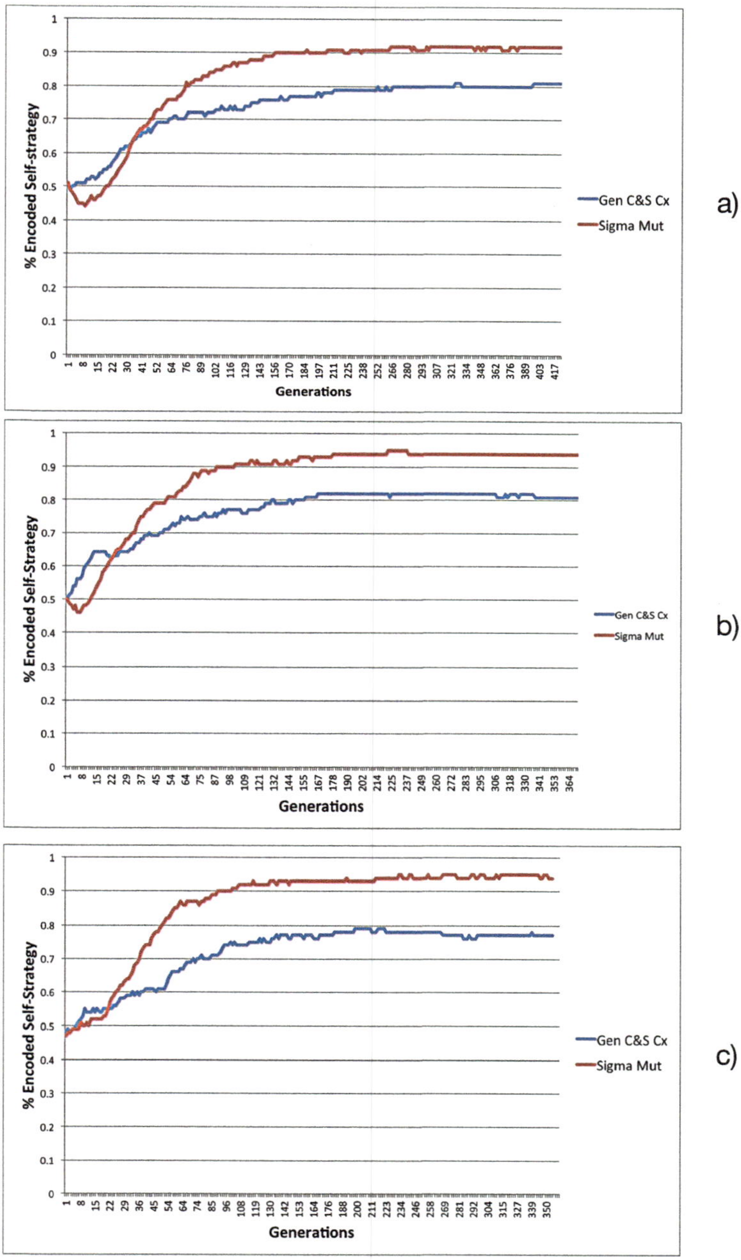

FIGURE 3.9 Evolution of the percentage of individuals encoding generalized C&S crossover and sigma mutation in the optimization of three Morse instances: (a) 43 atoms; (b) 68 atoms; and (c) 74 atoms.

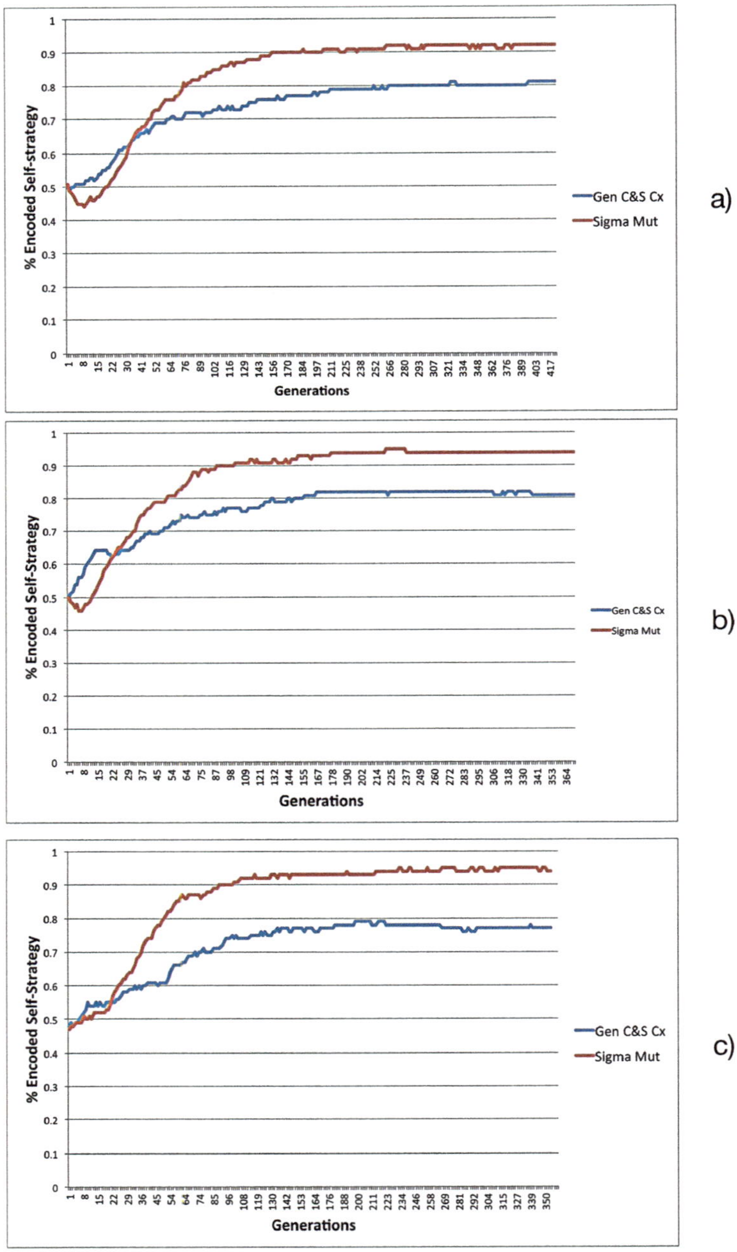

FIGURE 3.10 Evolution of the percentage of individuals encoding each of the distance measures in the optimization of three Morse instances: (a) 43 atoms; (b) 68 atoms; and (c) 74 atoms.

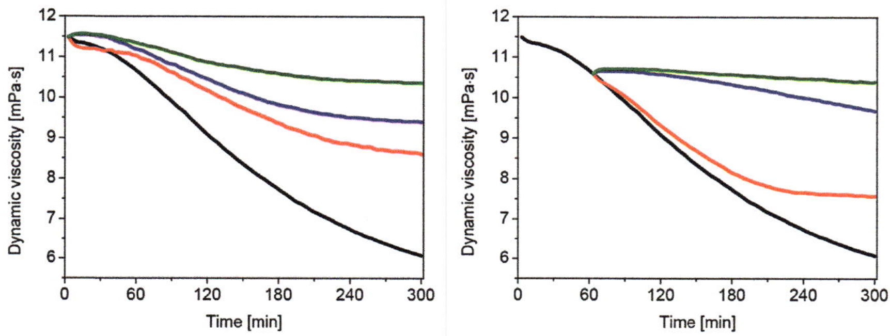

FIGURE 4.1 Time-dependent changes in dynamic viscosity of the hyaluronan solution in the presence of *A. indica* extract added to the reaction mixture just before the initiation of hyaluronan degradation (left panel) or 1 h later (right panel) in μg/ml: 0 (black), 5 (red), 25 (blue), and 125 (green).

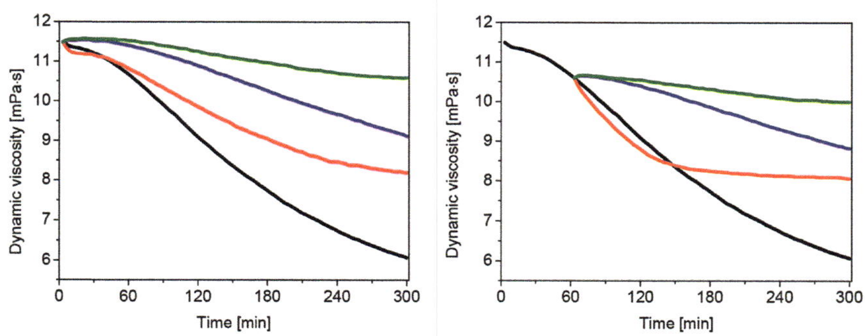

FIGURE 4.2 Time-dependent changes in dynamic viscosity of the hyaluronan solution in the presence of the extract of *O. tenuiflorum* added to the reaction mixture just before the initiation of hyaluronan degradation (left panel) or 1 h later (right panel) in μg/ml: 0 (black), 5 (red), 50 (blue), and 500 (green).

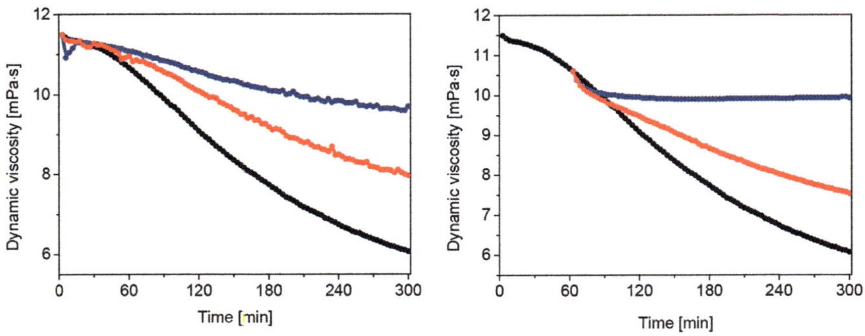

FIGURE 4.3 Time-dependent changes in dynamic viscosity of the hyaluronan solution in the presence of the extract of *W. somnifera* added to the reaction mixture just before the initiation of hyaluronan degradation (left panel) or 1 h later (right panel) in μg/ml: 0 (black), 5 (blue), and 25 (red).

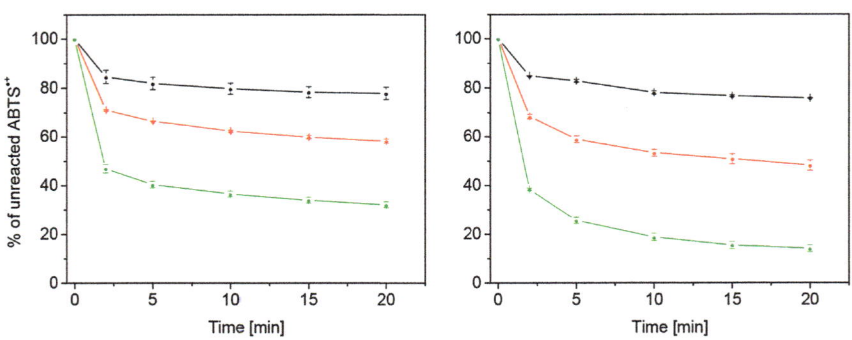

FIGURE 4.4 **(See color insert.)** Percentage of unreacted ABTS•+ after addition of the *O. tenuiflorum* extract (left panel) and the *A. indica* extract (right panel). Concentrations of the *O. tenuiflorum* extract in ABTS•+ solution were: 5 (black), 50 (red), and 65 μg/ml (green). Concentrations of the *A. indica* extract in ABTS•+ solution were: 25 (black), 35 (red), and 45 (green) μg/ml.

For practical confirmation of low hazard level of a medium filled with fire-extinguishing aerosol, the tests on animals were carried out. Overall, the tests reflect the presence of the risk of negative impact on living organisms. Methods in refs [11] and [12] were taken as the basis during the work.

White mice were used as the test animals. The exposure was 15 min long. Tests were conducted for:

- An aerosol with the basic content of solid dispersion phase, the concentration of solid fuel composition $q = 85$ g/m^3.
- An aerosol with low content of solid dispersion phase, the concentration of solid fuel composition $q = 175$ g/m^3.

Tests results are shown in Table 9.2.

TABLE 9.2 Results of the Experiment for Evaluating Acute Toxic Effect of the Aerosol.

	Unfiltered $q = 85$ g/m^3		**Porous powder material (PPM) filter $q = 175$ g/m^3**	
	CO_2, %vol.	O_2, %vol.	CO_2, %vol.	O_2, %vol.
Before exposure	~0.03	20.90–21.05	~0.03	20.90–21.05
During exposure	1.35–1.60	19.95–20.10	2.40–2.60	19.60–19.70
Died during exposure	None	None	None	None
Died during 10 days of observation	None	None	None	None

During exposure at both concentrations:

- There were no cases of deaths of test animals during exposure and 10 days after exposure.
- Oxygen concentration in the medium filled with the aerosol did not fall below 19%vol., and carbon dioxide concentration did not exceed 3.0%vol., which, according to refs[13,14] enables to be inside the protected premises for 20 min without risking functional changes in the body.

The obtained results indicate a low level of toxicity hazard of the medium filled with fire-extinguishing aerosol with both basic and low content of solid dispersion medium.

9.4.2 OPTICAL TRANSPARENCY OF THE MEDIUM

The most effective way of reducing the content of the solid dispersion phase is filtration of aerosols through different porous materials. The work comprised studies of various porous materials, which meet requirements for high thermal and erosion resistance and sufficient filtering capacity.

Filtering capacity was evaluated based on two parameters.

The efficiency of trapping solid-phase particles ($m_{\text{solid-phase total}}$) was evaluated from a change in filter mass ($\Delta m_{\text{trapped particles}}$). Efficiency ($KPD_{\text{filter}}$), which reflects the effectiveness of removing the solid phase, was determined:

$$KPD_{\text{filter}} = \frac{\Delta m_{\text{trapped particles}}}{m_{\text{solid-phase total}}} \times 100\ \% \tag{9.3}$$

Specific filtering area characterizes the filter surface area (S_i), used to filter the volume of aerosol, formed by mass of charge (M_{AFC}).

$$E = \frac{S_i}{M_{\text{AFC}}} \tag{9.4}$$

The total mass of the solid dispersion phase in the fire-extinguishing aerosol was determined during the studies. A complex-stage filter was developed to this end. The diagram of the filter is shown in Figure 9.4.

If no aerosol (white) trail, which is visible due to the presence of a solid phase, is observed during operation, the fraction of the trapped solid dispersion phase is close to 100%.

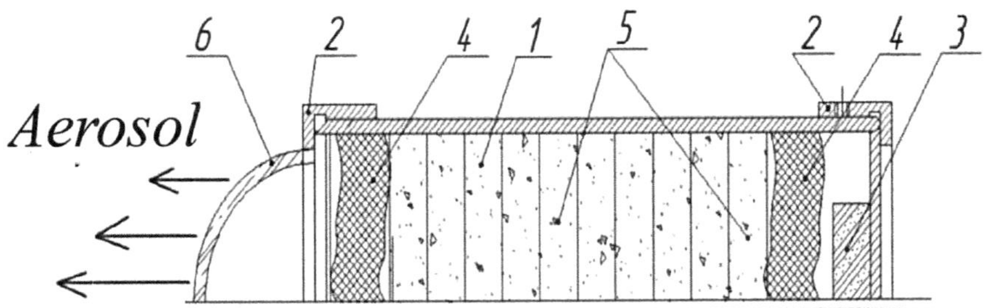

FIGURE 9.4 Diagram of the complex-stage filter, where 1 and 5—basalt needle felt, 2—filter body, 3—solid fuel charge, 4—metal stack, and 6—respirator (Petryanov cloth).

Results of the study showed that the aerosol contains about 14%wt. of the solid dispersion phase, the rest is a gas phase. Test results of filter materials are shown in Table 9.3.

TABLE 9.3 Comparison of Operating Parameters of Filter Materials.

Filter material	KPD, %	E, mm^2/g	%t.f.
Highly porous cellular material	~72.2	190	~3.9
PPM	~57.0	56.5	~6.0
Basalt needle felt	~60.1	78.5	~5.6

%t.f. is the remaining amount of solid dispersion phase in the aerosol.

The filter is based on highly porous cellular material (HPCM; nickel-based HPCM), disk configuration, ceramic porous powder material (PPM; titanium oxynitride-based PPM), tube configuration, and basalt needle felt (BNF), cloth configuration. The disk has an advantage of the presence of a large developed filtering surface.

HPCM was the most efficient, having trapped more than 70 %wt. of the solid phase from the aerosol. It has a large specific filtering area (*E*), which negatively affects cost and dimensions of the end product.

The PPM, with comparable efficiency with the BNF, has the smallest specific filtering area. This enables its use in a compact efficient filter. PPM was chosen as the basic version of the filter.

The aerosol particle counter Fluke 983 was used to determine particle size/fractional distribution of particles. Measurement results are shown for an aerosol with the basic content of solid dispersion-phase particles and filtered through PPM. Distribution curves of the solid dispersion-phase particles in particle size ranges are shown in Figure 9.5.

Particle distribution matches the normal distribution curve.

For an unfiltered aerosol, the basic weight of particles is in the size range of 1.0–10.0 μm, and 60%wt. particles—2.00–4.99 μm. Particles close to the nanosize range (0.30–0.49 μm) make up less than 5% of the total mass.

The use of the filter led to change in particle size. Results for the filtered aerosol show a shift of the particle mass distribution curve relative to the unfiltered aerosol. More than 50%wt. of the solid phase is in the particle size range of 1–2 μm.

The nature of particle distribution enables to extrapolate curves and shows the presence of a small weight fraction of nanoparticles $<1\%$.

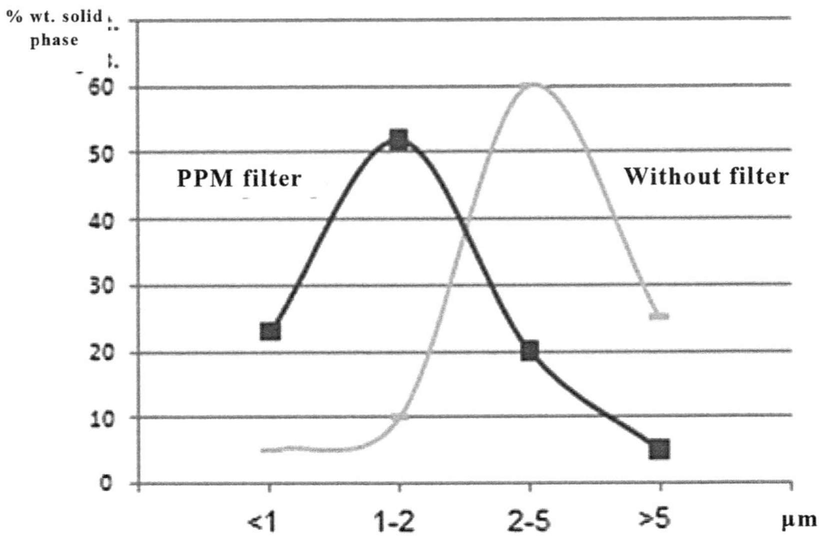

FIGURE 9.5 Distribution curves of solid dispersion-phase particles by mass for filtered and unfiltered aerosol.

Particle size for the aerosol with the basic content of particles was also measured.

The contactless optical profile meter "NewView 6300" showed the presence of particles of the solid dispersion phase with a particle size of 1 µm or higher (up to 12 µm). Up to 60% of the particles have a size of 3 µm, 10%—6 µm, and the remaining amount consists of larger particles.

The complex measurement system "NanoTest 600" detected particles with a size in the range of 1–12 µm.

Optical transparency of the medium was determined using a system consisting of a "Testo 540" lux meter and a red-light source—60 W Philips lamp. The lamp is mounted at a distance of 1 m from the lux meter.

Optical transparency of the medium is determined as the ratio of light flux values: N_i through a layer of the medium filled with aerosol, and N_0 through the transparent medium (without aerosol) (Eq. 9.5). The procedural principle is given as:[15]

$$T = \frac{N_i}{N_0} \tag{9.5}$$

The plotted curves of optical transparency versus aerosol concentration (Fig. 9.6) describe unfiltered aerosol, filtered aerosols through BNF, and PPM materials.

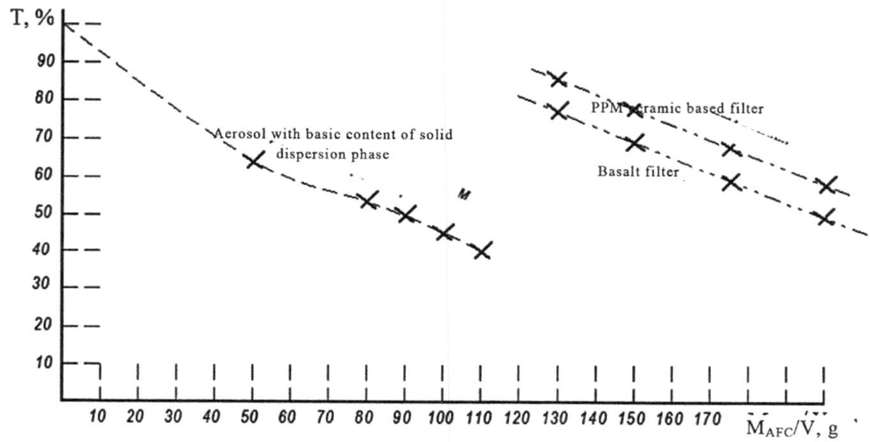

FIGURE 9.6 Optical transparency characteristics of the medium (T) filled with fire-extinguishing aerosols.

Results showed that light flux through a layer of aerosol filtered with PPM is higher than that of several experiments with a BNF filter. According to test results, BNF traps a larger solid phase.

Tests were conducted to evaluate the efficiency of fire-extinguishing sources, which showed that PPM aerosol is more effective than BNF with the same mass of the AFC (M_{AFC}).

This shows that a reduction of particle size toward the nano-range overall improves transparency of the medium filled with fire-extinguishing aerosols and increases the efficiency of fire-extinguishing sources.

Comparison of optical transparency and visibility distance of the light source was performed. Unfiltered aerosol and PPM-filtered aerosol were created in a specially prepared room. A light source in the form of an illuminated "Exit" sign of the "Blik S-12" was placed in the room. Distance from which light from the source is observable was determined for each aerosol.

The result of comparing optical transparency and visibility distance is shown in the diagram in Figure 9.7.

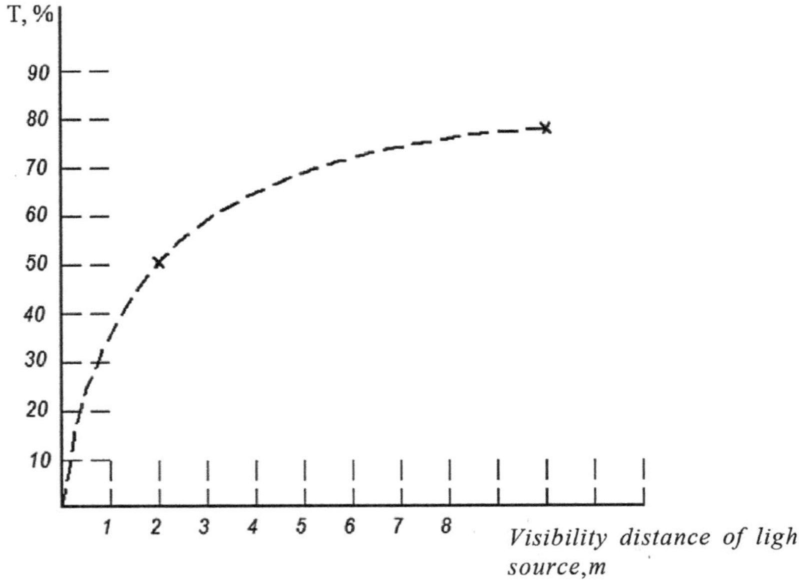

FIGURE 9.7 Comparison of optical transparency (T) and visibility distance of illuminated "Exit" sign.

In the aerosol after PPM filtering, the light source is visible at a distance of up to 10 m, and in the aerosol with the basic content of solid phase (unfiltered), a distance of up to 2 m.

9.5 CONCLUSIONS

1. The study of the gas composition was performed through experimental and calculation method concurrently, which enabled to compare the results.
2. The toxicity regulation was determined. The possibility of exposing humans to the aerosol medium for up to 10 min was demonstrated.
3. Tests were conducted on animals; survival rate of a special experimental group was 100% with 15 min exposure. Signs of toxic effect on external and internal organs were not detected.
4. A method of increasing optical transparency of the medium—aerosol filtration—was implemented.

5. A complex-stage filter was developed, which traps up to 100% solid dispersion phase from fire-extinguishing aerosol, which enabled to estimate the weight ratio of the solid and gas phases.
6. Several filter materials were tested. Based on the sum total of efficiency and specific filtering area, titanium nitride-based PPM was chosen.
7. The obtained PPM-filtered aerosol enables to recognize the "Exit" sign at a distance of up to 10 m that is at least five times greater than the same parameter for unfiltered aerosol.
8. Characteristics of the aerosol filtered through the titanium nitride-based PPM and the unfiltered aerosol were evaluated. More than 60% of the mass of the solid phase of the unfiltered aerosol lies in the 2–5 μm range. Up to 50% of the mass of the solid phase of the filtered aerosol lies in the 1–2 μm range.
9. The filtering result is a shift of particle size toward the nano-range. More effective extinguishing of fire sources with the filtered aerosol is observed.

On the basis of the conducted researches, a solid-fuel generator (optically transparent) of the fire-extinguishing aerosol "Nord" was created, which has no analogs.

ACKNOWLEDGMENT

The development was carried out during the implementation of the project "Fire-extinguishing gas-aerosol generator that generates an optically transparent medium," implemented in cofinancing with the government of the Perm Territory on the basis of the Agreement No. 2153 of 25.11.2010.

KEYWORDS

- **aerosols**
- **toxicity**
- **optical transparency**
- **micro-nanoparticle**

REFERENCES

1. Agafonov, V. V.; Kopylov, N. P. *Aerosol Fire Extinguishing Installations: Elements and characteristics, design, installation and operation;* VNIIPO: Moscow, 1999; p 232.
2. Bezrodny, I. F.; Merkulov, V. A.; Giletich, A. N. *Modern Fire Extinguishing Technologies;* Russian National Research Institute for Fire Protection of the Russian Interior Ministry: Moscow, 1997; pp 335–349.
3. Baratov, A. N.; Vogman, L. P. *Fire Extinguishing Powder Compositions;* Stroyizdat: Moscow, 1982; p 72.
4. Pivovarov, V. V.; Zhevlakov, A. F.; Karamnov, A. A. Selection of Fire Extinguishing Method for Premises with Electronic Equipment. In *Materials from the 26th International Scientific and Practical Conference on Fire Safety Issues;* VNIIPO: Moscow, 2012; pp 91–96.
5. Vakhrouchev, A. V.; Fedotov, A. Y.; Vakhrushev, A. A.; Golubchikov, V. B.; Givotkov, A. V. Multilevel Simulation of the Processes of Nanoaerosol Formation. Part 2. Numerical Investigation of the Processes of Nanoaerosol Formation for Suppression of Fires. Nanomech. *Sci. Technol. Int. J.* **2011,** *2*(3), 205–216.
6. Vakhrouchev, A. V.; Fedotov, A. Y.; Vakhrushev, A. A.; Golubchikov, V. B. Givotkov, A. V. Multilevel Simulation of the Processes of Nanoaerosol Formation. Part 1. Theory Foundation. *Nanomech. Sci. Technol. Int. J.* **2011,** *2*(2), 105–132
7. Vakhrushev, A. V.; Fedotov, A. Y.; Golubchikov, V. B.; Givotkov, A. V. Multilevel Modeling of the Processes of Condensation of a Molecular Mixture in Aerosol Fire Extinguishers. *Chem. Phys. Mesoscopy* **2011,** *13*(3), 340–350
8. Vakhrushev, A. V.; Fedotov, A. Y.; Golubchikov, V. B. Theoretical Bases of Modeling of Nanostructures Formed from the Gas Phase. *Int. J. Math. Comput. Simul.* **2016,** *10*, 192–201.
9. Vakhrushev, A. V.; Fedotov, A. Y.; Golubchikov, V. B. Research and Forecasting of Properties of Metallic Nanocomposites and Nanoaerosol Systems. In *Proceedings of the 6th International Conference of Nanotechnology*, Rome, Italy, November 7–9, 2015; pp 113–140.
10. Trusov, B. G. Terra Software System for Simulation of Phase and Chemical Equilibria in Plasma-Chemical Systems; N. E. Bauman MSTU Bulletin Series "Instrument-making", Moscow, pp 240–250, 2012.
11. Rakhmanin, Yu. A. *Evaluation of Toxic Hazard of Fire Extinguishing Gases and Aerosols Used for Saturation Fire Extinguishing—Study Guide;* VNIIPO: Moscow, 2005.
12. Environmental Protection Agency. Health Effect Test Guidelines. OPPTS 870.1300. Acute Inhalation Toxicity: United States Environmental Protection Agency, 1998.
13. Maximum Allowable Concentration of Hazardous Substances in Air and Water. Pub. 2nd, per. and sup.Chemistry: L, 1975; p 456.
14. Filov, V. A.; Ivin, B. A. *Harmful Chemical Substances. Inorganic Compounds of Group I–IV Elements;* Bandman, A. L., Ed., 1989; p 592.
15. Sharonov, V. V. Observation and Visibility; Military publication of the Ministry of Defense of the USSR: Moscow, 1953.

PART III

New Developments and Methods

CHAPTER 10

FAST QUALITATIVE INSPECTION OF DESIGNED EXPERIMENTS BY MEANS OF THE SUPERPOSING SIGNIFICANT INTERACTION RULES (SSIR) METHOD

EMILI BESALÚ[1,*], LIONELLO POGLIANI[2], and J. VICENTE JULIAN-ORTIZ[2]

[1]*Departament de Química, Institut de Química Computacional i Catàlisi (IQCC), Universitat de Girona, 17003 Girona, Catalonia, Spain*

[2]*Departamento de Química Física, Unidad de Investigación de Diseño de Fármacos y Conectividad Molecular, Facultad de Farmacia, Universitat de València, Burjassot, València, Spain and MOLware SL, Valencia, Spain, E-mail: liopo@uv.es, jejuor@uv.es*

**Corresponding author. E-mail: emili.besalu@udg.edu*

CONTENTS

ABSTRACT

After the presentation of the Superposing Significant Interaction Rules (SSIR) method in a previous volume of this AAP collection, this chapter shows how the procedure can be applied to inspect the data attached to an experimental design. This is possible due to two main reasons: first, the symbolic treatment of the data, that confers to it potential use in many fields; and second, because the rules being considered by the procedure are attached to events that are probabilistically evaluated. This chapter presents two example applications performed with data originally prepared for design of experiments. It is shown how the SSIR method is fast and helps to point to the correct direction of response optimization. The advantages of SSIR in terms of simplicity and the availability to deal with unbalanced experimental designs are discussed.

10.1 INTRODUCTION

For a longtime, both scientific and industrial processes faced the task to optimize the correct combinations of variables (factors) and their values (levels) in order to get optimal (maximal or minimal) responses for certain procedures. The correct way to proceed is not to optimize one variable at a time, as it is well established by the design of experiments (DoE) theory.[1] The DoE designs force the experimenter or engineer to prepare tests combining the levels of the variables in a very specific way, according to preestablished recipes. Despite those sometimes cumbersome prerequisites, DoE procedures have the merit to extract maximal amount of information from a minimal set of experiments. There is no doubt that the DoE field constitutes a well-established discipline which is able to grasp an optimal amount of quantitative and qualitative information from the designed experiments. Therefore, it is not the goal of this work to remove merit to the DoE theory. On the contrary, the intention is to provide a suitable independent tool to carry out the data treatments independently or even complementary. The Superposing Significant Interaction Rules (SSIR) procedure can be seen as an inspection tool and does not provide final data parameters such as DoE. However, despite that, SSIR constitutes a simple algorithm able to detect combinations of factors that, probabilistically speaking, can be responsible for variable effects. Further, SSIR does not need to prepare the data exactly as required by DoE (balanced and symmetric designs, homogeneous distribution of starred points, etc.). The method can handle unbalanced experiments (i.e., combined factors and levels in a quite arbitrary fashion) or even user-defined according to special (restrictive) needs. In many cases, the researcher faces restricted access to minimal data and this is maybe the main advantage of SSIR: it allows getting some clues about the optimization path that must be followed from the data available in a given moment.

10.2 GENERAL FRAMEWORK

The SSIR rationale and algorithmic procedure have been described in several places and also in this editorial book.[2–4] Here, the main characteristics of the method will be outlined focusing on the parallels with DoE procedures.

10.2.1 PARALLELS BETWEEN MOLECULAR TREATMENTS AND EXPERIMENTAL ARRANGEMENTS

The SSIR procedure was originally designed in the field of the molecular structure–activity relationships (SAR) for the congeneric molecular series.[2,3] In the same way that a congener molecular family can be described by a common compound skeleton and a series of possible molecular substitutions in n sites, a DoE can be seen as the combination of n factors (which have the role of substitution sites) at the corresponding levels (acting as particular residues attached to a substitution site in the SAR field) in order to describe and optimize the response of a system or experiment (the common molecular scaffold in congener SAR). Within the SAR field, a molecular response (physicochemical parameter, biological activity, etc.) for several compounds must be evaluated in order to construct the SAR model. Similarly, in the DoE field, the experimental response of a phenomenon must be determined by several experimental arrangements. In other words, because there is an isomorphism between the preparation of the congener molecular family (and the determination of the property value) and a designed experiment, SSIR can be applied in both cases. This is because the factors (sites) and levels (residues) are represented by arbitrary symbols in SSIR. To be more explicit, each factor in a DoE corresponds isomorphically to a molecular substitution site, and each factor level corresponds to a particular symbolic site residue. Each combination of factor levels (e.g., *A*, *B*, *C*, and *D*) needed in a DoE is parallel to a congeneric compound bearing the substitutions *A*, *B*, *C*, and *D* in the proper substitution sites. In both cases, the combination will be denoted here as *ABCD* (the position of each letter indicating the corresponding site). If in a DoE design *n* factors are studied and each bears *m* possible levels (for example, in a complete m^n factorial design), this opens the possibility to define m^n experiments. Similarly, the SSIR method was designed to study congener families that have *n* substitution sites, each one being able to accommodate up to *m* substituents and opening the possibility to deal with a total amount of m^n congener molecules. One of the most important points is that the exploratory nature of SSIR does not require the synthesis of all the m^n molecules nor a homogeneous fraction of n^{m-k} combinations. This opens the door for applying SSIR to explore the experiments without the requirement to follow a DoE recipe that consists of preparing a very specific set of predefined experimental combinations. The same considerations are applied when different numbers of levels (substituents) are combined along the factors (sites).

10.2.2 SSIR RULES, VOTES, AND RANKING

As described in the literature, SSIR generates combinatorial rules and selects statistically significant ones. For instance, within the molecular field, the *AXBX* rule represents all the molecules that have residue *A* at site number 1 and *simultaneously* present residue *B* at site number 3 (the symbol *X* is used as a wildcard). When translating this to the DoE field, the same symbolic rule stands for a set of experiments: all the available ones for which factor 1 is set to level *A* and, at the same time, factor 3 is set to its level *B*, regardless of the levels exhibited by the other factors involved in the experiments. A SSIR rule can also imply negative terms, as in the case $AX\bar{B}X$, which means the set of experiments (molecules) presenting the level (residue) *A* at factor (site) 1 and, at the same time, does not present the level (residue) *B* in factor (site) 3.

The rule *AXBX* is of order 2 because it specifies the two-factor levels. SSIR method virtually generates all the rules of a given order and selects the significant ones based on the available information.[2–4] Within the chemical field, this corresponds to assigning each rule probability only from the available set of training molecules that are consistent with that. The counterpart in the DoE field would take into account only the available experiments that fit the rule order. This is one of the main differences between SSIR and DoE. In general, DoE needs specific panels of experiments to be made available (for instance, the designed fractional factorial ones). SSIR will not only work with the same set of combinations available but also with others, even if they are less complete, nonhomogeneous, or asymmetric. The balanced, uniform, and homogeneous combinations defined by the DoE designs confer very desirable features to the final DoE results: they are quantitative, statistically based, providing confidence intervals, and so forth. The SSIR procedure does not support those properties but, on the other hand, confers freedom to the experimenter when preparing the experiments. Arbitrary combinations of factors and levels can be considered with SSIR. In some cases, the method is ready to be applied to the historical data the experimenter collected before the need to carry out the experiment arose.

The final SSIR model consists of a set of significant rules that help to "point" to the synthesis of new interesting molecules. The relevant rules are those that condense many molecules that have a significant ratio of compounds of interest. This concept is transferred to the experimental design process: the rules generated signal combinations of synergistic factor levels potentially capable of producing an optimal response to the experiment. SSIR requires a dichotomization of responses, namely in terms of interest or noninterest items (high or low values, membership of a particular group, etc.). The dichotomized responses allow SSIR to rank molecules or experiments during the model construction. The ranking is determined by the set of significant rules found. Each rule confers a cumulative positive or a negative vote to the experiments (or molecules) that agree with it. The quality of the series ranked by votes is measured with the area under the receiver operating characteristic (AU-ROC) curve.[5–8] It is assumed that the same rule combination obtained in training can be applied to any additional experiment. The application of several external entities will also lead to a ranked voted series and the top ranked experiments will be presumably linked to factors at levels that act synergistically to get an adequate response. Note that these new ranked experiments can be merely virtual. Therefore, SSIR helps to select new experiments which are going to be designed.

10.3 APPLICATION EXAMPLES

In this section, the information of two designed experiments have been taken from the literature and for each one, it is shown how the SSIR procedure is able to deal with the data and systematically lead to qualitative conclusions compatible with the findings of the literature. The idea we want to convey is that SSIR is able to extract qualitative or semiquantitative results even when dealing with incomplete data.

10.3.1 REACTION OPTIMIZATION FROM A TWO-LEVEL FULL FACTORIAL DESIGN

This example comes from the data reported in a study dealing with optimization within the design space of several 1,2,4-triazol-3-yl-azabicyclo[3.1.0]hexanes which are potent and selective modulators of D_3 dopamine receptors.[9] The main concern is related to the optimization of synthesis parameters affecting a key step of SN_2 displacement between the reagents 1,2,4-triazol-3-yl-halide derivative and variously substituted azabicyclo[3.1.0]hexanes. The approach was made with classical DoE. In Table 10.1 of the supplementary information, the investigation team provides the raw data of a central composite design[10] involving six factors that affect the reaction. The experiment consisted 77 runs but, in order to show the main features of SSIR, 13 experiments involving the starred points (2, 3, 10, 11, 28, 32, 46, 48, 54, 56, 63, and 75) and the center point (35) have been deleted. In this way, the design consists of 64 points, that is, a homogeneous 2^6 full factorial design. Table 10.1 summarizes the variables and levels involved in this calculation.

The two levels of each factor have been codified with the symbols L (lower) and H (higher) (see Table 10.1). Hence, each experiment is identified with a string composed of six symbols. For instance, the first experiment tabulated in the reference (employing the factor values 2.2, 2.0, 0.6, 80.0, 2.5, and 5.0, respectively) is simply encoded by the string "HLLHLH." It is useless to consider negative rule terms for factors involving only two levels[2–4] because the negative of the level L is simply the other level, H, and vice versa.

Five responses were evaluated in the original reference, but here the main, R_1, is analyzed. This response corresponds to the yield (in %) of the relevant products: 1,2,4-triazol-3-yl-azabicyclo[3.1.0]hexanes. As SSIR works with two-class responses, this dependent variable has been dichotomized. Response values of interest are defined as the top 15% percentile. This corresponds to yields of 96.6% or more (found in 10 out of the 64 experiments). The main goal of SSIR is to find probabilistic significant rules (according to a preestablished *p*-value threshold), to combine them and obtain a voting model[2–4] able to rank the series of dichotomized response. The evaluation of the ranking efficiency is measured by means of the AU-ROC obtained in the training. It is also possible to confirm the robustness of the model using a cross-validation procedure. In the examples shown here, the leave-one-out (L1O) process has been considered. Table 10.2 summarizes the main results.

TABLE 10.1 Variables and Levels Involved in the Experimental Design Described in the Supporting Information of Ref 9.

Factor no.	Factor	Design of experiments level and Superposing Significant Interaction Rules codifications	
		−1 (L)	+1 (H)
1	Equivalents of chloropropylthiotriazole	1.0	2.2
2	Equivalents of triethylamine	2.0	5.0
3	Equivalents of potassium iodide	0.6	2.0
4	Temperature (°C)	50.0	80.0
5	Volume of dimethyl sulfoxide (ml/g)	2.5	5.5
6	Ramp temperature (h)	1.0	5.0

For the maximization of the response R_1, the experiment designed by Massari et al. showed that some interactions of order 1 and 2 were relevant. SSIR results are consistent with that. Table 10.3 lists the first (i.e., most significant) rules of order 1 and 2 found for p-value $\leq 2\%$.

The presence of an only factor level in a rule is indicative of relevance. For example, the first rule in Table 10.3 (LXXXXX) explains that it is important to set the first factor at the low level. In this case, as the rule has negative vote, the setting of this level leads to decrease the response. Owing to the fact that this rule is of order 1 and implies a binary factor, both the rule and the vote can be reversed. The result is equivalent to saying that the rule HXXXXX bears a positive vote, that is, the high level has to be set up in order to maximize the response. The negative vote of rule LXXXXX will accumulate to all the experiments that present such a level.

TABLE 10.2 The Area Under the Receiver Operating Characteristic (AU-ROC) Values Obtained with Superposing Significant Interaction Rules (SSIR) for Several Threshold *p*-Values.

Threshold p-value (%)	Rule order (number of generated rules)				
	1 (6)	2 (60)	3 (160)	4 (240)	5 (192)
1	0.944	0.989	0.995	1.000	–[a]
	(2)	(5)	(4)	(5)	(–)
	0.944	0.896	0.848	0.956	–
2	0.944	0.995	0.993	1.000	–[a]
	(2)	(9)	(16)	(14)	(–)
	0.944	0.896	0.848	0.956	–
5	0.944	1.000	0.993	1.000	1.000
	(2)	(29)	(16)	(14)	(14)
	0.944	0.995	0.848	0.956	–[b]
10	0.944	1.000	0.993	1.000	1.000
	(2)	(29)	(16)	(14)	(14)
	0.839	0.995	0.888	0.955	–[b]

Each box shows the obtained AU-ROC for training, the number of significant rules found in parentheses, and the AU-ROC value attached to the leave-one-out (L1O) calculation. In the header of rule orders, the number in parentheses indicates the total number of rules generated and inspected.

[a]All the generated rules exceed the indicated threshold *p*-value.

[b]All the left out items receive zero votes during the L1O procedure.

TABLE 10.3 Significant Rules ($p \leq 2\%$) Selected by the SSIR Method when Generating Rules of Order 1 and 2 for the Maximization of the Response R_1.

Rule order	Rule no.	*p*-Value (%)	Rule	Vote
1	1	0.04	LXXXXX	−1
	2	0.04	XLXXXX	−1
	1	$5 \cdot 10^{-6}$	HHXXXX	+1
	2	0.14	XHXXLX	+1
	3	0.14	HXXXLX	+1
	4	0.14	XHHXXX	+1
2	5	0.14	HXHXXX	+1
	6	1.2	XHXLXX	+1
	7	1.2	XHXXXH	+1
	8	1.2	HXXXXH	+1
	9	1.2	HXXLXX	+1

Similar considerations can be made by reading the other selected rules. In the end, each experimental arrangement will bear a certain number of votes that rank it.

All the rules of order 2 are combining synergistic interactions. For instance, the most significant rule of order 2 (HHXXXX) involves the first and second factors, setting both to the highest level. The positive vote attached means that this partial experimental configuration helps to increase the response. In general, the combinations obtained indicate which factor and levels should be combined to obtain an optimal response. SSIR can use the mixtures of rules of several orders. For example, the complete model with rules of order 1 and 2 (2+5 rules with p-value $\leq 1\%$) gives AU-ROC of 0.989 in fitting and 0.952 for L1O.

An important feature of SSIR is that it can work with incomplete data sets. For instance, SSIR has been run by selecting rules of order 1 and 2 (p-value $\leq 1\%$) but constructing the training model by using only one-half (32) of the experiments. These 32 experiments were randomly chosen and, after the training was done, the significant rules selected were applied to the remaining 32 experiments (virtually simulating that these were not evaluated yet) and the rule votes were cumulated. This allowed to rank the test experiments using the partial knowledge of an inhomogeneous training set. Moreover, in order to evaluate the SSIR performance, this random calculation has been repeated 1000 times. In 942 occasions, a model was obtained, whereas 58 random combinations of training experiments did not report any significant rule. The AU-ROC values retrieved for the test (external) set of experiments ranged from 0.666 to 1.000. Figure 10.1 shows the histogram of the 942 AU-ROC values obtained for the external test set. Clearly, the information that SSIR extracts from each training set is applicable to the test 1 because the models are able to rank the training set with a mean value of 0.916 and most of the results are above of 0.9.

The ability of SSIR to transmit the rules to experiments that have not yet been prepared confers the possibility of predicting which one will presumably be placed at the beginning of the ranking (as in a surface response optimization procedure). This allows selecting the most promising experiments or the best combinations of factor levels. After a new set of 100 runs similar to the above randomized (but considering rules of orders 1–3), SSIR highly voted several experimental combinations that act as test arrangements. The following were the most voted: HHHLLH, HHHLLL, HHHLLH, HHHLHH, and HHLLLH. The general pattern coincides with the most voted experiment: HHHLLH. This points out to set the first three factors and the latter to the high level, whereas the fourth and fifth variables are recommended to be set to the lower level. This set of general preferences is in accordance with the signs of the coefficients of the multilinear model obtained by Massari and collaborators: positive for all the factors, but negative for the fourth and fifth. In fact, the experiment HHHLLH was prepared and gave one of the highest responses.

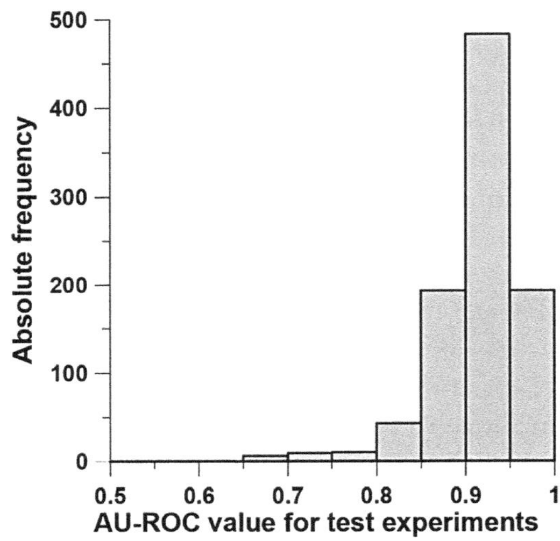

FIGURE 10.1 Histogram of the 942 area under the receiver operating characteristic (AU-ROC) values obtained for the external test experiments during the simulation of 1000 calculations randomly selecting the training experiments.

10.3.2 AN EXAMPLE WORKING WITH MORE THAN TWO LEVELS

In this section, an example[11] of designed experiment dealing with more than two qualitative levels is studied. The experiments are related to the study of parameters that affect the purification of curcumin by the process of crystallization by cooling. Details can be retrieved from the cited reference. Table 10.4 shows the encoding of the factor levels corresponding to the four variables involved in the crystallization conditions. Consider that all but one factors are categorical and that the two latter have three levels. Apart from replicates, a complete experiment will imply $2^2 3^2 = 36$ distinct arrangements. The reported DoE was designed using 30.

TABLE 10.4 Factors and Levels Involved in an Incomplete $2^2 3^2$ Design.

Factor no.	Factor	Levels (codifications)		
1	Stirring	Magnetic (m)	Overhead (o)	–
2	Cooling rate (°C/h)	8 (8)	40 (4)	–
3	Seeding	0 (0)	I (1)	II (2)
4	Step	First (f)	Second (s)	Third (t)

Several responses or properties where investigated: phase composition, mean particle size (d_m), coefficient in quartile variation (CQV), aspect ratio (and shape) of crystals, and percentages of yield and purity (curcumin purity). Of these, the phase composition and the shape are categorical. This example has been chosen not only for the multilevel structure but also for the presence of these categorical variables that will be treated as responses here. Table 10.5 lists the SSIR encoded experiments (according to the codes of Table 10.4) and the responses. All responses have been dichotomized and the asterisks stand for values of interest (i.e., high value or being a member of a particular class). When possible, for continuous variables, top five cases (adding ties if any) have been selected as being the experiments of interest (see the entry at the bottom of Table 10.5).

At the lower part of Table 10.5, the main results of SSIR are shown. For the sake of simplicity, all calculations have been done using the same parameters. Only rules of order 2 have been inspected and negative rule terms were allowed for the three-level factors. In this case, 88 rules were generated. Only the rules having p-values equal to or less than 5% were considered significant. The number of significant rules is specified at the bottom of Table 10.5. The last two file entries in Table 10.5 correspond to AU-ROC values, one for the training process and the second, more relevant and indicative, for the L1O cross-validation calculation. Many of the L1O values are greater than 0.9. The worse values of 0.611 and 0.333 for the shapes N (four cases) or S (two instances) were possibly due to the fact that only a very few experiments belonged to these classes. When these two classes were merged into one, the AU-ROC value improved substantially to 0.889.

One of the most important responses reported is the yield. The AU-ROC for L1O was 0.908 and the ROC curve is depicted in Figure 10.2.

TABLE 10.5 SSIR Encodings of the Experiments and the Response or Descriptor Values for the Optimization Example of Crystallization Conditions.

Experiment no.	Encoded experiment				Phase						Shape			
		I	II	I+II	d_m	CQV	AR	R	N	S	N, S	RS	Y	CUR
1	o80f	*						*						
2	o80s	*			*			*						
3	o80t	*			*			*						
4	o81f	*						*						
5	o81s	*				*		*						
6	o81t	*					*		*		*			*
7	o82f		*		*	*				*	*			
8	o82s		*		*	*				*	*			
9	o82t			*	*	*	*					*		*
10	o40f	*				*		*						
11	o40s	*						*						
12	o40t	*					*		*		*			
13	m80f	*						*						
14	m80s	*						*						
15	m80t	*					*		*		*		*	
16	m81f	*						*						
17	m81s	*						*						
18	m81t	*						*					*	*
19	m82f		*									*		

TABLE 10.5 *(Continued)*

Experiment no.	Encoded experiment	Phase						Shape						
		I	II	I+II	d_m	CQV	AR	R	N	S	N, S	RS	Y	CUR
20	m82s		*									*		
21	m82t			*								*	*	*
22	m40f	*						*						
23	m40s	*						*						
24	m40t	*					*		*		*			
25	m41f	*						*						
26	m41s	*						*						
27	m41t	*						*					*	
28	m42f		*									*		
29	m42s		*									*		
30	m42t			*								*	*	
No. of items of interest		21	6	3	5	5	5	17	4	2	6	7	5	5
No. of significant rules (5%)		20	12	5	7	6	7	17	7	4	5	14	7	10
AU-ROC for training		1.000	1.000	1.000	1.000	0.952	0.960	0.993	0.966	1.000	0.948	1.000	1.000	0.996
AU-ROC for L1O		1.000	1.000	1.000	1.000	0.840	0.944	0.982	0.611	0.333	0.889	1.000	0.908	0.968

*Mean value of interest or experiment belonging to a specified class. Property values are: phase composition (I, II, or I+II), d_m (particle size), coefficient of quartile variation (CQV), aspect ratio (AR), shape distribution (R, N, S, the mix N or S, and the RS case), yield (Y), and curcumin purity (CUR).

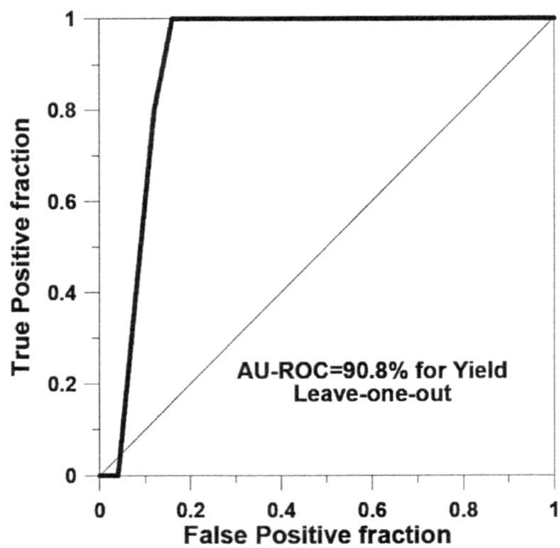

FIGURE 10.2 The ROC curve for the leave-one-out calculation that predicts higher yields. The integrated area under the curve (AU-ROC) is 0.980 (see Table 10.5).

As expected, the most efficient yields are those related to the third crystallization because these three steps were performed consecutively. It is worth noting that the SSIR rules automatically reveal this clear trend: all the significant rules associated to a positive vote bear the rule term XXXt (imposition of the third crystallization step) and, at the same time, the rules with negative vote bear the term XXX|t, that is, the negation of the third step. Table 10.6 lists the seven significant rules selected and the corresponding *p*-values and votes.

TABLE 10.6 Significant Rules ($p \leq 5\%$) Selected by SSIR Method when Generating Order 2 Rules to Maximize the Yield of Curcumin.

Rule no.	*p*-Value (%)	Rule	Vote
1	0.0042	mXXt	+1
2	0.26	XX|0t	+1
3	0.56	mXX|f	+1
4	0.56	mXX|s	+1
5	3.1	XX“2”t	−1
6	3.1	XX“1”t	−1
7	4.1	X8Xt	+1

The symbol “|” stands for the negation term.

10.4 CONCLUSIONS

SSIR is an exploratory tool originally designed to deal with congeneric molecular families. Here, it has been shown how it can also be used as a companion to carry out experimental optimizations. The examples shown demonstrate how the SSIR technique is able to qualitatively determine the most relevant combination of factors and levels to be established to optimize the results of a conducted experiment. This tool can be used to make pretreatments among the data and give further insight into more focused experiments.

ACKNOWLEDGMENTS

The author acknowledges the Generalitat de Catalunya (Departament d'Innovació, Universitats i Empresa) for the financial support to the research group Química teòrica i Modelatge i Enginyeria Molecular of the University of Girona (code 2014-SGR-1202).

KEYWORDS

- **superposing significant interaction rules method**
- **design of experiments**
- **QSAR**
- **ranking procedures**

REFERENCES

1. Eriksson, L.; Johansson, E.; Kettaneh-Wold, N.; Wikström, C.; Wold, S. *Design of Experiments: Principles and Applications*; Umetrics Academy: Sweden, 2000.
2. Besalú, E. Fast Modeling of Binding Affinities by Means of Superposing Significant Interaction Rules (SSIR) Method. *Int. J. Mol. Sci.* **2016,** *17*(827). DOI: 10.3390/ijms17060827.
3. Besalú, E.; Pogliani, L.; de Julián-Ortiz, J. V. Superposing Significant Interaction Rules (SSIR) Method: A Simple Procedure for Rapid Ranking of Congeneric Compounds. *Croat. Chem. Acta* **2016,** *89*(4) 481–492. DOI: 10.5562/cca3027.

4. Besalú, E.; Pogliani, L.; de Julián-Ortiz, J. V. The Superposing Significant Interaction Rules (SSIR) method. In *Applied Chemistry and Chemical Engineering, Volume 4 (Experimental Techniques and Methodical Developments)*; Haghi, A. K., Pogliani, L., Castro, E. A., Balköse, D., Mukbaniani, O. V., Chia, C. H., Eds.; Apple Academic Press: Waretown, New Jersey, 2017.
5. Egan, J. P. *Signal Detection Theory and ROC Analysis;* Academic Press: New York, 1975.
6. Besalú, E.; De Julián Ortiz, J. V.; Pogliani, L. On Plots in QSAR/QSPR Methodologies. In *Quantum Frontiers of Atoms and Molecules;* Putz, M. V., Ed; NOVA Publishing Inc: New York, 2010; pp 589–605.
7. Forlay-Frick, P.; Van Gyseghem, E.; Héberger, K.; Vander Heyden, Y. Selection of Orthogonal Chromatographic Systems Based on Parametric and Non-Parametric Statistical Tests. *Anal. Chim. Acta.* **2005,** *539,* 1–10.
8. Mason, S. J.; Graham, N. E. Areas Beneath the Relative Operating Characteristics (ROC) and Relative Operating Levels (ROL) Curves: Statistical Significance and Interpretation. *Q. J. R. Meteorol. Soc.* **2002,** *128,* 2145–2166.
9. Massari, L.; Panelli, L.; Hughes, M.; Stazi, F.; Maton, W.; Westerduin, P.; Scaravelli, F.; Bacchi, S. A Mechanistic Insight into a Simple C-N Bond Formation via SN_2 Displacement: A Synergistic Kinetics and Design of Experiment Approach. *Org. Process Res. Dev.* **2010,** *14,* 1364–1372. DOI: 10.1021/op100176u.
10. Box, G. E. P.; Hunter W. G.; Hunter, J. S. *Statistics for Experimenters. An Introduction to Design, Data Analysis, and Model Building;* John Wiley & Sons, Inc: New York, 1978.
11. Ukrainczyk, M.; Kieran Hodnett, B.; Rasmuson, Å. C. Process Parameters in the Purification of Curcumin by Cooling Crystallization. *Org. Process Res. Dev.* **2016,** *20*, 1593–1602. DOI: 10.1021/acs.oprd.6b00153.

CHAPTER 11

OPTIMUM SELECTION OF SYSTEM OF A GAS CLEANING WITH AN ECONOMIC EFFICIENCY ESTIMATION

R. R. USMANOVA[1] and G. E. ZAIKOV[2]

[1]*Ufa State Technical University of Aviation, Ufa 450000, Bashkortostan, Russia, E-mail: Usmanovarr@mail.ru*

[2]*N. M. Emanuel Institute of Biochemical Physics, Russian Academy of Sciences, Moscow 119991, Russia, E-mail: chembio@chph.ras.ru*

**Corresponding author. E-mail: GEZaikov@yahoo.com*

CONTENTS

ABSTRACT

In this chapter, the reliable method for selection of a gas cleaning system with economic efficiency estimation is discussed and reviewed in detail.

11.1 INTRODUCTION

Apparatuses of cyclonic type are widely used in chemical engineering. It is enough to specify such widespread devices as centrifugal machines, apparatuses with the rabble, cyclone separators, scrubbers of centrifugal act, vortex tubes, whirlwind compressors, and many other things. All these apparatuses are merged by the general principle: their functioning is based on centrifugal force used. In the scientific and technical literature, traffic in regional zones which are usually expelled from the analysis, as a rule, is not observed, and the sticking condition as value of a peripheral velocity on a wall is accepted nonzero is not satisfied, and conditions of loss of stability of an eddy flow in the presence of a viscous radial stream are not analyzed. Now, the theory of the twirled currents in connection with aspiration to explain the nature of whirlwind effect which remains till now not revealed intensively develops. Therefore, works in this direction represent scientific interest.

The application of apparatuses with the twirled traffic of phases in gas cleaning systems is especially perspective. Thereupon, it is necessary to notice that engineering protection of a circumambient is based on well-developed chemical engineering.

Necessity and importance of the solution of a problem of raise of the efficiency of the gas cleaning apparatuses based on functional features with the twirled traffic of phases define an urgency of the given research.

11.2 DUST COLLECTOR DESIGN

In dust collection equipment, most or all of the collection mechanisms may be operating simultaneously, their relative importance is determined by the particle and gas characteristics, the geometry of the equipment, and the fluid-flow pattern. Although the general case is exceedingly complex, it is usually possible in specific instances to determine which mechanism or mechanisms may be controlling. Nevertheless, the difficulty of theoretical treatment of dust collection phenomena has made necessary simplifying assumptions with the introduction of corresponding uncertainties. Theoretical studies have been hampered by a lack of adequate experimental techniques for verification of predictions. Although theoretical treatment of collector performance has been greatly expanded in the period since 1960, a few of the resulting performance models have received adequate experimental confirmation because of experimental limitations.[1]

The best-established models of collector performance are those for fibrous filters and fixed-bed granular filters, in which the structures and fluid-flow patterns are reasonably well defined. These devices are also adapted to small-scale testing under controlled laboratory conditions. Realistic modeling of full-scale electrostatic precipitators and scrubbers is incomparably more difficult. Confirmation of the models has been further limited by a lack of monodisperse aerosols that can be generated on a scale suitable for testing equipment of substantial sizes. When a polydisperse test dust is used, the particle-size distributions of the dust both entering and leaving a collector must be determined with extreme precision to avoid serious errors in the determination of the collection efficiency for a given particle size.

The design of industrial-scale collectors still rests essentially on empirical or semiempirical methods, although it is increasingly guided by concepts derived from theory. Existing theoretical models frequently embody constants that must be evaluated by experiment and that may actually compensate for deficiencies in the models.

The basic critical bucklings of the cyclone separator are resulted in Figure 11.1. It has diameter of cyclone separator D (or its radius R) which is usually accepted in the capacity of a characteristic size of the cyclone separator; altitude of a conic part of the case of the cyclone separator is $h_к$, altitude of its cylindrical part $h_ц$, and full altitude of cyclone separator H. In most cases, all geometrical sizes are expressed in shares of diameter of the cyclone separator and therefore usually observe the relative sizes which have been referred to its characteristic size.

In TABLE 11.1, our recommendations for sampling of design data are presented. Till now in a scope of apparatuses for a gas cleaning, there is no uniform nomenclature, the uniform inoculated rows of apparatuses are not developed, and the general approach is not developed for an estimation of the basic technological parameters of apparatuses.

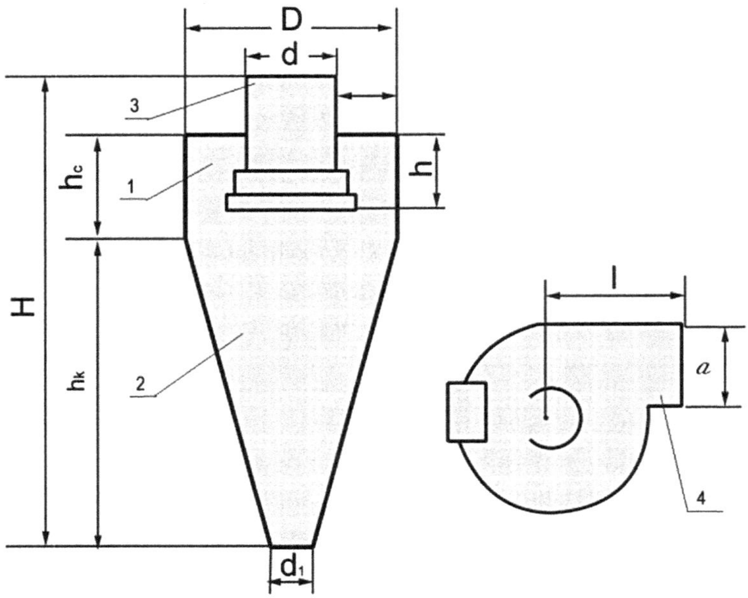

FIGURE 11.1 The Constructive circuit design of a cyclonic deduster: 1—a cylindrical shell; 2—a conic shell; 3—a pipe exhaust; and 4—the tangential upstream end.

TABLE 11.1 Recommendations for Sampling of Design Data.

Parameter	Aspect	Value
Relative width of the upstream end	b/D	0.05÷0.35
Relative diameter of the upstream end	d/D	0.15÷0.75
Relationship of sizes of the upstream end	h/b	1÷6
Relationship of the square of an entry and exit	A_{en}/A_{ex}	0.6÷2.5
Relative length of the exhaust tube	L_{ex}/D	0.5÷1.8
Relative length of the case of the apparatus	l/D	1.5÷5.5
The relation of a conic part to length of the case	L_{con}/l	0÷1

Depending on the predominance of cylindrical or conic parts, cyclone separators are conditionally divided as cylindrical and conic. The altitude of a cylindrical part at conic cyclone separators is usually equal to input channel altitude. Conic cyclone separators, at equal productivity with cylindrical, differ from the last big gabarits and consequently usually are not applied in group modification.[2]

Speaking about conditions of feeding into of a stream, it is necessary to note that the majority of industrial cyclone separators, unlike fire chambers and furnaces, are characterized by rather big relative square of an entry.

Industrial implementation was gained by following basic designs of feeding of a gas stream in the cyclone separator: simple tangential feeding (Fig. 11.2a), tangential feeding into with a screw overhead part (Fig. 11.2c), simple spiral feeding (Fig. 11.2b), and spiral feeding into with a screw overhead part (Fig. 11.2d).

Simple tangential feeding is applied at the big dustiness of a stream, sizes of corpuscles, and speed of gas. As a rule, the channel has a squared shape. The round form of an input channel in cyclonic apparatuses meets less often than in hydrocyclone separators.

Simple spiral feeding (Fig. 11.2b) provides high separation efficiency of gases and is used in modern effective conic and cylindrical cyclone separators for fine-dust branch, for example, active black, a cement powder. The similar design of feeding can be applied at any dustiness of a stream.

It is expedient to use spiral screwed feeding at small concentration of a dust (Fig. 11.2d).

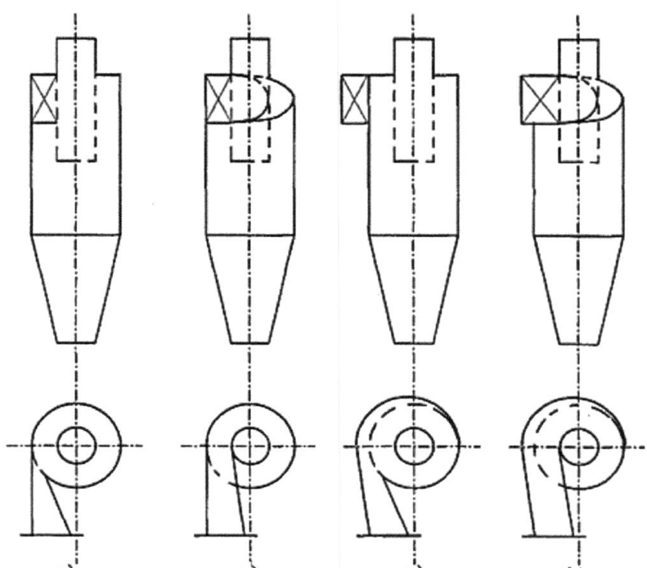

FIGURE 11.2 Designs of knot of feeding into of a stream in cyclone separators.

The basic technical data of the cyclone separator characterizing it as the dust removal apparatus are:

d_{50}—diameter of the corpuscles trapped with probability of 50%, a micron;

η_p—fractional factor of clearing:

$$\eta_p = \frac{M_2}{M_1}$$

where M_2—weight of the trapped corpuscles of the set fraction of a dust;

M_1—weight of all corpuscles of this fraction on an entry in the cyclone separator;

ζ—water resistance factor:

$$\xi = \frac{2\Delta P}{\rho_g v^2}$$

where ΔP—a difference of inlet pressures and a cyclone separator exit;

v—average speed of a stream in cyclone separator cross section;

ρ_g—gas density.

As a rule, these parameters are defined experimentally. The overall performance of cyclone separators is sized up on their factor of clearing η_p. Quality of clearing of gases from a dust depends on its properties, first of all sizes of corpuscles.

11.3 DUST COLLECTION EQUIPMENT

The gravity settling chamber is probably the simplest and earliest type of dust collection equipment, consisting of a chamber in which the gas velocity is reduced to enable dust to settle out by the action of gravity. Its simplicity lends it to almost any type of construction. Practically, its industrial utility is limited to removing particles larger than 325 mesh (43-fim diameter). For removing smaller particles, the required chamber size is generally excessive.

Gravity collectors are generally built in the form of long, empty, horizontal, rectangular chambers with an inlet at one end and an outlet at the side or top of the other end. By assuming a low degree of turbulence relative to the settling velocity of the dust particle as question, the performance of a gravity settling chamber is given by ref 2.

$$\eta_y = \frac{2}{1-m^2}\int_m^1 t_1\left(1-e^{-\lambda \cdot T}\right)dt$$

The height need is made only large enough so that the gas velocity V in the chamber is not so high as to cause reentrainment of separated dust. Generally, V should not exceed about 3 m/s (10 ft/s).

This equation is a result of the residence time theory of particle collection. In this theory, the time that it takes for a particle to reach v is balanced by the time that a particle spends in the cyclone. The particle size that makes it to the wall by the time that it exits the cyclone is the particle size collected at 50% collection efficiency, Dpj/Dpt.

When consistent units are used, the particle size calculated by the above equation will be in either meters or feet. The equation contains effects of cyclone size, gas velocity, gas viscosity, gas density, and particle density of the solids. In practice, a design curve as given in Figure 11.3 uses Dpj/Dpt as the size at which 50% of solids of a given size are collected by the cyclone.

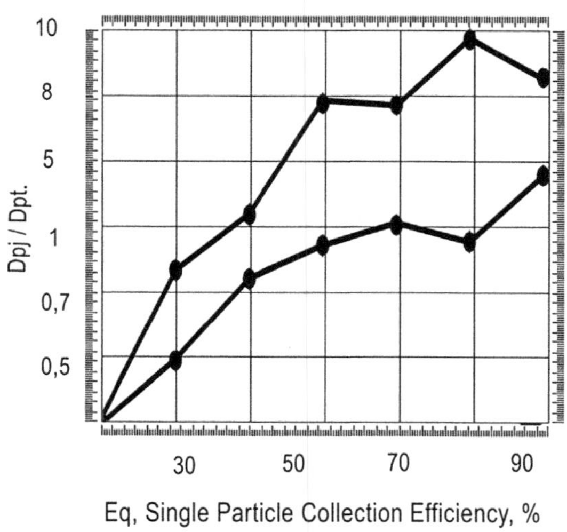

FIGURE 11.3 Single-particle collection efficiency curve.

The material entering the cyclone is divided into fractional sizes, and the collection efficiency for each size is determined. The total efficiency of collection is the sum of the collection efficiencies of the cuts.

The above applies for very dilute systems, usually on the order of 1 gr/ft^3, or 2.3 g/m^3 where 1 gr (1/7000) lb. When denser flows of solids are present in the inlet gas, cyclone efficiency increases dramatically. This is thought to be due to the coarse particles colliding with fine ones as they move to the wall, which carry a large percentage of the finer particles along with them. Other explanations are that the solids have a lower drag coefficient or tend to agglomerate in multiparticle environments, thus effectively becoming larger particles. At very high inlet solids loadings, it is believed that the gas cannot simply hold that much solid material in suspension at high centrifugal forces, and the bulk of the solids simply "condenses" out of the gas stream.

The initial efficiency of a particle-size cut is found on the chart, and the parametric line is followed by the proper overall solids loading. The efficiency for that cut size is then read from the graph.

A single cyclone can sometimes give sufficient gas–solid separation for a particular process or application. However, solids collection efficiency can usually be enhanced by placing cyclones in series. Cyclones in series are typically necessary for most processes to minimize particulate emissions or to minimize the loss of expensive solid reactant or catalyst. Two cyclones in series are most common, but sometimes three cyclones in series are used. Series cyclones can be very efficient. In fluidized catalytic cracking regenerators, two stages of cyclones can give efficiencies of up to and even greater than 99.999%.

The pressure drop through a settling chamber is small, consisting primarily of entrance and exit losses. Because low gas velocities are used, the chamber is not subject to abrasion and may therefore be used as a precleaner to remove very coarse particles and thus minimize abrasion on subsequent equipment.

Cyclones work by using centrifugal force to increase the gravity field experienced by the solids. They then move to the wall under the influence of their effectively increased weight. The movement to the wall is improved as the path the solids traverse under centrifugal flow is increased. This path is equated with the number of spirals the solids make in the cyclone barrel. Figure 11.4 gives the number of spirals N_s as a function of the maximum velocity in the cyclone. The maximum velocity may be either the inlet or the outlet velocity depending on the design. The equation for Dpj, the theoretical size particle removed by the cyclone at 50% collection efficiency, is

$$d_E = \sqrt{\frac{\mu\, R_z^3 G^{0,6}}{\rho\; G\varphi R^{1,2}\vartheta^{0,6}}},$$

where p—gas density and p_v—particle density. For a given volumetric airflow rate, collection efficiency depends on the total plan cross section of the chamber and is independent of the height.

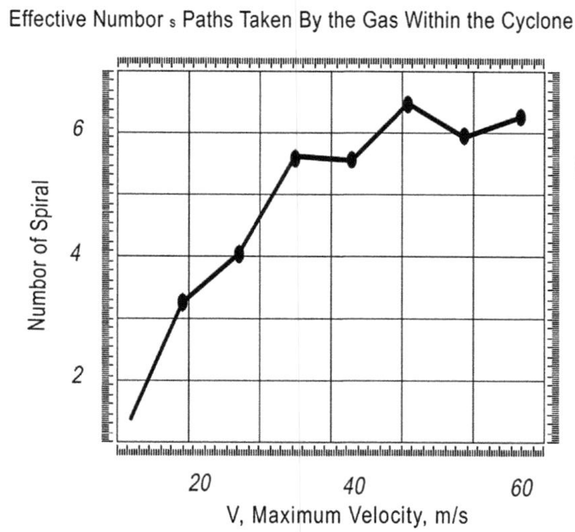

FIGURE 11.4 Effective number of spiral paths taken by the gas within the body of a cyclone.

Fields of Application: Within the range of their performance capabilities, cyclone collectors offer one of the least expensive means of dust collection from the standpoint of both investment and operation. Their major limitation is that their efficiency is low for the collection of particles smaller than 5–10 μm. Although cyclones may be used to collect particles larger than 200 μm, gravity settling chambers or simple inertial separators (such as gas reversal chambers) are usually satisfactory for this size of the particle and are less subject to abrasion. In special cases in which the dust is highly agglomerated or high dust concentrations (over 230 g/m^3

or 100 gr/ft^3) are encountered, cyclones will remove dusts having small particle sizes. In certain instances, efficiencies as high as 98% have been attained on dusts having ultimate particle sizes of 0.1–2.0 μm because of the predominant effect of particle agglomeration due to high interparticle forces. Cyclones are used to remove both solids and liquids from gases and have been operated at temperatures as high as 1200°C and pressures as high as 50,700 kPa (500 atm). Cyclones can be very small or very large. The smallest cyclones range from approximately 1 to 2 cm in diameter and the largest up to about 10 m in diameter. The number of cyclones used for a single fluidized bed can vary from 1 to 22 sets of first-stage and second-stage cyclones (44 cyclones in total).[1]

The inner vortex (often called the core of the vortex) rotates at a much higher velocity than the outer vortex. In the absence of solids, the radius of this inner vortex has been measured to be 0.4–0.8 r. With axial inlet cyclones, the inner core vortex is aligned with the axis of the gas outlet tube. With tangential or volute cyclone inlets, the vortex is not exactly aligned with the axis. The nonsymmetrical entry of the tangential or volute inlet causes the axis of the vortex to be slightly eccentric from the axis of the cyclone. This means that the bottom of the vortex is displaced some distance from the axis and can "pluck off" and reentrain dust from the solids sliding down the cyclone cone if the vortex gets too close to the wall of the cyclone cone.

At the bottom of the vortex, there is substantial turbulence as the gas flow reverses and flows up the middle of the cyclone into the gas outlet tube. As indicated above, if this region is too close to the wall of the cone, substantial reentrainment of the separated solids can occur. Therefore, it is very important that cyclone design take this into account.

The vortex of a cyclone will process (or wobble) about the center axis of the cyclone. This motion can bring the vortex into close proximity to the wall of the cone of the cyclone and "pluck" off and reentrain the collected solids flowing down along the wall of the cone. The vortex may also cause erosion of the cone if it touches the cone wall. Sometimes an inverted cone or a similar device is added to the bottom of the cyclone in the vicinity of the cone and dipleg to stabilize and "fix" the vortex. If it is placed correctly, the vortex will be attached to the cone and the vortex movement will be stabilized, thus minimizing the efficiency loss due to cyclone cone (Fig. 11.5).

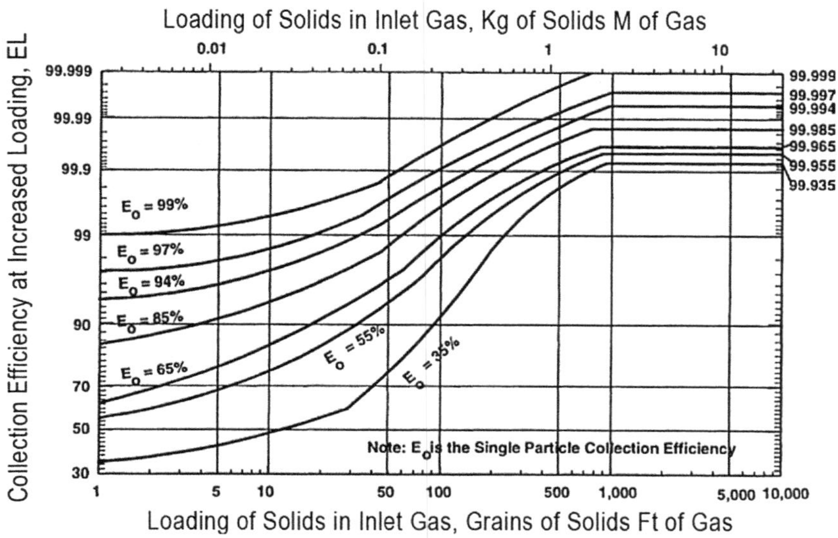

FIGURE 11.5 Effect of inlet loading on collection efficiency for Celdart group A and group C particles.

Typically, first-stage cyclones will have an inlet gas velocity less than that of second-stage cyclones. The lower inlet velocity of first-stage cyclones results in lower particle attrition rates and lower wall erosion rates. After most of the solids are collected in the first stage, a higher velocity is generally used in second-stage cyclones to increase the centrifugal force on the solids and collection efficiency. Inlet erosion rates are generally low in the second stage because of the vastly reduced flux of solids into the second-stage cyclone. However, cone erosion rates in second-stage cyclones are much greater than in first-stage cyclones.

11.4 AGENCY OF CRITICAL BUCKLINGS OF THE CYCLONE SEPARATOR ON ITS EFFICIENCY

In most cyclonic apparatuses being used for gas cleaning, the cleaning factor is the most important parameter to be considered. We will observe agency of design data of the cyclone separator on its parameters, first of all, on its efficiency.

Sampling of a rational slope of the upstream end allows to raise the efficiency of clearing of gases and to lower an apparatus aerodynamic resistance a little. Agency of an angle of feeding into a stream on the aerodynamics of cylindrical cyclone separators is investigated in process.[2]

Application of tangential feeding allows to carry out smooth feeding into a stream in the cyclone separator that promotes separation martempering. Various intermediate alternatives between simple tangential feeding and screw[3] are often applied.

Sampling of value of the dimensionless square of an entry at designing of cyclone separators is defined by necessity of maintenance of demanded efficiency of clearing, hence, enough high level of tangential speeds in a swept volume; maintenance of high carrying capacity on gas; and exclusions of driving down by a dust of a surface of the inlet branch.

11.4.1 CYCLONE DESIGN FACTORS

Cyclones are sometimes designed to meet specified pressure drop limitations. For ordinary installations, operating at approximately atmospheric pressure, fan limitations generally dictate a maximum allowable pressure drop corresponding to a cyclone inlet velocity in the range of 8–30 m/s (25–100 ft/s). Consequently, cyclones are usually designed for an inlet velocity of 15–20 m/s (50–65 ft/s), although this need not be strictly adhered to.

Because of the relatively high gas velocities at the inlet of cyclones, particle attrition in fluidized-bed systems is generally dominated by the attrition produced in the cyclone. In some catalytic systems with very expensive catalysts, the economics of the process can be dependent on low attrition losses. In such cases, reducing the inlet velocity of the cyclone will significantly reduce the attrition losses in the process. To compensate for the reduction in inlet velocity, the exit gas velocity will generally be increased (by reducing the diameter of the outlet tube) in order to maintain high cyclone efficiencies. Reducing the outlet tube diameter increases the outlet gas velocity and the velocity in the vortex of the cyclone—increasing collection efficiency. However, as the vortex velocity is increased, its length is also increased. Therefore, care must be taken to ensure that the cyclone is long enough to contain the increased vortex length. If it is not, the vortex can extend far into the cone and can entrain solids, lowing on the sides of the cone as it comes near them.[1]

11.4.2 CYCLONE INLETS

The design of the cyclone inlet can greatly affect cyclone performance. It is generally desired to have the width of the inlet B as narrow as possible so that the entering solids will be as close as possible to the cyclone wall where they can be collected. However, narrow inlet widths require that the height of the inlet II be very long in order to give an inlet area required for the desired inlet gas velocities. Therefore, a balance between narrow inlet widths and the length of the inlet height has to be struck. Typically, low-loading cyclones (cyclones with inlet loadings less than approximately 2–5 kg/m^3) have height/width ratios H/B_c of between 2.5 and 3.0. For high-loading cyclones, this inlet aspect ratio can be increased to as high as 7 or so with the correct design. Such high inlet aspect ratios require that the cyclone barrel length increases.

A common cyclone inlet is a rectangular tangential inlet with a constant area along its length. This type of inlet is satisfactory for many cyclones, especially those operating at low solids loadings. However, a better type of inlet is one in which the inner wall of the inlet is angled toward the outer cyclone wall at the cyclone inlet. This induces solids momentum toward the outer wall of the cyclone. The bottom wall of the inlet is angled downward so that the area decreases along the inlet. Flow path is not too rapid and acceleration is controlled. In addition, the entire inlet can be angled slightly downward to give enhanced efficiencies. This type of inlet is superior to the constant-area tangential inlet, especially for higher solids loadings (greater than 2–5 kg/m^3).

The experimental results obtained describe the complexity for efficiency measurements in cyclone separators. It speaks complexity and coherence of the processes proceeding in the cyclone separator.

11.5 SAMPLING OF SYSTEM OF A GAS CLEANING

In this study, we have to consider the ecological concepts and economic efficiency provisions. The problem is put to inject into calculation of a damage to a circumambient D_t operational parameters of the given clearing installation to pass to relative magnitudes that will allow to divide out number of the factors which are not influencing functioning of system to reduce criteria of ecological efficiency to technological to develop methods of calculation of relative efficiency of the treatment facilities, giving the chance to choose the most rational approaches and the equipment of systems of trapping of harmful atmospheric emissions. In the most general case, the damage D_t caused by atmospheric emissions can be computed[3]:

$$D_t = B \cdot M$$

The reduced mass of emission switches on N of components in an aspect

$$M = \sum_{i=1}^{N} A_i \cdot m_i,$$

and the weight of emission m_i is proportional to an overshoot through system

$$m_i = (1 - \eta_i) \cdot m_{oi}.$$

In an industrial practice, it is usually set or the share of concrete pollution in departing gas is known C_{oi}. We will consider gas diluted enough so its density ρ does not depend on the presence of impurity

$$m_{oi} = C_{oi} \cdot \rho \cdot Q.$$

Let us compute a damage caused to an aerosphere per unit weight of trapped pollution D_m

$$D_m = \frac{B \cdot \sum_{i=1}^{N} A_{i\,i} \cdot (1-\eta_i) \cdot C}{\sum_{i=1}^{N} \eta_i \cdot C_{oi}}.$$

To size up gas cleaning system on average parameters, $\eta_i = \eta$

$$A = \frac{1}{N} \cdot \sum_{i=1}^{N} A_i \cdot C_{oi},$$

that

$$D_m = \frac{B \cdot A \cdot (1-\eta)}{\eta}.$$

Let us formulate a principle of the ecological efficiency of nature protection provisions, at least a damage put to a circumambient. Purpose function in this case will appear in the form of $D_m \rightarrow \min$. Magnitude D_m diminishes with value growth.

$$E = \frac{\sum_{i=1}^{N} \eta_{ii} \cdot \tilde{N}}{B \cdot \sum_{i=1}^{N} A_i \cdot (1-\eta_i) \cdot C_i}.$$

We will consider magnitude E as criterion of ecological efficiency of nature protection provisions. The criterion of relative ecological efficiency θ is representable in the form of the relation of values E computed for compared alternative E_1 and base E_0 accepted in the capacity of E_0:

$$\theta = E_1/E_0.$$

In case of the single-component pollution value of criterion of relative ecological efficiency, we will find θ as

$$\theta = \frac{\eta_1}{\eta_i} \cdot \frac{1-\eta_i}{1-\eta_1}.$$

Thus, for two systems of a gas cleaning of the concrete manufacture, different in separation extent, $\eta_1 \neq \eta_0$, the relative ecological system effectiveness is sized up by technological parameter $\theta \rightarrow \max$. The prevented damage D_n is computed as a difference between economic losses of two competing alternatives.

$$D_n = D_o - D_1.$$

We will be restricted to a case of comparison of two alternatives of the gas cleaning intended for the same manufacture with fixed level of technological perfection. In the capacity of base alternative D_0, we will accept the greatest possible damage atmospheric emissions of the manufacture which technological circuit design does not provide a clearing stage, $\eta_{io}=0$. For the fixed technological circuit design of manufacture efficiency of a stage of clearing, we will size up in shares from the maximum damage

$$E_n = D_n/D_0$$

$$E_i = \frac{D_i}{D_0} = \frac{\sum_{i=1}^{N} A_{ii} \cdot C_i \cdot \eta}{\sum_{i=1}^{N} A_{ii} \cdot C}.$$

To consider that all components of harmful emission with aggression average indexes are trapped in equal extents ($A_i = A$, $\eta_i = \eta$), we come to $E_n = \eta$. Thus, the widespread extent of trapping η is a special case of criterion of ecological efficiency E_n computed for one-parametric pollution or for emission with average characteristics.[3] At sampling of te system of a gas cleaning, it is necessary to give preference to the installation providing higher values of criterion E_n.

11.6 CONCLUSIONS

The analysis of known designs of cyclone separators has allowed to reveal major factors (loading on a liquid phase, a stream monkey wall, and discharge connection sizes), working upon the aerodynamic structure of a stream. It is experimentally installed that the slope of blades makes strong impact on stream structure. Design data of intensity and the twisting gained for various types of cyclone separators are observed. For the characteristic of the intensity of a twisting of an air stream, it is offered to use a mean of the speed of the stream, defined on geometrical characteristics of the apparatus.

On the basis of a method of an estimation of economic efficiency of the carried out nature protection provisions, relationships for calculation of the damage put to a circumambient by atmospheric emissions of manufacture are gained. Transition to relative parameters has allowed to divide out number of the factors which are not influencing the process of a gas cleaning and to produce criteria of the ecological efficiency of systems the gas cleanings allowing on a design stage to make their rational sampling.

KEYWORDS

- **boundary problem**
- **inertial devices**
- **mathematical model**
- **computational grid**
- **nature protection provisions**
- **criterion**
- **ecological efficiency**

REFERENCES

1. Pell, M.; Dunson, J. B.; Knowlton, T. M. Gas-Solid Operations and Equipment. In *Perry's Chemical Engineers' Handbook*; The McGraw-Hill Companies: USA, 2008, Inc.67p.
2. Usmanova, R. R.; Zaikov, G. E. *Clearing of Industrial Gas Emissions: Theory, Calculation, and Practice;* AAP: U.S., Canada, 2015.
3. Usmanova, R. R; Zaikov, G. E. Ecological Safety of the Production Technology of Synthetic Rubber. *Elastomery* **2015,** *1*, 9–11.

CHAPTER 12

GENETIC ALGORITHMS AND THEIR APPLICATIONS IN DIFFERENT AREAS OF ENGINEERING AND SCIENCE

SUKANCHAN PALIT[1,2]

[1]*Department of Chemical Engineering, University of Petroleum and Energy Studies, Bidholi, via Premnagar, Dehradun, Uttarakhand 248007, India, Tel.: +91-8958728093*

[2]*43, Judges Bagan, Haridevpur, Kolkata 700082, India*

[*]*Corresponding author. E-mail: sukanchan68@gmail.com, sukanchan92@gmail.com.*

CONTENTS

ABSTRACT

Science and technology in present day human civilization are moving at a rapid pace. The world of challenges and the vision to move forward are immense and far-reaching. The futuristic vision of applied mathematics and chemical process engineering today are surpassing visionary scientific frontiers. Genetic algorithm (GA) is a branch of applied mathematics that needs to be reenvisioned and restructured in today's research pursuit in science and technology. Petroleum engineering science and chemical engineering are the two branches of scientific endeavor where there are massive applications of GA. GAs are also used in areas of multi-objective optimization which involves modeling, simulation, and optimization. In this treatise, the author rigorously points out toward the futuristic vision in the application of GA in different areas of engineering and science. Mankind's immense scientific prowess, the vast technological vision, and the deep scientific introspection will lead a long and visionary way in the true realization of optimization science and GAs. The areas covered in this chapter are basic fundamentals of GA, the scientific success, and the vast scientific potential behind GA applications. Scientific vision, scientific revelation, and the deep scientific introspection are the forerunners toward a newer scientific emancipation in the field of GA. This treatise unfolds the hidden intricacies of GA applications with the sole objective of the furtherance of science and engineering.

12.1 INTRODUCTION

Technology and engineering science in today's world are moving at a rapid and drastic pace. Human civilization and human scientific endeavor today stands in the midst of a deep scientific hope and unimaginable scientific vision. Scientific profundity and scientific forbearance are the cornerstones of scientific research pursuit in our present day human civilization. Science today is a gigantic colossus with a vast vision of its own. Provision of basic human needs are the scientific imperatives today. Food, energy, and water are the pivotal components of scientific revelation and scientific sagacity today. GA, multi-objective optimization, and applied mathematics tools are the vision of tomorrow. Science has few answers to the intricacies of scientific endeavor in petroleum engineering science and water technology. In this chapter, the author rigorously points out the scientific success, the vast technological vision, and profundity in the application areas of GA in petroleum engineering science, petroleum refining, water treatment, and energy engineering.[12,13] the technologies of petroleum refining and petroleum engineering are at a deep stake with the depletion of fossil fuel resources. Alternate energy resources and renewable energy are the cornerstones of scientific endeavor today. This treatise targets the immense success of GA toward the furtherance of science and engineering. Energy sustainability is the other visionary avenue of this comprehensive review. Energy sustainability and petroleum energy are the two opposite sides of the scientific visionary coin today. GA and other tools of applied mathematics will veritably lead a long and visionary way in the true scientific emancipation and the true realization of holistic sustainable development today.

12.2 THE AIM AND OBJECTIVE OF THE STUDY

The world of engineering and science is rapidly moving at a drastic pace. Technology and engineering science needs to be reenvisioned and revamped as mankind trudges a weary path toward a newer scientific destiny and a newer vision.[12,13] Mathematical tools such as GAs, differential evolution, and multi-objective optimization are the vital scientific imperatives toward a newer visionary eon today. GA is a next generation mathematical tool that has immense applications in designing of petroleum refining units today. The success, the targets, and the vision today are replete with deep scientific imagination. Human scientific endeavor and engineering science today stands in the midst of scientific introspection and wide vision. In this treatise, the author pointedly focuses on the wide applications of GA in designing of petroleum refining units and the wide area of water technology. The vision and the challenge of this treatise lies in the hands of deep scientific introspection in the field of multi-objective optimization and GA. The author also lucidly delineates the field of multi-objective simulated annealing. Multi-objective simulated annealing is another broad area of scientific research pursuit today. Technological marvels, scientific endurance and the broad scientific girth are the forerunners toward a newer visionary era in the field of chemical process engineering and petroleum engineering science.

12.3 WHAT DO YOU MEAN BY GENETIC ALGORITHM?

GA is an area of frontier science and technology. Scientific vision in the application of GA in multi-objective optimization is gaining immense scientific heights as human civilization treads a visionary road toward a newer eon. The widespread use of GA for problem-solving and scientific realization is not new. The pioneering work of J. H. Holland in the 1970s proved to be a significant contribution for scientific and engineering applications.[12,13] Technology has travelled a vast and visionary path since then. Since 1970, the output of research work in this area of scientific endeavor has grown exponentially although the successful contributions have been, and are largely initiated, from academic institutions worldwide.[12,13] Only in recent times research work appreciation and true realization of GA applications are truly opening up new areas of scientific revelation and wide vision. The concept of GA still stands in the midst of scientific introspection. This vast concept is still bleak and replete with scientific imagination. However, the major obstacle that drives engineers and computer scientists away from using GA is the major and vexing difficulty of speeding up the computational process as well as the intrinsic nature of randomness. Nonetheless, GA development has been visionary and scientifically stoic. Technological profundity and scientific maturity are the other cornerstones toward a scientific revolution in differential evolution and GA today.[12,13]

12.4 SCIENTIFIC DOCTRINE BEHIND MULTI-OBJECTIVE OPTIMIZATION

Multi-objective optimization and multi-objective simulated annealing are gaining new heights in this century in order to gear forward towards a newer vision in computer science. Multi-objective optimization today has immense applications in chemical process engineering, petroleum engineering science and diverse branches in engineering science. The vision and the challenge of optimization need to be readdressed and reenvisioned. Scientific doctrine, scientific cognizance, and the needs of the human society are the pallbearers toward a newer realization of optimization science today. Energy, water, and food stands as major pillars in the furtherance of science and engineering today. Science of sustainability today needs to be reenvisioned and restructured with the passage of scientific history, scientific vision, and time. Multi-objective optimization and application of GA are today restructuring the vast scientific landscape of applied mathematics, chemical process engineering, and petroleum engineering science. The need for energy sustainability is a veritable challenge to the mankind. In this treatise, the author pointedly focuses on the scientific success, the deep scientific doctrine, and the futuristic vision of application of multi-objective optimization for designing petroleum engineering and chemical engineering systems. Technological vision and vast scientific objectives and profundity are of utmost need in the research pursuit and scientific forays.[12,13]

12.5 SCIENTIFIC VISION AND SCIENTIFIC DOCTRINE IN GA APPLICATIONS

Scientific vision and deep scientific discernment are the imperatives of scientific endeavor today. The challenges of energy sustainability are changing the deep scientific fabric of emancipation and contemplation. GA applications today span in diverse branches of scientific endeavor. Petroleum engineering operations today are veritably in the need of advanced modeling and simulation. Design of the fluid catalytic cracking unit (FCCU) and other important units in petroleum refining are today linked with the wonders of improved mathematical tools such as GA. Multi-objective optimization as well as multi-objective simulated annealing are the visionary eons of tomorrow's scientific research pursuit. In this treatise, the author pointedly focuses on the scientific and academic rigor behind GA applications. The wide scientific success, the real world application of computer science, and the applications of GA are the forerunners towards a greater visionary paradigm in engineering science.

12.6 APPLICATION OF GA IN DIVERSE BRANCHES IN SCIENCE AND ENGINEERING

Application of GA in vast areas of science and engineering are creating new waves of scientific regeneration and deep scientific forbearance. Technological difficulties and intricate barriers of optimization science are readdressing the whole domain of GA. Applied mathematics and chemical process engineering are today in a state of immense scientific upheaval and reconstruction. Scientific girth, scientific determination, and deep scientific contemplation today are changing the path of human civilization. Nowadays, GA encompasses diverse branches of research endeavor. The vast challenges, the deep intricacies, and the scientific girth will today lead a long and visionary way in the true emancipation of science and engineering. GA can be applied in designing chemical and petroleum engineering systems. Technological candor, scientific contemplation, and the wide research questions are the forerunners toward a newer and a true realization of chemical process engineering and petroleum engineering. Mankind's immense scientific prowess today is at stake due to the ever-growing concern of fossil fuel depletion and the frequent environmental catastrophes. This treatise vastly addresses the wide scientific success, the scientific potential, and the scientific challenges in GA applications in chemical process engineering, petroleum engineering science, and water technology. Thus, technology will not remain behind as scientific validation and scientific comprehension veritably gain visionary heights.

12.7 PETROLEUM REFINING AND PETROCHEMICALS

Petroleum refining and petrochemicals are progressing towards scientific regeneration. Human scientific endeavor and the vast scientific and academic rigor of petroleum engineering are opening up new challenges and a newer vision in the field of petroleum refining. The depletion of fossil fuel resources has plunged the human civilization in a deep abyss. In such a crucial juxtaposition of scientific history and time, GAs and other mathematical tools are leading a long and visionary way in the true realization and true scientific emancipation of petroleum refining technology. Petrochemicals and petroleum refining products are the energy needs of the present day human civilization. In this treatise, the author pointedly focuses on the scientific success, the scientific fortitude, and the scientific forbearance in the true application of petroleum engineering to human society. Petroleum refining and petrochemicals are witnessing immense scientific intricacies and deep challenges. Application of mathematical tools, the crisis of fossil fuel depletion, and the futuristic vision are the torchbearers toward a greater emancipation of petroleum refining, and modeling and simulation of petroleum refining units. Here GA is gaining immense scientific heights along with the sole vision of application and scientific realization.

12.8 RECENT SCIENTIFIC ENDEAVOR IN THE FIELD OF GA APPLICATIONS

Scientific endeavor in the field of GA applications are far-reaching and surpassing visionary frontiers. Mankind's immense scientific determination, the technology revelation and the immense need of petroleum engineering revamping will all lead a long and visionary way in the true realization of mathematical tools such as GA and multi-objective optimization. GA is robust and groundbreaking with the passage of scientific history, scientific forbearance, and time.

Man et al. (1996)[1] deeply discussed with cogent insight concepts and applications in the GA. Scientific vision, scientific validation, and deep scientific forays are the challenges of human civilization and human scientific endeavor. This chapter introduces GA as a complete entity, in which the knowledge of this emerging technology can be integrated together to form the framework of an effective design tool for industrial engineers. In many ways, industrial engineering and computer science are linked by an unsevered umbilical cord. The various facets of the application of GA tools are elucidated in this well-researched treatise.[1] The use of GA for problem-solving is vast and visionary and not latent. The pioneering work of J. H. Holland in the 1970s proved to be a watershed and significant contribution for the scientific and engineering applications.[1] Since then, the vast output of research pursuit in this field has grown exponentially, although the contributions have been largely initiated from academic institutions worldwide.[1] The vast vision of science and the targets of engineering science are today culminating and evolving into new knowledge dimensions in the GA applications in this century.[1] Applied science and applied mathematics are today witnessing newer scientific regeneration and vast scientific reenvisioning with the passage of visionary time frame. The concept of GA is still not understood and scientific intricacies are immense.[1] The obvious obstacle that may drive engineers and technologists away from using GA is the vast difficulty of speeding up the computational process as well as the intrinsic nature of the randomness that veritably leads to a problem of performance assurance.[1] Engineering science research pursuit, the success of mathematical tools, and the wide vision of technology applications are all today leading a long and visionary way in the true realization of GA science today. Validation of science and technology and the true realization of sustainability are the utmost need of the hour.[1] This treatise redefines the vision of GA applications with the sole purpose of enhancing scientific vision and discernment.[1] GA is inspired by the mechanism of natural selection, a biological process in which the stronger individuals are likely be the winners in the competitive environment.[1] Scientific discernment and scientific sagacity are in the process of immense restructuring in this century. Applied science and applied mathematical tools are the imperatives of scientific research pursuit today. Human scientific vision, the veritable needs of human society, and the sustainability science will definitely lead toward the vast emancipation of applied mathematics, applied mathematical tools, and differential evolution science today. GA uses a direct analogy of such natural evolution.[1] It presumes that the potential solution of a problem is an individual and can be represented by a set of parameters. The scientific endeavor and the deep scientific vision of GA today are opening new areas of challenges and deep scientific fortitude in decades to come.[1]

Mardle et al. (1998)[2] deeply pondered upon the vast scientific success of GAs in a well-researched treatise. This chapter investigates GAs for the optimization of multi-objective fisheries bio-economic models.[2] The use of GAs for optimization problems offer a promising and an alternative approach to the traditional solution methods. Scientific profundity, scientific intellect, and a deep scientific cognizance are the pillars of this well-researched paper. GA follows the concept of solution evolution, by stochastically developing generations of solution populations using a given fitness statistic.[2] GA is the future generation mathematical tool and needs to be reenvisioned and readdressed with the passage of scientific history, scientific vision, and time.[2] In this paper, a nonlinear goal program of the North Sea demersal fishery is used to develop a GA for optimization.[2] Optimization science today stands in the midst of deep scientific introspection and clarity.[2] This treatise widely opens the windows of scientific innovation, scientific intellect, and vast scientific vision in the field of GA.[2]

Vasconcelos et al., (2001)[3] deeply pondered on the concept of improvements in GAs. The deep scientific vision, scientific profundity, and the farsightedness of science will veritably open up new avenues of endeavor in GA in years to come.[3] This paper presents an exhaustive study of the simple genetic algorithm (SGA), the steady state genetic algorithm (SSGA), and the replacement genetic algorithm (RGA).[3] The performance and the outcome of each of the method is analyzed in relation to several operator types of crossovers, selection and mutation as well as in relation to the probabilities of crossover and mutation with and without dynamic change of its values during the optimization process. Validation of science and technology are today replete with vision and forbearance.[3] The scientific intellect, the far-sight of science, and the definite targets and vision of research will all lead a long and visionary way in the true realization of GA application in science and engineering.[3]

Mitchell (1996),[4] in a phenomenal treatise comprehended, with deep insight, the world of GA.[4] The author in this book has covered the overview of GA, GAs in problem-solving, GAs in scientific modeling, and the deep theoretical foundations of GA.[4] Slowly and steadily, the science of optimization and GA are moving toward a newer scientific evolution and scientific justification.[4] This book is a watershed text in the field of optimization and GA.[4] In the 1950s and 1960s, several computer scientists delved deep into the field of evolutionary systems with the sole objective of merger of evolutionary methods with optimization science. GAs were discovered by John Holland in the 1960s and were developed by his students and colleagues at the University of Michigan, United States of America in the 1960s and 1970s.[4] Technological profundity and scientific forbearance of optimization applications in the designing of chemical engineering and petroleum engineering systems are vast and far-reaching. Technology and engineering science today stand in the midst of a deep crisis with the ever-growing concerns of fossil fuel depletion and the frequent environmental catastrophes.[4] This treatise gives a holistic view of the scientific success and scientific destiny in genetic engineering applications.[4]

During the past decade, several methods have been proposed for handling and adjusting constraints by GAs for numerical optimization problems. It is a visionary and common knowledge that a successful implementation of an evolutionary technique for a particular real-world problem requires additional heuristics. Today GA is opening a new scientific evolution and new scientific strategies with the sole purpose of furtherance of scientific discernment and scientific evolution. The challenge and the vision of research fundamentals in chemical process engineering and petroleum engineering systems are veritably opening new doors of optimization science and GA. Technology of GA is today far-reaching and surpassing visionary boundaries. The author of this text repeatedly pronounces upon the vast scientific potential and the wide scientific genesis in the field of GA applications.

Coley (1999)[5] dealt with immense lucidity the scientific vision, the deep scientific profundity and the futuristic vision of application of GA in engineering and scientific applications.[5] GAs are numerical optimization algorithms inspired by both natural selection and natural genetics. The vision of engineering science, the immense prowess in research pursuit and the wide scientific rigor will all lead a long and visionary way in the true realization of applied mathematics and optimization science.[5] This method is a general one, capable of being applied in different knowledge dimensions.[5] The science of GA needs to be reenvisioned and restructured with the passage of scientific history and time. The algorithm is simple to understand and required an easy to write computer code. This is a concept of natural selection and natural genetics which has not been tried previously. However GAs have been invented by one man—Dr. John Holland—in the 1960s.[5] The technology is highly evolved and highly advanced.[5] Human scientific research pursuit is in a state of immense scientific regeneration and deep introspection. His 1975 book, "Adaptation in Natural and Artificial Systems" is particularly worth reading, and today his treatise needs to be reenvisioned and justified again with the course of scientific history and vision.[5]

12.9 RECENT SCIENTIFIC ENDEAVOR IN MULTI-OBJECTIVE OPTIMIZATION

Multi-objective optimization and multi-objective simulated-annealing are today in the path of newer scientific regeneration and deep scientific cognizance. Multi-objective optimization can be applied in design problems in petroleum engineering and water technology. Water science and water technology are today evolving into a newer field of visionary scientific endeavor. The author redefines the scientific research pursuit of GA giving two visionary angles such as, the design of FCCU and water technology. Scientific revelation and scientific sagacity are reframing the world of technology and the engineering science of GA, and multi-objective optimization. The sagacity of science is slowly evolving.

Multiple, often conflicting, objectives arise naturally in most real-world optimization scenarios. As evolutionary algorithms possess several characteristics that are important for this problem, this class of search strategies has been used for multi-objective optimization for more than a decade. Science and technology of GA and multi-objective optimization are witnessing drastic challenges and deep upheavals.

Diakonikolas (2011),[6] in a doctoral thesis propounded the approximation of multi-objective optimization problems. Optimization problems with multiple objectives are pervasive across many diverse disciplines—in economics, engineering, health care, and biology to name a few. The author in this treatise devised efficient algorithms for the succinct approximation of the Pareto set for a large class of multi-objective problems.[6] Scientific clarity and scientific determination are the pillars of this doctoral thesis.[6] Multi-objective optimization has witnessed remarkable changes over the years. Efficiency, robustness, and enhanced computational paradigm are the torchbearers toward a greater visionary era in the field of applied mathematics and optimization science.[6] The vision of science, the challenges of technology, and the futuristic vision of optimization science will all lead a long and visionary way in the true emancipation of engineering science.[6] The author presents the scientific subtleties which encompass a wide background and introduction, succinct approximate Pareto sets, approximate convex Pareto sets, the Chord algorithm, and a comprehensive conclusion.[6] Multi-objective optimization is a branch of mathematics that stands in the interface of operations research and microeconomics. This area has been under intense study since the 1950s.[6] Technology validation and enshrinement are the challenges of research pursuit. In that aspect, the author deeply contemplates the intricacies of optimization applications in diverse areas.[6] This type of multicriteria or multi-objective problems arise across diverse disciplines such as engineering, economics, health care, biology, and manufacturing. In multi-objective optimization, usually one cares not only in single optimal solution but in a more complicated object, the set of Pareto-optimal solutions or Pareto set.[6] These are the solutions that are not dominated by other solutions, that is to say, a solution is Pareto optimal if there does not exist another solution that is simultaneously better in all criteria.[6] The entire thesis delves deep into the wide domain of Pareto set and its scientific intricacies.[6]

Zitzler (1999)[7] described methods and applications in evolutionary algorithms for multi-objective optimization. Scientific vision and technological objectives are the new visionary domains of the future. Human civilization's scientific prowess, the futuristic endeavor, and the challenges of technology are today the torchbearers toward a newer visionary eon in the field of optimization.[7] Many real-world problems involve two types of problem difficulties: (1) multiple conflicting objectives, and (2) a highly complex search space.[7] Scientific vision and motivation are the forerunners toward a newer world of endeavor and research pursuit. In the pursuit of multi-objective optimization, on the one hand instead of a single optimal solution, competing goals give rise to a set of compromise solutions, generally denoted as Pareto optimal. In the absence of a preference know-how, none of the corresponding trade-offs can be said to be better than the others. On the other hand, the search space can be too large and too complex to be computed mathematically. Here comes the importance of efficient optimization strategies. The challenge is enormous and pragmatic as science and engineering move toward a newer era. The subject of this work is the comparison and the improvement of existing multi-objective evolutionary algorithms and their vast application in system design problems in computer science. The author in this thesis pointedly focuses on the intricacies of multi-objective optimization and computer science as a greater scientific understanding in the field of engineering science.[7]

KanGAL Report Number 2,009,005 (2011)[8] delineated an interactive evolutionary optimization method based on progressively approximated value functions. The challenge and the vision of this report go beyond scientific imagination.[8] This paper suggests a preference-based methodology, which incorporates an evolutionary multi-objective optimization algorithm to lead a decision-maker to the most preferred solution of his or her choice.[8] Scientific vision, scientific sagacity, and the world of challenges in evolutionary computation will definitely lead a long and visionary way in the true realization of optimization science and applied mathematics.[8]

12.10 PETROLEUM REFINING AND GA

Petroleum refining and the application of GA, and other mathematical tools are the two opposite sides of the visionary scientific coin. Human scientific endeavor, the needs of energy sustainability, and the vast scientific vision are today leading a long and visionary way in the true realization of scientific forbearance in mathematical tools today. Technology is highly advanced today. GA and other applied mathematical tools are changing the scientific frontiers. The vision of science, validation of engineering science and the futuristic vision of energy sustainability are the forerunners towards a newer eon of deep positivity in energy sustainability. Energy and environmental sustainability today are in the midst of deep scientific comprehension and vision. Petroleum engineering and the true realization of renewable technology are the visionary facets of Sustainable Development Goals of this century. The status of petroleum refining today is vast and versatile. In order to achieve sustainable development goals as propounded by the United Nations, the intense need and the intense target should be towards the scientific success and vision for energy self-sufficiency, energy vision, and renewable and nonrenewable energy technology.

12.11 APPLICATION OF GA IN MODELING AND SIMULATION OF PETROLEUM REFINING UNITS

Modeling and simulation of petroleum refining units is entering into a new era of scientific revelation and deep scientific fortitude. Today human scientific endeavor stands in the midst of deep scientific comprehension and vast contemplation. The question of energy sustainability comes into the scientific horizon as human civilization moves from one scientific paradigm over another.

Khandalekar (1993)[9] in a well-researched postgraduate thesis elucidated on the control and optimization of fluidized catalytic cracking process. Technology motivation and scientific objectives are the challenges of human civilization and human scientific endeavor today. The author focuses in these research domains. The authors delineate on the literature survey, process description and dynamic simulation, nonlinear process model-based control, constraint control, and FCCU optimization.[9] The challenge and the wide vision of scientific research pursuit go beyond scientific imagination and scientific forbearance. Dynamic simulation today stands in the midst of deep scientific comprehension and wide contemplation.[9] The FCCU receives multiple feeds consisting of high boiling components from several other refining units and cracks these streams into lighter and more valuable components.[9] The author deeply comprehends the scientific success, the vast scientific vision, and the deep scientific introspection in the application of multi-objective optimization and GA in the design of an FCCU.[9] The world of challenges, the veritable needs of human society, and the vital need of energy sustainability are the torchbearers toward a newer era of design, modeling, and simulation domain. Fluidized catalytic cracking is one of the most difficult processes to control in the petroleum refinery. The wide vision, the veritable research question of energy sustainability, and the targets of catalytic cracking will lead a long and visionary way in true realization of petroleum engineering science.[9] The author in this treatise deeply comprehends the vast scientific vision and the scientific forbearance behind modeling, control and optimization of an FCCU with the sole vision toward furtherance of science and engineering.[9]

Ruqiang et al. (2008)[10] discussed with a deep and cogent insight a novel close-loop strategy for integrating the process operations of an FCCU with production planning optimization. Technological prowess, scientific intellect, and the vast futuristic vision of petroleum engineering will all lead a long and visionary way in the true emancipation of science and engineering.[10] This treatise reflects the challenges in the process operations of an FCCU.[10] Production planning models generated by common modeling systems do not involve constraints for process operations and a solution optimized by these models is veritably called quasi-optimal plan.[10] In order to determine a practically feasible optimal plan and the corresponding operating conditions of an FCCU, a novel close-loop integrated strategy, including determination of a quasi-optimal plan, search of operating conditions of FCCU, and the revision of the production planning model was proposed in the model.[10] Technology and engineering science of process integration and control are entering a newer visionary era as human civilization and human scientific endeavor crosses vast and versatile scientific boundaries. This treatise is a challenge to scientific efforts and scientific fortitude.[10] The author rigorously points towards the application of GA in the operation of an FCCU and pinpoints the scientific intricacies and the scientific justification behind the operation.[10]

12.12 WATER SCIENCE AND TECHNOLOGY AND GA

At present, water science and technology are important needs of the human society. Drinking water treatment, industrial wastewater treatment and water purification are the visionary scientific endeavors of tomorrow. Here GA is also playing a decisive role toward greater scientific emancipation and true realization of energy engineering today. A water distribution network is a system containing pipes, reservoirs, pumps, and valves of different types, which are connected to each other to provide water to customers. Here comes the application of GA and multi-objective optimization. In the case of design of a pipe network, the optimization problem can be stated as follows: minimize the cost of the network components subject to the satisfactory performance of the water distribution system. This leads to an unfavorably constrained, combinatorial optimization problem. The author in this treatise addresses the water science interface with optimization science. The technological motivation, the challenges of applied mathematics, and the scientific success of water distribution network are the forerunners toward a veritable and true emancipation of optimization science today.

Guc (2006)[11] in his postgraduate thesis deeply elucidated optimization of water distribution networks using GA. The vast scientific vision and the scientific challenges are slowly unfolding and opening up a new era in the application of GA in water technology.[11] This study gives an intense description about a computer model, RealPipe, which relates GA to the well-known problem of the least cost design of water distribution network. GA is an evolutionary process which needs deep scientific introspection.[11] GA is essentially an efficient search method basically for nonlinear optimization cases. The immense scientific and academic rigor behind GA is the utmost need of deep comprehension today.[11] The vision and the targets of research pursuit in water technology and water distribution networks need to be reenvisioned and revamped with the passage of scientific history and time. GA optimization is well suited for optimization of water distribution systems, especially large and complex systems. RealPipe optimizes given water network distribution systems by considering capital cost of pipes only. Five definite operators are involved in the program algorithm.[11] These operators are generation, selection, elitism, crossover, and mutation.[11] Optimum population size is found to be between 30–70 depending on the size of the network (i.e., pipe number) and number of commercially available pipe size.[11] Technology of optimization is far-reaching today with the passage of scientific history, scientific forbearance and time. Optimization science in this treatise widely opens up new vistas of scientific research pursuit in decades to come.[11]

12.13 FUTURE FRONTIERS AND FUTURE FLOW OF THOUGHTS

At present, GA is on the path of immense scientific regeneration and wide scientific rejuvenation. This area of scientific endeavor is latent today and yet far-reaching. The needs of human civilization such as food, energy, and water are urging the scientific community to gear towards newer innovations and true realization of sustainability. Green technology and green engineering are the visionary paths of endeavor and scientific research pursuit today. Environmental and energy sustainability are the true need of the hour.[14,15,16] Petroleum engineering science and water technology today stand in the midst of deep scientific fortitude and forbearance. The design of a petroleum refining unit or an industrial wastewater/drinking water system needs greater scientific revamping. Here comes the need of GA, differential evolution, and multi-objective optimization. The success of engineering science today is at stake. The future frontiers need to be readdressed as regards true emancipation of energy sustainability and petroleum engineering science. Future targets and future flow of thoughts need to be reenvisioned and revamped with the passage of scientific history and time. The vision of science and the world of engineering science need to be revamped as human civilization trudges a weary and definite path towards energy sustainability.[12,13]

12.14 FUTURE RESEARCH TRENDS

Engineering and science are today in a state of immense catastrophe with the growing concerns of sustainable development. Energy sustainability and environmental engineering science are in a state of crisis with the evergrowing concern of depletion of fossil fuel resources, the burgeoning water crisis, and global water shortage. Technology has few answers to the ravages of water, environment, and energy of the present day human civilization. This treatise repeatedly pronounces the success of mathematical tools in designing petroleum refinery and water treatment units. GA is the future generation mathematical procedure. The future research trends pointedly focus on the scientific success, scientific revelation, and the wide scientific vision in the application of evolutionary computation and GA in designing chemical engineering, petroleum engineering, and environmental engineering systems. Research trends in applied mathematics and optimization are targeted towards the success of computational time and the efficiency of the process. Human scientific endeavor today stands in the midst of deep scientific revelation and intellectual prowess. This treatise pinpoints the wide vision and imagination of science in applied mathematics and optimization science. Technological masterpieces, scientific marvels, and scientific profundity of GA will veritably open new knowledge dimensions in years to come.[14,15,16]

12.15 SUMMARY, CONCLUSION AND FUTURE PERSPECTIVES

Science and the vision of scientific research pursuit are today surpassing the frontiers and intricacies of endeavor and intellect. Future perspectives of GA, mathematical tools, and differential evolution are changing the scientific landscape. In this treatise the author repeatedly points out the scientific excellence and scientific genesis of GA and the multi-objective optimization in designing chemical engineering and petroleum engineering systems. The scientific struggle and the scientific destiny of GA are veritably changing the paradigm of human research pursuit. The future perspectives of engineering and the science of GA remains challenged and needs to be reenvisioned and readdressed. This is a century of immense scientific introspection. Applied science and engineering science are in the midst of deep concern such as environmental crisis and depletion of fossil fuels. Here comes the veritable need of computer technology and applied mathematics. GA and multi-objective optimization are today changing the future of design of chemical engineering and petroleum engineering systems. The author deeply elucidates on the human genesis and related scientific success in the huge branch of applied mathematics and computational science. The technology needs to be newly reenvisioned with the passage of scientific vision, forbearance, and the timeframe.

12.16 ACKNOWLEDGMENT

The author with great respect acknowledges the contribution of his late father Shri Subimal Palit, an eminent textile engineer from India who taught the author the rudiments of chemical engineering.

KEYWORDS

- **optimization**
- **multi-objective**
- **evolution**
- **vision**
- **genetic**
- **algorithm**

REFERENCES

1. Man, K. F.; Tang, K. S.; Kwong, S. Genetic Algorithms: Concepts and Applications. *IEEE Trans. Ind. Electron.* **1996,** *43*(5), 519–534.
2. Mardle, S.; Pascoe, S.; Tamiz, M. An Investigation of Genetic Algorithms for the Optimization of Multi-Objective Fisheries Bioeconomic Models. CEMARE Research Paper 136, University of Portsmouth, United Kingdom, 1998.
3. Vasconcelos, J. A.; Ramirez, J. A.; Takahashi, R. H. C.; Saldanha, R. R. Improvements in Genetic Algorithm. *IEEE Trans. Magn.* **2001,** *37*(5), 3414–3417.
4. Mitchell, M. *An Introduction to Genetic Algorithms*; MIT Press: Cambridge, 1996.
5. Coley, D. A. *An Introduction to Genetic Algorithms for Scientists and Engineers;* World Scientific Publishing Co. Pte. Ltd.: Singapore, 1999.
6. Diakonikolas. I. Approximation of Multi-Objective Problems. Ph.D.Thesis, Columbia University, USA, 2011.
7. Zitzler. E. Evolutionary Algorithms for Multi-Objective Optimization: Methods and Applications. PhD. Thesis, Swiss Federal Institute of Technology, Zurich, Switzerland.
8. Deb, K.; Sinha, A.; Korhonen, P.; Wallenius, J. *An Interactive Evolutionary Multi-Objective Optimization Method Based on Progressively Approximated Value Functions*; Kanpur Genetic Algorithm Laboratory Report No.2009005: IIT Kanpur, India, 2011.
9. Khandalekar, P. Control and Optimization of Fluidized Catalytic Cracking Process. Master's Degree thesis, Texas Technological University, USA, 1993.
10. Ruqiang, W.; Chufu, L.; Xiaorong, H. E.; Bingzhen, C. A Novel Close-Loop Strategy for Integrating Process Operations of Fluidized Catalytic Cracking Unit with Production Planning Optimization. *Chin. J. Chem. Eng.* **2008,** *16*(6), 909–915.
11. Guc, G. Optimization of Water Distribution Networks using Genetic Algorithm. Master's Degree Thesis, Middle East Technical University, Turkey, 2006.
12. www.google.com (accessed Oct 1, 2017).
13. www.wikipedia.com (accessed Oct 1, 2017).
14. Bandyopadhyay, S.; Saha, S.; Maulik, U.; Deb, K. A Simulated Annealing-Based Multiobjective Optimization Algorithm: AMOSA, (2008). *IEEE Trans. Evol. Comput.* **2008,** *12*(3), 269–283.
15. Deb, K. *Multi-Objective Optimization Using Evolutionary Algorithms: An introduction*; Kanpur Genetic Algorithm Report No. 2011003, 2011.
16. Ramteke, M.; Gupta, S. K. Kinetic Modeling and Reactor Simulation and Optimization of Industrial Important Polymerization Processes: A Perspective. *Int. J. Chem. React. Eng.* **2011,** *9*, R1.

CHAPTER 13

DEHYDRATION KINETICS OF ZINC BORATES ($2ZnO \cdot 3B_2O_3 \cdot 3H_2O$ AND $4ZnO \cdot B_2O_3 \cdot H_2O$) BY THERMAL ANALYSIS

MEHMET GÖNEN[1,*], DEVRIM BALKÖSE[2], and SEMRA ÜLKÜ[2]

[1]*Department of Chemical Engineering, Engineering Faculty, Süleyman Demirel University, Batı Yerleşkesi, 32260 Isparta, Turkey*

[2]*Department of Chemical Engineering, Izmir Institute of Technology, Gülbahçe Köyü—Urla, 35430 Izmir, Turkey*

[*]*Corresponding author. E-mail: mehmetgonen@sdu.edu.tr*

CONTENTS

ABSTRACT

The kinetic parameters for the dehydration of zinc borates, $2ZnO{\cdot}3B_2O_3{\cdot}3H_2O$ (ZnB-1) and $4ZnO{\cdot}B_2O_3{\cdot}H_2O$ (ZnB-2), were determined using dynamic thermal gravimetric (TG) technique. Kissinger method was used in the analysis of the TG and derivative TG data for determination of dehydration kinetics. Zinc borate (ZnB-1) has started to loose mass at 290°C and its thermal dehydration consists of mainly two consecutive kinetic steps related to the OH group interactions as inferred from derivative TG curves. The dehydration activation energies of ZnB-1 were calculated as 183.5 and 556.1 kJ mol^{-1} for the first and second kinetic steps, respectively. On the other hand, ZnB-2 began to dehydrate at 450°C and dehydration took place only in one step and the dehydration activation energy was calculated as 201.7 kJ mol^{-1}.

13.1 INTRODUCTION

Metal borates occurring in nature as mineral or produced synthetically can be classified into two main groups, hydrated and anhydrous. Most of the synthetic metal borates look like minerals in chemical structure, containing isolated polyborate anions or complex polyborate rings, chains, sheets, or networks. Hydrated borates consist of boron, oxygen, metal, and hydrogen atoms. The building blocks in borates are trihedral (BO_3) and tetrahedral (BO_4) boron–oxygen units. Various types of borate anions, such as metaborate, triborate, tetraborate, and pentaborate are formed due to the higher order of boron–oxygen coordination. A water molecule in hydrated borates could be present as a water of crystallization (interstitial water) or hydroxyl groups attached to boron atoms (B–OH). Borate anions interact with cations in complex ways and often display extensive hydrogen bonding integrating their structures.[19] Two types of hydrogen bonding, stronger O–H and weaker O–H bonds could be in the structure of hydrated borates.[26]

Zinc borates are an important branch of inorganic hydrated borates and there is series of zinc borates ($4ZnO{\cdot}B_2O_3{\cdot}H_2O$, $ZnO{\cdot}B_2O_3{\cdot}{\sim}1.12H_2O$, $ZnO{\cdot}B_2O_3{\cdot}{\sim}2H_2O$, $6ZnO{\cdot}5B_2O_3{\cdot}3H_2O$, $2ZnO{\cdot}3B_2O_3{\cdot}7H_2O$, $2ZnO{\cdot}3B_2O_3{\cdot}3H_2O$, $3ZnO{\cdot}5B_2O_3{\cdot}14H_2O$, $ZnO{\cdot}5B_2O_3{\cdot}4.5H_2O$) that have been developed since the 1940s.[20] The properties of the zinc borates are determined by B_2O_3/ZnO molar ratio and the amount of water in its structure. They have been used in a wide range of applications, such as a flame retardant in polymers and coatings,[8] preservative in wood products,[7] an additive in lubricants,[4] and a corrosion inhibitor in coatings.[3] Zinc borates of $2ZnO{\cdot}3B_2O_3{\cdot}3H_2O$ and $4ZnO{\cdot}B_2O_3{\cdot}H_2O$ are the most widely used types in the polymer industry since they have moderately high dehydration temperatures.[2,21] It is expected that zinc borates used as fire retardant in polymers will not decompose during the processing of the polymer in an extruder but should dehydrate in the case of fire. Thermal decomposition of hydrated borate takes place in the following order: dehydration of product results in amorphous product and recrystallization or solid-state reaction of remaining constituents. The formation of an amorphous phase of zinc borate, $2ZnO{\cdot}3B_2O_3$ after the dehydration of $2ZnO{\cdot}3B_2O_3{\cdot}3H_2O$ was pointed out in the literature.[16] $Zn_4B_4O_7$ and $Zn_4B_6O_{13}$ crystals were present in $2ZnO{\cdot}3B_2O_3{\cdot}3H_2O$ samples heated at 800°C.

Two independent stages occur in the decomposition of hydrated borates as determined by Waclawska.[24] First, the formation of free water molecules from interstitial water or condensation of hydroxyl groups takes place and second, water formed transfers from the interior of the solid matrix to the outside. The removal of interstitial water occurs at relatively low temperature compared with the removal of water formed by condensation of hydroxyl groups.[24] For instance, borax decahydrate ($Na_2B_4O_5(OH)_4{\cdot}8H_2O$) starts to release 8 mol of water of crystallization at 330 K and 2 mol of water in the form of hydroxyl groups attached to the borate anion at 415 K.[5] The thermal behavior of zinc borates is substantially important, especially when they are used as a flame retardant in polymers. It is expected that they must be stable while the polymer is being processed and they should function when they are exposed to high temperature during the fire. Hydrated metal borates function not only by releasing their hydration water to dilute both fuel vapor and oxygen in the combustion zone but also by forming an inorganic glassy layer, which acts as a barrier for both mass and heat transfers. The entrapped material beneath the glaze layer increases the formation of char. The release of water from interstitial water or condensation of hydroxyl groups absorbs heat from the environment, as well.[13] Fire retardancy of metal borates in polymers is usually controlled by these two phenomena. Since dehydration kinetics of metal borates plays a crucial role in providing fire retardancy, it should be examined and the parameters need to be determined for zinc borates $2ZnO{\cdot}3B_2O_3{\cdot}3H_2O$ and $4ZnO{\cdot}B_2O_3{\cdot}H_2O$. Thermal dehydration kinetics of zinc borate ($Zn_3B_6O_{12}3.5H_2O$), which was synthesized from boric acid and zinc salts (zinc sulfate heptahydrate and zinc chloride) was investigated by Kipcak et al. (2015).[11] They reported that thermal decomposition of zinc borate occurred in two stages. In the Coats–Redfern method, activation energies for the first and second steps were calculated as 225.40 and 570.63 $kJ \cdot mol^{-1}$, respectively. The dehydration behaviors of two different hydrated zinc borate species, $Zn[B_3O_3\ (OH)_5] \cdot H_2O$ and $Zn[B_3O_4(OH)_3]$, which are industrially important flame retardants, were investigated by thermal gravimetric (TG) analysis and in situ diffuse reflectance infrared Fourier transform spectroscopy.[1] Dehydration onset temperatures of Zn $[B_3O_3(OH)_5] \cdot H_2O$ and Zn $[B_3O_4(OH)_3]$ were 129 and 320°C, respectively, at a $10°C \cdot min^{-1}$ ramp rate. Average activation energies for dehydration of Zn $[B_3O_3(OH)_5] \cdot H_2O$ and Zn $[B_3O_4(OH)_3]$ were calculated as 54.9 and 166.2 $kJ \cdot mol^{-1}$ by using Ozawa method.

After the clarification of the $2ZnO \cdot 3B_2O_3 \cdot 3H_2O$ structure by Schubert et al. (2003),[20] different dehydration temperatures were reported for this product by different researchers.[6,23] Mergen et al. (2012)[14] studied the production of nano-sized zinc borate ($4ZnO \cdot B_2O_3 \cdot H_2O$) using zinc borate of $2ZnO \cdot 3B_2O_3 \cdot 3.0–3.5H_2O$ and they investigated the polyvinyl chloride (PVC) thermal stability in the presence of $4ZnO \cdot B_2O_3 \cdot H_2O$ (). The addition of zinc borate into PVC (1 and 5 %wt.) has enhanced the PVC flame retardancy by increasing limiting oxygen index of virgin PVC from 41 to 47% and to 54%, respectively. On the other hand, the addition of nano zinc borate into PVC did not affect inversely the mechanical properties of zinc borate–PVC composites even at the high amount of 5 %wt. There is also no detailed thermal analysis for the recently developed zinc borate species $4ZnO \cdot B_2O_3 \cdot H_2O$ in the literature. It is expected that knowing the decomposition parameters of the aforementioned zinc borates will provide important data for the formulation of fire retardants in polymers.

13.1.1 KINETIC METHODS

A kinetic parameter of interest was the apparent activation energy (E_a) relating to decomposition of hydrated borates. There are numerous techniques available in the literature to calculate E_a.[9] The explicit derivations of these methods can be found in the literature. The Ozawa and the Kissinger methods are generally used in the determination of apparent activation energy for decomposition of most inorganic and polymeric materials.[12,15]

In this study, the Kissinger method was used to determine E_a, A, and k of zinc borates. The ultimate operative equations used in this chapter are as follows:

$$\frac{d\alpha}{dt} = \beta\left(\frac{d\alpha}{dT}\right) = k(T)f(\alpha), \tag{13.1}$$

where β is the heating rate (°C·min^{-1}), α is the conversion, $k(T)$ *is* the rate constant function of temperature, and T is the temperature (K).

α, conversion or fraction decomposed, is defined as:

$$\alpha = \frac{W_o - W_t}{W_o - W_f} \tag{13.2}$$

where W_o is the initial mass of the sample, W_t is the mass of the sample at temperature t and W_f is the final mass at a temperature at which the mass loss is approximately not changed.

Applying *n*th order reaction for the dehydration of zinc borates to obtain

$$f(\alpha) = (1 - \alpha)^n \tag{13.3}$$

The temperature dependence of the rate constant of dehydration is expressed by Arrhenius as given below:

$$k(T) = A\exp(-\frac{E_a}{RT}) \tag{13.4}$$

Inserting Equations (13.3) and (13.4) into Equation (13.1) gives:

$$\frac{d\alpha}{dt} = \beta\frac{d\alpha}{dT} = (1-\alpha)^n A\exp(-\frac{E_a}{RT}) \tag{13.5}$$

where A is the pre-exponential factor (min^{-1}), E_a is the activation energy of reaction (J·mol^{-1}), and R is the gas constant (8.314 J·mol^{-1}·K^{-1}). Equation (13.5) is the basic analytical equation for the TG data.

13.1.1.1 KISSINGER METHOD

In this technique, the decomposition rate gradually increases to a maximum value and then decreases to zero as the material is consumed. The maximum decomposition rate is reached when the change of decomposition rate with respect to time is zero (*d/dt* [*dα/dt*] = *0*). When Equation 13.5 was differentiated and set to zero, the following equation was obtained:

$$\frac{\beta E}{RT_m^2} = A.n(1-\alpha)_m^{n-1}\exp(-E_a / RT_m) \tag{13.6}$$

where T_m represents the temperature at maximum decomposition rate. The Kissinger method assumes that the product $n(1-\alpha)_m^{n-1}$ is independent of β and close to 1.0; the following expression can be derived:

$$\ln\left\{\frac{\beta}{T_m^2}\right\} = \left\{\ln\frac{AR}{E_a} + \ln\left[n(1-\alpha)_m^{n-1}\right]\right\} - \frac{E}{RT_m} \tag{13.7}$$

Derivative TG data obtained at different heating rates are used to determine T_m and then activation energy and frequency factor are calculated using slope and intercept of the line in $\ln(\beta / T_m^2)$ versus $1/T_m$ plots, respectively.

In the present study, the reaction processes of thermal dehydration of $2ZnO \cdot 3B_2O_3 \cdot 3H_2O$ (ZnB-1) and $4ZnO \cdot B_2O_3 \cdot H_2O$ (ZnB-2) were investigated using TG analysis. The Kissinger method was used in order to obtain fundamental data, activation energy, and frequency factor.

13.2 EXPERIMENTAL

13.2.1 MATERIALS

Commercial zinc borate samples, Firebrake 2335, $2ZnO \cdot 3B_2O_3 \cdot 3H_2O$ (ZnB-1) and Firebrake 415, $4ZnO \cdot B_2O_3 \cdot H_2O$ (ZnB-2) from U.S. Borax Inc. were used in the experiments. They were characterized in order to determine their properties by X-ray crystallography (XRD), Fourier–transform infrared (FTIR) spectroscopy, and scanning electron microscope (SEM). Thermal characterization of those species and dehydration kinetic study were performed by using TG analysis.

13.2.2 CHARACTERIZATIONTECHNIQUESOFZINCBORATES

The crystalline structure and purity of the zinc borate samples were determined using X-ray powder diffractometer (Philips Xpert-Pro) with CuKα radiation at 45 kV and 40 mA. The scattering data were collected in the 5–70° 2θ range. SEM (Philips XL30 SFEG) was used for the identification of particle size and morphology. The transmission spectra of KBr pellets prepared by mixing 4.0 mg of zinc borate and 196 mg of KBr in an agate mortar and pressing the mixture under 10 tons were obtained using FTIR spectrophotometer (Shimadzu 8601).

13.2.3 THERMOGRAVIMETRIC (TG) ANALYSIS

Dynamic TG runs were obtained with TGA (Setaram Labsys) thermobalance in nitrogen atmosphere at a flow rate of 40 $cm^3 \cdot min^{-1}$. The initial mass of the samples was about 10–12 mg. The dynamic runs were carried out from ambient temperature to 600°C at rates of 5, 10, 15, 20°C·min^{-1}. Raw TG data were processed to obtain derivative TG data.

13.3 RESULTS AND DISCUSSION

13.3.1 CHARACTERIZATION OF ZINC BORATES

FTIR spectra of zinc borates of ZnB-1 and ZnB-2 are shown in Figure 13.1. The spectrum in Figure 13.1a indicates that zinc borate species has identical spectrum with the spectrum given for $2ZnO \cdot 3B_2O_3 \cdot 3H_2O$.[6] All hydrated borate salts have a broad peak at 3000–3600 cm^{-1} resulting from stretching vibrations of OH groups in their structure.[10] $2ZnO \cdot 3B_2O_3 \cdot 3H_2O$ has two major sharp peaks at 3210 and 3470 cm^{-1} in this region, which is accepted as characteristics of $2ZnO \cdot 3B_2O_3 \cdot 3H_2O$.[6] The three hydrogen atoms of $2ZnO \cdot 3B_2O_3 \cdot 3H_2O$ are hydrogen bonded with acceptor O atoms of B–O–B groups. They have distinctly different H-bonding strengths as determined by nuclear magnetic resonance spectroscopy.[20] Another distinguishing peak emerging at 2520 cm^{-1} belongs to hydrogen-bonded OH groups stretching. Thus, FTIR also indicated three OH groups with different H-bonding strength existed in $2ZnO \cdot 3B_2O_3 \cdot 3H_2O$. The chemical structure of zinc borate (ZnB-1) has been recently updated as $(Zn\,[B_3O_4(OH)_3])$, which has a polymeric network structure. Although the structure of $2ZnO \cdot 3B_2O_3 \cdot 3.5H_2O$ was revised as $2ZnO \cdot 3B_2O_3 \cdot 3H_2O$, it is still known by its previous oxide formula in trade literature. The other peaks between 1400 and 400 cm^{-1} wavenumbers belong to various B–O coordination that was defined in literature.[10,25] Figure 13.1b shows the FTIR spectrum of $4ZnO{\cdot}B_2O_3{\cdot}H_2O$ in which major peaks are 3500, 3400, 1346, 1250, 1028, 717, 530, 474, 400 cm^{-1} wavenumbers. The bands at 3500, 3400 cm^{-1} belong to H-bonded OH vibrations. The bands at 1346 cm^{-1} and 1028 cm^{-1} are due to the asymmetric stretching of BO_3 and BO_4 structures, respectively. The bands at 1250 cm^{-1} and 717 cm^{-1} could be due to the in-plane bending of B–O–H and out-of-plane bending of BO_3, respectively. The band at 530 cm^{-1} is caused by symmetric pulse vibration of pentaborate anion $[B_5O_6(OH)_4]^-$ and the band at 474 cm^{-1} could be due to bending of BO_4 and the band at 400 cm^{-1} can be due to ZnO coordination. Another important feature that was observed in infrared spectra of both zinc borates is that they do not have water of crystallization in their structure since there is no peak at 1635 cm^{-1} related to bending vibrations of H–O–H structure. However, zinc borate of $4ZnO{\cdot}B_2O_3{\cdot}H_2O$ produced by boiling the aqueous NH_3 solution in the presence $2ZnO{\cdot}3B_2O_3{\cdot}3H_2O$ has crystalline water according to the FTIR analysis as pointed out in the literature (Mergen et al., 2015).

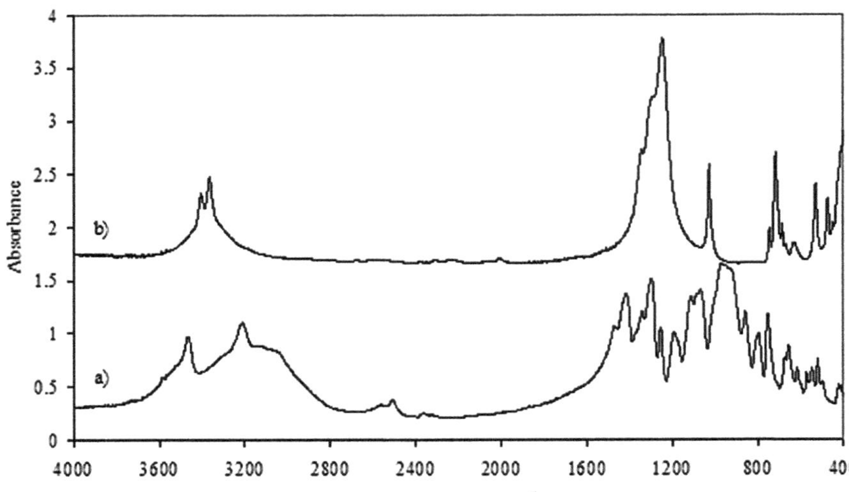

FIGURE 13.1 **(See color insert.)** Fourier-transform infrared (FTIR) spectra of zinc borate species (a) $2ZnO{\cdot}3B_2O_3{\cdot}3H_2O$ and (b) $4ZnO{\cdot}B_2O_3{\cdot}H_2O$.

The presence of sharp peaks in those XRD patterns shown in Figure 13.2 indicates that both zinc borate species have crystalline structure. The most important characteristic of this zinc borate, $2ZnO{\cdot}3B_2O_3{\cdot}3H_2O$, is that the major peak begins to occur at 17.9° 2θ value, and other major peaks in Figure 13.2a are observed at about 20.5, 21.7, 23.6, 25.7, 28.6, 30.0, and 36.7° 2θ values. These are consistent with the XRD pattern of $2ZnO{\cdot}3B_2O_3{\cdot}3H_2O$, which was reported by Sawada et al. (2004).[17] The observed major peaks in Figure 13.2b, 18.6, 21.7, 28.3, 31.5, 32.5, 36.1, 37.4 and 40.5° 2θ values, are consistent with the peaks which were defined for zinc borate of $4ZnO{\cdot}B_2O_3{\cdot}H_2O$.[18,22] Types of zinc borates regarding to the literature were verified.

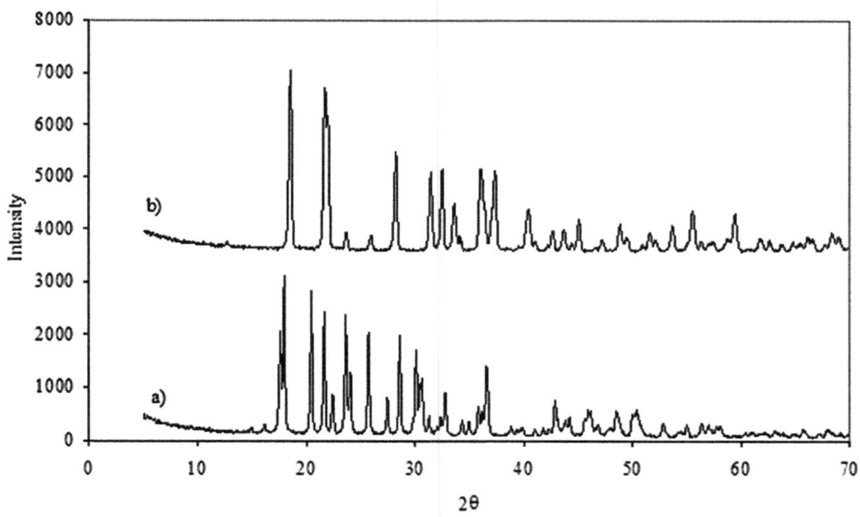

FIGURE 13.2 (See color insert.) X-ray crystallography (XRD) patterns of zinc borate species (a) $2ZnO{\cdot}3B_2O_3{\cdot})\ 3H_2O$ and (b) $4ZnO{\cdot}B_2O_3{\cdot})_2O$.

Figure 13.3 and 13.4 show SEM microphotographs of zinc borates of $2ZnO{\cdot}3B_2O_3{\cdot}3H_2O$ and $4ZnO{\cdot}B_2O_3{\cdot}H_2O$, respectively. It is clearly observed that smaller particles with 0.4–1 μm particle size have smooth surfaces and random shapes were agglomerated and formed larger particles as seen in Figure 13.3. The morphology of $4ZnO{\cdot}B_2O_3{\cdot}H_2O$ was determined as plate-like as seen in Figure 13.4. Those particles are about 10 μm in length and 2–4 μm in width and less than 1 μm in thickness. No agglomeration of those particles was observed in Figure 13.4.

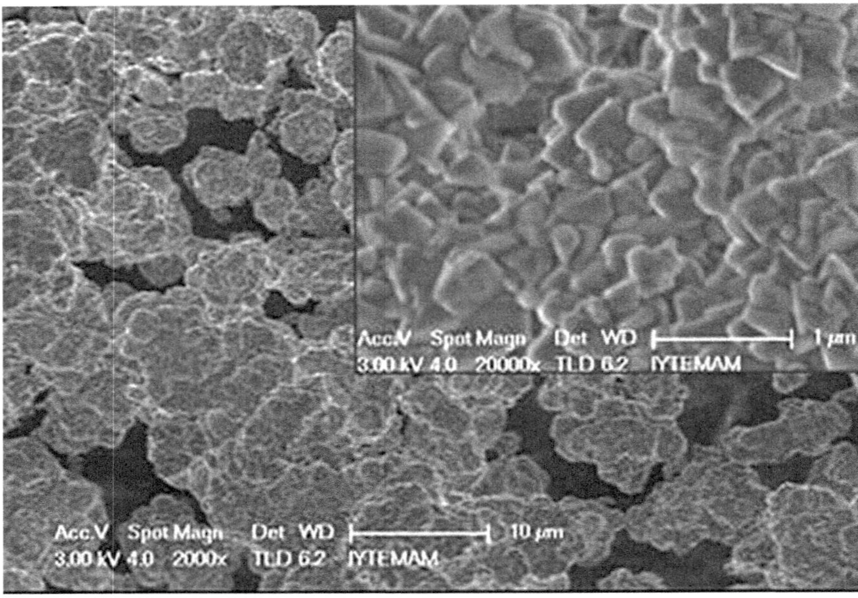

FIGURE 13.3 Scanning electron microscope (SEM) microphotographs of zinc borate species of $2ZnO{\cdot}3B_2O_3{\cdot}3H_2O$.

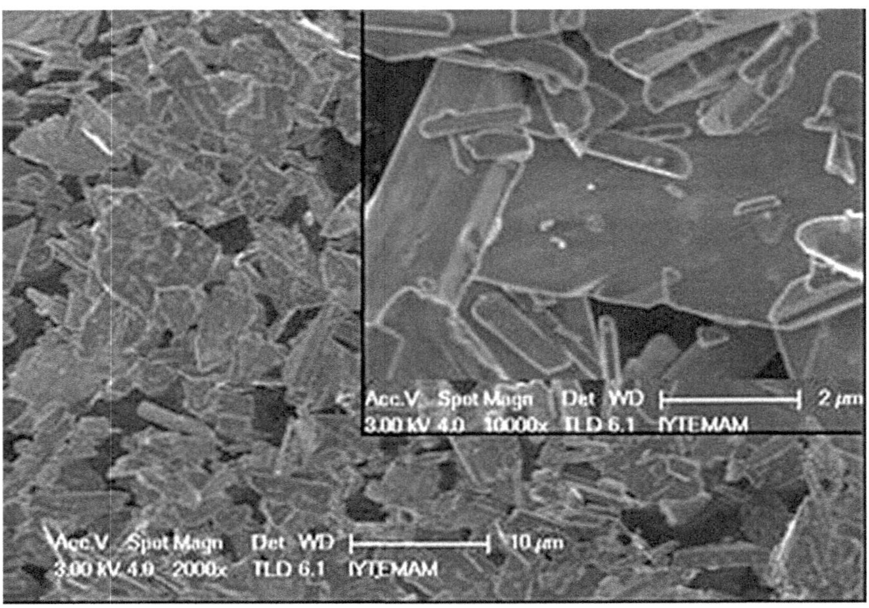

FIGURE 13.4 SEM microphotographs of zinc borate species of $4ZnO{\cdot}B_2O_3{\cdot}H_2O$.

13.3.2 DEHYDRATION KINETICS OF ZINC BORATES

TG curve obtained at 10°C·min^{-1} heating rate and derivative TG curve of zinc borate (ZnB-1) are shown in Figure 13.5. The thermal decomposition of ZnB-1 starts at 290°C and ends at 550°C with a mass loss of 12.5% corresponding to theoretical H_2O content value of 12.69%. Thermal dehydration of $2ZnO{\cdot}3B_2O_3{\cdot}3H_2O$ consists of two consecutive kinetic steps as observed in derivative TG curve (obtained at 10°C·min^{-1}) with maxima of 387.6°C and 435.9°C. Total mass losses at these maxima are 6 and 12.5%, respectively. The dehydration of $2ZnO{\cdot}3B_2O_3{\cdot}3H_2O$ in two steps was reported previously but no further discussion was made for this phenomenon.[11,16] It is thought that the removal of water molecule formed from weakly coordinated O–H groups takes place in the former steps and the release of water from the condensation of more stable O–H groups occurs in the last step. In other words, rate of condensation reaction changes according to the number of hydroxyl groups bonded to the borate main structure. For instance, if there is more than one OH group attached to boron atom, condensation reaction takes place more easily; or if there is only one OH group bonded to boron atom, it should interact with OH groups of another boron atom to form water molecule. Thermal dehydration of hydrated zinc borate (ZnB-1) has occurred according to the representative reaction given in Equation 13.8.

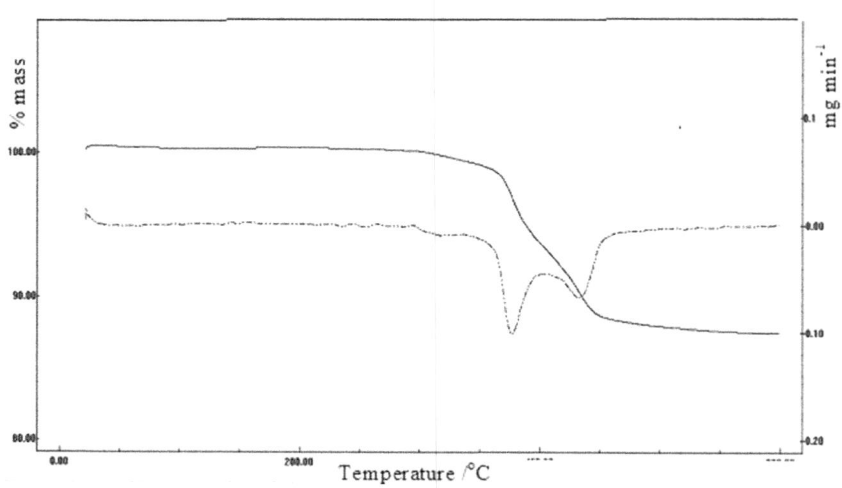

FIGURE 13.5 TG and derivative TG curves of zinc borate species of $2ZnO{\cdot}3B_2O_3{\cdot}3H_2O$ obtained at 10°C·min^{-1}.

$$2ZnO{\cdot}3B_2O_3{\cdot}3H_2O \rightarrow 2ZnO{\cdot}3B_2O_3 + 3H_2O \quad (13.8)$$

The temperature ranges of the two steps overlap each other. Second step starts without completion of the first step.

TG analysis and derivative TG data for $4ZnO{\cdot}B_2O_3{\cdot}H_2O$ are reported in Figure 13.6. ZnB-2 started to decompose at around 450°C and ended at 600°C. It had lost 4.0% of its initial mass, which corresponds to theoretical water content of 4.35%. The decomposition occurs according to Equation 13.9.

$$4ZnO{\cdot}B_2O_3{\cdot}H_2O \rightarrow 4ZnO{\cdot}B_2O_3 + H_2O \quad (13.9)$$

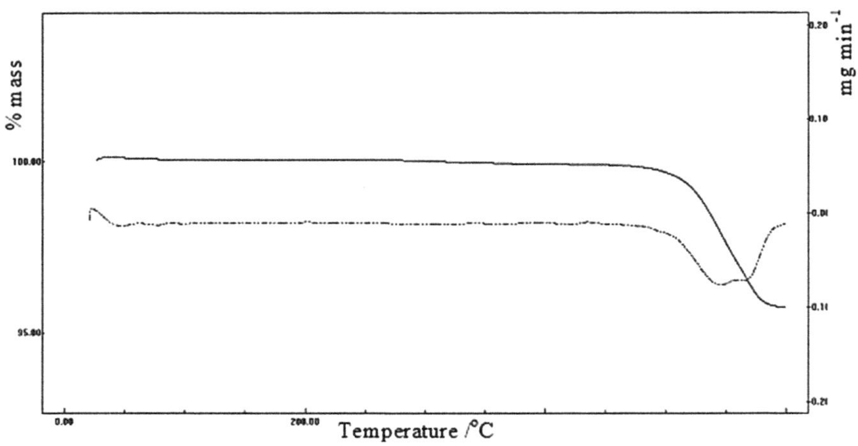

FIGURE 13.6 TG and derivative TG curves of zinc borate species of $4ZnO{\cdot}B_2O_3{\cdot}H_2O$ obtained at $10°C{\cdot}min^{-1}$.

Unlike the ZnB-1 dehydration behavior, ZnB-2 dehydration occurred in a single step with a broad maximum at 556°C, in which the removal of water formed from strongly bonded O–H groups' condensation took place. Since there is no water of crystallization in the structure of above zinc borates, ZnB-1 and ZnB-2, dehydration reactions onset at relatively higher temperatures of 290 and 450°C, respectively. The first step of ZnB-1 dehydration peak was neglected since it was a broad and small peak. Dehydration peak temperatures of ZnB-1 and ZnB-2 at different heating rates were reported in Table 13.1. In the calculation of activation energy, those temperatures were utilized in the Kissinger method.

TABLE 13.1 Thermal Decomposition Kinetics' Parameters for Zinc Borates.

Parameter		Peak temperature in dTG curves (°C)		
		ZnB-1 ($2ZnO{\cdot}3B_2O_3{\cdot}3H_2O$)		ZnB-2 ($4ZnO{\cdot}B_2O_3{\cdot}H_2O$)
		First step	Second step	One step
Heating rate ($°C{\cdot}min^{-1}$)	5	376.6	431.5	535.2
	10	387.6	435.9	549.5
	15	395.2	438.6	565.9
	20	402.9	441.9	568.7
Slope of Kissinger plot		−22.076	−66.888	−24.261
Intercept of Kissinger plot		22.7	83.5	18.3
Kissinger plot R^2		0.99	0.98	0.97
E_a ($kJ{\cdot}mol^{-1}$)		183.5	556.1	201.7
$A{\cdot}s^{-1}$		4.4×10^{10}	3.4×10^{37}	5.6×10^{8}

TG curves of zinc borates (ZnB-1 and ZnB-2) are obtained at different heating rates. Derivative TG data were obtained from those TG thermograms. The Kissinger method utilizing derivative TG curves was used in the determination of decomposition activation energy and frequency factor for zinc borates. As shown in Equation 13.7, the $\ln\left(\beta / T_m^2\right)$ versus $1/T_m$ plot gives straight line. The Kissinger plots obtained using the peak temperature of dehydration steps in the derivative TG curves are shown in Figure 13.7 and Figure 13.8. Since there are two steps in derivative TG curves of ZnB-1, two Kissinger plots were obtained as straight line as shown in Figure 13.7. Thus, two slopes were determined considering this fact: the former for initial dehydration step and the latter for second dehydration step. The Kissinger plot of ZnB-2 shown in Figure 13.8 was used in activation energy calculations. The activation energies of zinc borates (ZnB-1 and ZnB-2) were calculated from the slope of the lines, and the frequency factors were obtained from the intercept of those lines. Slopes and intercept of the Kissinger plots for ZnB-1 and ZnB-2 are given in Table 13.1.

The activation energy for ZnB-1 was calculated as 183.5 and 556.1 $kJ{\cdot}mol^{-1}$ for the first step and the second step, respectively, which are in good agreement with values in the literature.[11] The activation energy of ZnB-1 calculated for the second step is greater than the one in the first step. The distinction in activation energy might be attributed to the condensation reaction of hydroxyl groups, which are coordinated differently. The activation energy for the dehydration of ZnB-2 was determined as 201.7 kJ·mol^{-1}.

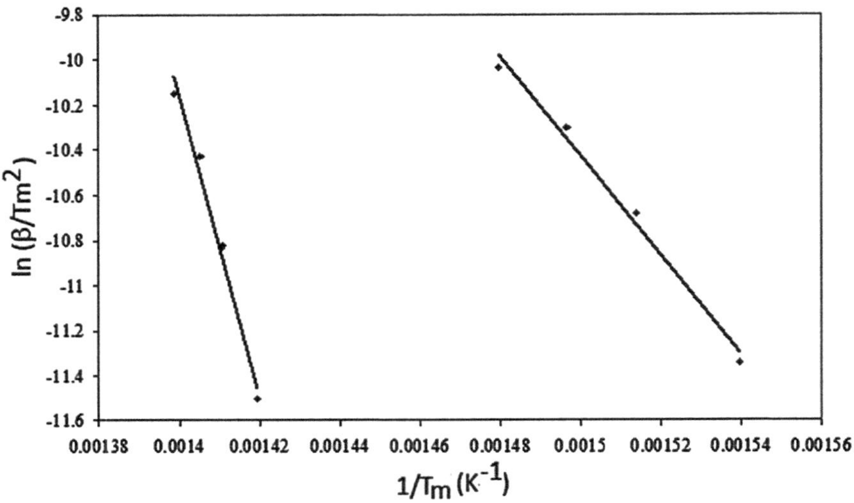

FIGURE 13.7 Kissinger plots for $2ZnO{\cdot}3B_2O_3{\cdot}3H_2O$ dehydration.

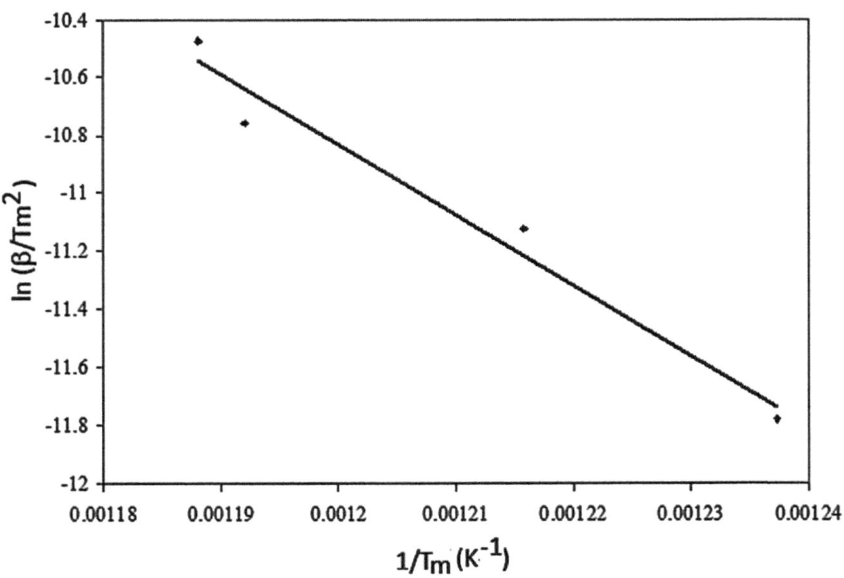

FIGURE 13.8 Kissinger plot for $4ZnO{\cdot}B_2O_3{\cdot}H_2O$ dehydration.

13.4 CONCLUSIONS

In this work, the decomposition kinetics of zinc borates has been investigated using dynamic TG analysis with the help of the Kissinger method. The decomposition process for zinc borate of ZnB-1 started at about 290°C and reached mass loss of 12.44% and zinc borate (ZnB-2) dehydration began at about 450°C and reached mass loss of 4.0% at 600°C. It was concluded that zinc borate of ZnB-1 dehydrated mainly within two-step processes as it was determined from its derivative TG data. Since the coordination of OH groups in zinc borate structure is different, dehydration process occurred in two kinetic steps based on the OH groups' interactions. The dehydration activation energy of ZnB-1 was calculated as 183.5 and 556.1 $kJ \cdot mol^{-1}$ for the first and second steps, respectively. On the other hand, zinc borate of ZnB-2 decomposed in a single step as observed in derivative TG curve and the dehydration activation energy of ZnB-2 was determined as 201.7 $kJ \cdot mol^{-1}$.

ACKNOWLEDGMENTS

The authors acknowledge the financial support from The Scientific and Technical Research Council of Turkey (TÜBİTAK) (project number: 105M358) and Dr. Saied Kochesfahani from U.S. Borax Inc. for providing commercial zinc borate samples. The authors also acknowledge the support of Burcu Alp for TG analysis of the samples.

KEYWORDS

- **dehydration kinetics**
- **activation energy**
- **hydrated zinc borates**
- **fire retardant**
- **Kissinger method**

REFERENCES

1. Alp, B.; Gönen M., Savrık, S. A.; Balköse, D.; Ülkü, S. Dehydration, Water Vapor Adsorption and Desorption Behavior of Zn B_3O_3 $(OH)_5$. H_2O and Zn B_3O_4 $(OH)_3$. *Drying Technol.* **2012,** *30*(14), 1610–1620.
2. Bourbigot, S.; Bras, M. L.; Leeuwendal, R.; Shen, K. K.; Schubert, D. Recent Advances in the Use of Zinc Borates in Flame Retardancy of EVA. *Polym. Degrad. Stab.* **1999,** *64*, 419–425.
3. Brooman, E. V. Modifying Organic Coatings to Provide Corrosion Resistance: Part II-Inorganic Additives and Inhibitors. *Met. Finish.* **2002,** *100* (5), 42.
4. Dong, J. X.; Hu, Z. S. A Study of the Anti-wear and Friction-reducing Properties of the Lubricant Additive, Nanometer Zinc Borate. *Tribol. Int.* **1998,** *31*, 219–223.
5. Ekmekyapar, A.; Baysar, A.; Künkül, A. Dehydration Kinetics of Tincal and Borax by Thermal Analysis. *Ind. Eng. Chem. Res.* **1997,** *36*, 3487–3490.
6. Eltepe, H. E.; Balköse, D.; Ülkü, S. Effect of Temperature and Time on Zinc Borate Speices Formed from Zinc Oxide and Boric Acid in Aqueous Medium. *Ind. Eng. Chem. Res.* **2007,** *46*, 2367–2371.
7. Garba, B. Effect of Zinc Borate as Flame Retardant Formulation on Some Tropical Woods. *Polym. Degrad. Stab.* **1999,** *64*, 517–522.
8. Giudice, C. A.; Benitez, J. C. Zinc Borate as Flame-Retardant Pigments in Chlorine-Containing Coatings. *Prog. Org. Coat.* **2001,** *42*, 82–88.
9. Hatakeyama, T.; Quinn, F. X. *Thermal Analysis Fundamentals and Applications to Polymer Science,* 2nd ed.; J. Wiley & Sons: Chichester, 1999.
10. Jun, L.; Shupping, X.; Shiyang, G. FT-IR and Raman Spectroscopic Study of Hydrated Borates. *Spectrochim. Acta.* **1995,** *51A*(4), 519–532.
11. Kipcak, A. S.; Senberber, F. T.; Derun, E. M.; Tugrul, N.; Piskin, S. Characterization and Thermal Dehydration Kinetics of Zinc Borates Synthesized from Zinc Sulfate and Zinc Chloride. *Res. Chem. Intermed.* **2015,** *41*(11), 9129–9143.
12. Kissinger, H. Reaction Kinetics in Differential Thermal Analysis. *Anal. Chem.* **1957,** *29*, 1702–1706.
13. Lewin, M.; Weil, E. D. *Fire Retardant Materials*; Horrocks, A. R., Price, D., Eds.; CRC Press: Boca Raton, 2001; p 55.
14. Mergen, A.; Ipek, Y.; Bölek, H.; Öksüz, M. Production of Nano Zinc Borate ($4ZnO{\cdot}B_2O_3{\cdot}H_2O$) and its Effect on PVC. *J. Eur. Ceram. Soc.* **2012,** *32*, 2001–2005.
15. Ozawa, T. A New Method of Analyzing Thermogravimetric Data. *Bull. Chem. Soc. Jpn.* **1965,** *38*, 1881–1886.
16. Samyn, F.; Bourbigot, S.; Duquesne, S.; Delobel, R. Effect of Zinc Borate on the Thermal Degradation of Ammonium Polyphosphate. *Thermochim. Acta* **2007,** *456*, 134–144.
17. Sawada, H.; Igarashi, H.; Sakao, K. Zinc Borate, and Production Method and Use Thereof. U. S. Patent 6,780,913 B2, August 24, 2004.
18. Schubert, D. M. Zinc Borate. U. S. Patent 5,472,644, December 5, 1995.
19. Schubert, D. M. Borates in Industrial Use. Group 13 Chemistry III Industrial Applications. In *Structure and Bonding;* Roesky, H. W., Atwood, D. A., Eds.; Springer: Berlin, 105, 2003; pp 1–40.

20. Schubert, D. M.; Alam, F.; Visi, M. Z.; Knobler, C. B. Structural Characterization and Chemistry of Industrially Important Zinc Borate Zn $[B_3O_4(OH)_3]$. *Chem. Mater.* **2003,** *15*, 866–871.
21. Shen, K. K.; Kochesfahani, S.; Jouffret, F. Zinc Borates as Multifunctional Polymer Additives. *Polym. Adv. Technol.* **2008,** *19*, 469–474.
22. Shi X.; Yuan L.; Sun X.; Chang C.; Sun J. Controllable Synthesis of $4ZnO.B_2O_3.H_2O$ Nano/Microstructures with Different Morphologies: Influence of Hydrothermal Reaction Parameters and Formation Mechanism. *J. Phys. Chem. C.* **2008,** *112*, 3558–3567.
23. Tian, Y.; Guo, Y.; Jiang, M.; Sheng, Y.; Hari, B.; Zhang, G.; Jiang, Y.; Zhou, B.; Zhu, Y.; Wang, Z. Synthesis of Hydrophobic Zinc Borate Nanodiscs for Lubrication. *Mater. Lett.* **2006,** *60*, 2511–2515.
24. Waclawska, I. Controlled Rate Thermal Analysis of Hydrated Borates. *J. Therm. Anal.* **1998,** *53*, 519–532.
25. Yongzhong, J.; Shiyang, G.; Shuping, X.; Jun, L. FT-IR Spectroscopy of Supersaturated Aqueous Solutions of Magnesium Borates. *Spectrochim. Acta Part A* **2000,** *56*, 1291–1297.
26. Yu, D.; Xue, D.; Ratajczak, H. Microscopic Characteristics of Hydrogen Bonds of Hydrated Borates. *Phys. B.* **2006,** *371*, 170–176.

CHAPTER 14

CATALYTIC CRACKING AND PETROLEUM REFINING: SCIENTIFIC VALIDATION AND A VISION FOR THE FUTURE

SUKANCHAN PALIT[1,2,*]

[1]*Department of Chemical Engineering, University of Petroleum and Energy Studies, Bidholi, via Premnagar, Dehradun, Uttarakhand 248007, India, Tel.: +918958728093*

[2]*43, Judges Bagan, Haridevpur, Kolkata 700082, India*

[*]*Corresponding author. E-mail: sukanchan68@gmail.com, sukanchan92@gmail.com*

CONTENTS

ABSTRACT

The science and engineering of petroleum technology is moving forward at a rapid and drastic pace. Petroleum refining today stands in the midst of scientific vision and deep scientific introspection. The immense challenges and the definitive vision of petroleum engineering science today are slowly evolving. Depletion of fossil fuel resources is challenging the wide scientific fabric of science and technology. In such a crucial juncture of scientific history and time, petroleum chemistry, the intricacies of petroleum refining, and the wide world of energy engineering needs to be reenvisioned and revamped. Sustainable development is the utmost need of the hour. Energy and environmental sustainability are challenging the scientific scenario today. Here comes the immense importance of petroleum chemistry and catalytic cracking. Mankind's immense scientific prowess, the true vision of fossil fuel science, and the needs of alternate energy sources will all lead a long and visionary way in the true emancipation of energy sustainability today. In this treatise, the author deeply comprehends the wide scientific success and the vast scientific potential of petroleum refining units particularly Fluid Catalytic Cracking (FCC) Unit. In such a scientific fervor, catalytic cracking assumes vast importance. The author in this treatise investigates the intricacies of catalytic cracking chemistry and its importance in petroleum refining. This treatise veritably opens up a new chapter in the field of catalytic cracking and FCC in particular.

14.1 INTRODUCTION

The world of petroleum engineering science and energy technology are moving toward a visionary realm in today's human civilization. Technological barriers are surpassed and scientific vision realized. Depletion of fossil fuel resources is a vexing issue of our present day human civilization. In this treatise, the author elucidates on the vision of catalytic cracking and chemical kinetics which stands as an important component of petroleum refining. The chemical kinetics of catalytic cracking is immensely intricate and recent research pursuit is far-reaching. Petroleum engineering science today stands in the midst of wide scientific vision and deep scientific comprehension. In this century, challenges and vision of science are immense and the fruits of human scientific endeavor groundbreaking. Petroleum engineering science and petroleum engineering are in a state of immense scientific quagmire. This treatise is a well-researched and well-viewed piece of work which surpasses scientific imagination. Today, there are questions of application of nonrenewable energy technology to human scientific progress. In such a situation, the need for alternate energy sources comes into prominence.[1–3]

14.2 THE VISION OF THIS STUDY

Technology of petroleum engineering science is moving toward newer knowledge dimensions. Catalytic cracking and fluid catalytic sracking (FCC) are veritably changing the scientific landscape. Energy sustainability and recent research trends in sustainability science are the other objectives of this treatise. The major issue lies in the domain of catalytic cracking and its chemical kinetics. Energy security and energy efficiency domain needs to be reenvisioned and readdressed with the passage of scientific vision, scientific history, and time frame. In this treatise, the author pointedly focuses on the deep scientific success, scientific potential, and the wide scientific forbearance of petroleum refining. The challenge and vision of petroleum technology are veritably challenging the vast scientific landscape. This study deeply elucidates the immense intricacies of petroleum engineering science especially the vision behind catalytic cracking and FCC Unit. The aim and objective of this paper is replete with vision and fortitude as science and engineering today moves forward in this 21st century. FCC operation and design today stands in the midst of deep scientific understanding and visionary discernment. The challenge of modeling, simulation, optimization, and control of FCC Unit are the other facets of this comprehensive review. Monitoring of a petrochemicals plant and a petroleum refinery are the other vast avenues of scientific endeavor. The author rigorously points out toward the immense success and the vast vision behind cracking chemistry and the intricacies linked with it. This treatise will surely open up new avenues of research endeavor in areas of petroleum chemistry in decades to come.[1–6,13,14]

14.3 NEED AND THE RATIONALE OF THIS STUDY

Science and technology advancements, the futuristic vision and the needs of the human society are the forerunners toward a newer visionary era in the field of petroleum engineering science today. Science today is a huge colossus with a definite vision and willpower of its own. Scientific and technological profundities are changing the very vision and mindset of a petroleum engineer today. Depletion of fossil fuel resources and the need for renewable energy technology are challenging and changing the wide scientific panorama of research pursuit in petroleum refining. Here comes the immense need of this comprehensive review. The rationale of this study goes beyond scientific imagination and scientific forbearance. Technology and science has few answers to the intricacies of petroleum refining and catalytic cracking. Cornerstones of science and technology today are the basic advancements and basic fundamentals of petroleum refining and petroleum engineering science. The need for this study is immense and evergrowing. Scientific and technological profundities are gaining immense heights in the field of energy and environmental sustainability.[4–9]

14.4 ENERGY SUSTAINABILITY AND PETROLEUM REFINING

Energy and environmental sustainability today are ushering in a new eon in the human scientific progress and deep scientific rigor. Technology and engineering science of petroleum refining stands in the midst of deep scientific vision and vast scientific profundity. FCC and catalytic cracking, in particular, needs to be readdressed and revalidated as science and engineering of petroleum refining crosses vast scientific frontiers. Scientific vision, scientific forbearance, and deep scientific cognizance are the utmost need of the hour as the depletion of fossil fuel resources destroys the scientific landscape of petroleum engineering science. Mankind today is in the grip of unimaginable environmental catastrophes and environmental issues. Sustainability and global Sustainable Development Goals as defined by the former Prime Minister of Norway, Dr. Gro Harlem Brundtland, need to be redefined and justified again. The energy crisis is evergrowing as human civilization treads a weary path toward a newer scientific destiny. Today Sustainable Development Goals are redefined as United Nations reframed the entire concept of sustainability. In such a crucial juncture of deep scientific progress, research trends should be toward alternate energy sources and renewable energy. Wind energy, solar energy, wave energy, and the energy from biomass are the utmost need of the hour as mankind trudges forward. The scientific potential and the scientific success of alternate energy sources are revamping the entire energy scenario.[1–3,13,14]

Sustainable development is the organizing principle for meeting human development goals while at the same time sustaining the ability of natural systems to provide the natural resources and ecosystem services upon which the economy and society depends. Scientific vision, technological motivation, and the futuristic vision of petroleum engineering are all today leading a long way in the true realization and true emancipation of human research endeavor. The desirable conclusion of energy and environmental sustainability is a state of human society where living conditions and resource use continue to meet human needs without undermining the integrity and stability of the natural systems. Energy sustainability is the need of the hour whether it is a developed or a developing nation and today, scientific endeavor in sustainability science is replete with vision and technological fervor. Petroleum refining and science of energy sustainability are two opposite sides of the visionary coin. Mankind's immense scientific intellect and prowess and the futuristic vision of petroleum engineering science will go a long and visionary way in the true realization of energy sustainability today. In a similar manner, environmental sustainability and water technology assumes immense importance with the passage of human scientific history and visionary time frame. In this treatise, the author pointedly focuses on the vast scientific potential, the scientific regeneration, and the scientific sagacity behind energy sustainability and petroleum engineering science.[13,14]

14.5 WHAT DO YOU MEAN BY CATALYTIC CRACKING?

Catalytic cracking stands as an important component in petroleum refining and petroleum engineering science. The vision and the challenge of petroleum refining encompass catalytic cracking or FCC Unit. Petroleum engineering science today stands as a vital parameter in the progress of human civilization. Depletion of fossil fuel resources and the need for alternate energy resources are issues of deep concern in our present day human civilization. Research trends and research profundity are today targeted toward the success of the application of alternate energy resources and renewable energy technology.

Catalytic cracking is the premier conversion process in a petroleum refining. Nearly, 20% of all distilled crude oil is processed in catalytic cracking units. An average size-cracking unit processes 50,000 barrels a day, although some units exceed 120,000 barrels per day.[13,14] These units process an extremely wide variety of feedstocks.[13,14] The feeds used include molecules with carbon numbers from simple C_7–C_8 molecules to complex structures of 100 or more carbon atoms.[13,14] Molecular types predominantly present are paraffins, cycloparaffins (naphthenes), and aromatics. Many of the aromatic rings contain multiple fused rings (rings directly joined). Catalytic cracking is extremely intricate and veritably complex. The easiest feeds to crack consist of more paraffins and naphthenes, with minimal quantities of aromatics. Feedstock quality determines the value of products that come from catalytic cracking to a very large degree.[13,14] Today the research pursuit in the field of catalytic cracking is surpassing visionary frontiers. Products from the unit are fuel gases, liquefied petroleum gas, and gasoline range naphtha. They also include diesel range light cycle oil (LCO), heavy cycle oil (HCO), and coke.[1–3,13,14]

Catalytic cracking is endothermic, meaning heat is absorbed by the reactions. The temperature of the reaction mixture declines as the reactions proceed. Heat to drive the process comes from combustion of coke formed in the process.[13,14] Coke is a necessary product of cracking. It is a solid, black material that is rich in carbon and low in hydrogen; chemists call this condition "highly unsaturated."[13,14] Coke forms on the surface and in the pores of the catalyst during the cracking process, covering the active sites and deactivating the catalyst. During regeneration, this coke is burnt off from the catalyst to restore its activity. Similar to all combustion processes, regeneration is exothermic, liberating heat. Many improvements have been envisioned to enhance the unit mechanical reliability and its ability to crack heavier, lower-value feedstocks. Technological and scientific validations are of utmost importance with the progress of scientific and academic rigor today. The scientific challenge today goes beyond scientific imagination.[13,14]

The FCC unit utilizes a microspherodial catalyst which fluidizes when properly aerated. The main purpose of the unit is to convert high-boiling petroleum fractions called gas oil to high-value, high octane gasoline and heating oil. Gas oil is the portion of crude oil that boils in the range of 650–1050°F and contains a diverse mixture of paraffins, naphthenes, aromatics, and olefins.[4–6,13,14]

14.6 SCIENTIFIC DOCTRINE AND SCIENTIFIC COGNIZANCE BEHIND CHEMICAL KINETICS OF CATALYTIC CRACKING

Chemical reaction engineering stands as a decisive parameter in the catalytic cracking process today. The scientific doctrine and the vast scientific cognizance of petroleum refining units are today ushering in an era of scientific endeavor. Chemical process engineering and petroleum engineering science are two opposite sides of the visionary coin. Chemical kinetics and chemical reaction engineering together are the heart of chemical process engineering paradigm today. The deep scientific vision, the scientific cognizance, and the futuristic vision of science and technology will all lead a long and visionary way in the true realization of chemical process engineering today. Chemical process engineering has a wide technological vision and is surpassing vast and versatile visionary scientific frontiers. Technology is advancing at a rapid pace today. Petroleum engineering science and chemical process engineering are the forerunners of a greater realization of energy sustainability. Catalytic cracking and FCC Unit are major components of petroleum refining. Petroleum refining and its vast vision need to be readdressed and revamped with the passage of scientific history, scientific fortitude, and the visionary time frame. Human civilization and human scientific endeavor are today in a state of immense scientific regeneration and deep introspection. Depletion of fossil fuel resources and their grave technical concern has urged the scientific community to gear toward newer vision and newer scientific innovations. Technology has few answers toward the intricacies of scientific research pursuit in petroleum engineering science. Scientific revelation and deep scientific genesis need to be reframed with the course of scientific history and time.[7–9]

14.7 TECHNOLOGICAL VISION BEHIND PETROLEUM REFINING TODAY

Petroleum refining and petroleum engineering science stand in the midst of deep scientific rejuvenation and deep introspection. Today scientific revelation and scientific forbearance in the research pursuit of petroleum refining are crossing visionary boundaries. Today technology has few answers to the scientific success and scientific potential in the path toward greater emancipation and greater scientific destiny. Refining and petrochemicals also stand in the midst of deep scientific research questions and vast contemplation. Petroleum refining and engineering science today need to be reenvisioned and restructured with the evergrowing crisis of fossil fuel depletion and environmental catastrophes. Energy, food, and water are the utmost needs for the success of human civilization today. In a similar manner, energy and environmental sustainability are in the path of immense scientific regeneration. Each unit in petroleum refining today stands in the crossroads of scientific discernment and deep vision.[10–12]

Human civilization and human scientific endeavor are today in the path of scientific rejuvenation. Technology has few answers to the vexing issue of depletion of fossil fuel resources. Mankind's immense scientific prowess and intellect, the vast technological vision, and the needs of sustainable development will all lead a long and visionary way in the true realization of energy and environmental sustainability. Global energy scenario is at stake today. Petroleum engineering science and petroleum refining today stand at the crossroads of deep scientific introspection and vast research queries. In this treatise, the author pointedly focuses on the scientific success, the definite vision for the future, and the need for scientific research endeavor in the field of design and operation of petroleum refining units such as catalytic cracking or FCC unit.

14.8 RECENT SCIENTIFIC ADVANCES IN CATALYTIC CRACKING

Catalytic cracking and FCC today plays vital roles in scientific realization and emancipation of human scientific endeavor today. FCC is a revolutionary and promising domain of scientific research pursuit in the present day human civilization. The vast vision, scientific challenges, and scientific determination are the torchbearers toward a newer eon of technological divinity in the decades to come. In this treatise, the author pointedly focuses on the scientific forays, the deep scientific sagacity, and the vast scientific landscape of catalytic cracking and FCC in petroleum engineering science. The challenge, the vision, and the potential of science and engineering are immense in today's scientific and technological forays in engineering science. This treatise opens newer avenues of scientific sagacity in the field of catalytic cracking chemistry and also petroleum chemistry in the decades to come. The targets of petroleum engineering science are vast and versatile. Depletion of fossil fuel resources are of major concern globally. Concerns of nonrenewable energy technology and its efficacies are challenging the scientific landscape. Thus, there is need for renewable energy technologies such as solar energy, wind energy, and biomass energy. In this section, the author deeply comprehends the success of FCC in meeting energy needs and greater realization of energy sustainability.

Occelli (2000)[6] discussed with cogent and lucid insight advances in FCC with special emphasis on testing, characterization, and environmental regulations.[6] Scientific vision, scientific profundity, and deep scientific forbearance are the pillars of this vast scientific research pursuit. The author treads a difficult path into the intricacies of FCC process. Technology of FCC needs to be revamped and restructured as science marches forward into this century.[6] The treatise begins with the technology evaluation of maximizing FCC LCO by HCO recycling.[6] Then it discusses new catalytic process approach, catalyst evaluation, pilot unit test of residue type catalysts, novel FCC catalysts, improvement of profitability of FCC unit, troubleshooting, advanced artificial troubleshooting, coke characterization, FCC emission reduction technologies, and the vast domain of environmental restrictions.[6] Science of catalytic cracking is entering into a new phase of scientific vision and profundity. The challenges and the vision are immense and surpass scientific imagination. The vast and versatile technological advancements and scientific achievements of catalytic cracking and petroleum refining today are changing the scientific fabric.[6] The question of energy sustainability and the wide vision of science of sustainability are the forerunners toward a newer visionary eon in the field of energy engineering. The scientific success, the scientific intricacies and the scientific challenges need to be restructured with the passage of history and time. FCC design and operation is a vexing issue and a challenging domain of research endeavor. In this treatise, the author rigorously points toward the immense success, the deep scientific and technological vision and the scientific barriers behind the design and operation of Catalytic cracking unit.[6]

Sadeghbeigi (2000)[3] discussed with deep and cogent insight design, operation, and troubleshooting of FCC unit in a comprehensive treatise. The deep scientific revelation, the challenge, and the vision of science and the vast need of energy sustainability are the torchbearers toward a newer era of petroleum refining and petroleum engineering science today.[3] The author of this treatise dealt lucidly with the process description, FCC feed characterization, the importance of FCC catalysts, chemistry of FCC reactions, unit monitoring and control, FCC products and economics, project management and hardware design, troubleshooting, optimization, and finally emerging trends in FCC.[3] Vision of petroleum engineering science and petroleum refining are moving from one definite paradigm toward another.[3] The scientific success, the scientific genesis, and vast scientific progeny are the pallbearers toward a newer visionary eon in the field of petroleum refining. The author in this treatise deeply comprehended issues related to the FCC process and provides practical and proven recommendations to largely improve the performance and the vast reliability of FCC Unit operations. The technological vision and the scientific profundity in the design and operation of FCCU have challenged the scientific domain today. FCCU continues to play a decisive role in an integrated refinery in the primary conversion process.[3] The catalytic cracker is the key to profitability to check whether or not the refiner can be competitive in the global energy market. Science of sustainability is the other side of the visionary scientific coin today. Since the first operation of FCC unit in 1942, many drastic changes have taken place. Mankind's immense scientific justification, the scientific intellect and the challenges of engineering science will all lead a long and visionary way in the true realization of energy engineering. Science of FCC is moving at a rapid pace today. Human scientific research pursuit in the field of catalytic cracking is gaining immense importance in the present day human civilization. The FCC unit uses a microspheroidal catalyst, which behaves like a liquid when properly aerated by gas. The main purpose of the unit is to convert high-boiling petroleum fractions called gas oil to high-value, high octane gasoline, and heating oil. Technological advancements and scientific profundity are in a state of immense scientific regeneration. The case of catalytic cracking process is similar.[3] Gas oil is the portion of crude oil that commonly boils at 650+ to 1050+°F (330–550°C).[3] Before going further, it is highly prudent to investigate how a catalytic cracker fits into the refinery process.[3] A petroleum refinery comprises of several processing units that convert raw crude oil into usable products such as gasoline, diesel, and jet oil.[3] The crude unit is the first major unit in the petroleum refinery. Here, the raw crude is distilled into several intermediate products: naphtha, kerosene, diesel, and gas oil. The heaviest portion of the crude oil, which cannot be distilled in the atmospheric tower, is heated and sent to the vacuum tower where it is split into gas oil and tar. Scientific regeneration, deep scientific vision and the vast challenges will veritably open up windows of instinct and innovation in petroleum engineering science in the years to come. The tar from the vacuum tower is sent to be further processed in a delayed coker, deasphalting unit or a visbreaker or is sold as fuel oil or road asphalt.[3] The FCC process is extremely complex. For scientific vision and scientific clarity the process can be divided into six separate sections:

1. Feed preheat[3]
2. Riser-reactor-stripper
3. Regenerator and heat recovery
4. Main fractionators
5. Gas plant and treating facilities[3]

In this book, the author rigorously points toward the vast scientific success, the visionary aisles and the scientific imagination behind FCC Unit design and operation. Today, human scientific research pursuit stands in the midst of deep vision and technological intellect of highest order.[3] The vast academic and scientific rigor in modeling, simulation, optimization, and control of FCC Unit are presented in minute details and with deep scientific precision. Technology and engineering science are today in the path of immense scientific rejuvenation.[3] This treatise is an eye-opener toward a newer visionary scientific girth and determination in the field of operation of FCC unit. The authors in this treatise also elucidated in minute details feed preheat, riser–reactor–stripper, catalyst separation, stripping section, regenerator heat, and energy recovery, the entire valve system of the FCC unit, and the vast technological vision behind the chemistry of catalytic cracking.[3]

Ray Chaudhuri (2011)[12] discussed with lucid and cogent details fundamentals of petroleum and petrochemical engineering. The vast scientific vision, the immense scientific profundity and the cause of fossil fuel depletion are today leading a long and visionary way in the true realization of petroleum engineering science. Human scientific endeavor today is in the crossroads of deep scientific vision and scientific fortitude.[12] The challenge, the vision, and the targets of science need to be restructured and reorganized with the passage of scientific history, forbearance, and visionary timeframe.[12] The author in this book presents an introduction to petroleum engineering and petrochemicals, then the crude petroleum characteristics, petroleum products, and test methods, the scientific success of various types of fuels, processing operations in a petroleum refinery, the vast domain of lubricating oil and grease, the scientific frontiers of petrochemicals, off-site facilities and utilities, and the scientific vision of the operation of various accessories.[12] The author also elucidates upon heat exchangers and pipe still furnaces, distillation and stripping, extraction and various unit operations of chemical process engineering, and petroleum engineering science. Petroleum is a fossilized mass that accumulates below the earth's surface from ages. The vast technological profundity and the deep scientific motivation are the pillars of scientific truth and wide vision in this book. The hydrocarbon industry is basically divided into three distinct classifications—petroleum production, petroleum refining, and petrochemical manufacture. The vast challenges and the scientific imagination of research pursuit in petroleum engineering are changing the scientific landscape. This treatise opens up new scientific frontiers and challenges the scientific canons and scientific truth in the domain of petroleum engineering science and petrochemicals in particular.[12]

Khandalekar (1993)[7] discussed the control and optimization of FCC process with deep and cogent insight. Design of the FCC process is slowly moving from one scientific vision toward another.[7] In the vast domain of chemical process engineering, modeling and simulation stands as a vital pillar toward the greater emancipation of science and engineering. The scientific success, the scientific imagination and the world of challenges will all lead a long and visionary way in the true realization of energy sustainability and petroleum engineering.[7] Modeling, simulation, optimization, and control stand as an important pillar toward furtherance in science and engineering today.[7,10–12] The scientific potential, the scientific success, and the deep scientific profundity in optimization of petroleum refining processes are the forerunners toward a greater visionary eon in the field of petroleum engineering. Human scientific endeavor and march of human civilization are changing the scientific landscape and evolving into a newer engineering emancipation.[7] Technological challenges in modeling, simulation, and optimization of FCC unit are immense and far-reaching. In this treatise, the author rigorously points toward the vast scientific potential and scientific revelation behind chemical process technology and petroleum engineering. The author in this treatise clearly defines the objectives of an ideal FCC Unit control system. The literature review comprises of catalytic cracking and coke formation, regeneration of catalyst, dynamic simulation and control of FCC unit, and the vast domain of multivariable process control.[7,10–12] Then the author delves deep into the dynamic simulation, the operation of wet gas compressor, regenerator, air blowers, and other operating units of FCCU.[7] The author also elucidates on the nonlinear process model-based control. The challenge of FCCU is slowly changing the scientific landscape and surpassing visionary frontiers. This text is a veritable eye-opener to the success of modeling, simulation, and process control domain of chemical process engineering, particularly petroleum refining. The FCC is an important scientific endeavor in petroleum refining. Gasoline is the major product. Owing to its large throughput and ability to produce gasoline which is more valuable, economic operation and greater emancipation of process control needs to be envisioned and reenvisaged.[7] Fluidized Catalytic Cracker poses a multivariable, strongly interacting, and a highly nonlinear control process. Technological challenges are numerous and versatile. Operating constraints are vast and numerous. Here comes the question of robust dynamic simulation. The author pointedly focuses on the scientific intricacies, the scientific barriers and the vast scientific foresight in the furtherance of science of petroleum refining.[7,10–12]

14.9 ENGINEERING CHALLENGES IN CATALYTIC CRACKING

Technological challenges and scientific vision are groundbreaking in present-day human civilization. Science today is a huge colossus with a vast vision of its own. Catalytic cracking needs to be reenvisioned and scientifically restructured. Energy sustainability and holistic sustainable development are the forerunners toward a newer visionary era in the field of petroleum refining and petroleum engineering science. The immense scientific prowess of human civilization, the vast scientific vision and the scientific motivation will all lead a long and visionary path toward the greater emancipation of engineering science today. Catalytic cracking and FCC are the visionary scientific processes of tomorrow.[10–12] Engineering challenges and technological profundity are veritably changing the scientific landscape of scientific validation, chemical process engineering, and petroleum engineering science. Environmental sustainability and the need for zero-discharge norms are the two sides of the visionary coin. Catalytic cracking is a scientific imperative today in petroleum refining as mankind and scientific research pursuit delves deep into the murky depths of scientific intricacies. The challenges and the vision of petroleum engineering today are slowly unfolding the real scientific fabric of energy sustainability. Renewable energy technology such as solar power, wind power, and biomass energy are the utmost need of the hour. Efficiency of the process, the robustness, and the futuristic vision of FCC are the scientific imperatives and scientific necessities of human scientific endeavor. Depletion of fossil fuel resources is veritably challenging the global energy scenario. Nonrenewable energy is the scientific blunder of today. This brings about the need for energy sustainability and the wide emancipation of energy engineering. Today chemical process engineering and energy engineering are linked with each other by an unsevered umbilical cord. Catalytic cracking is a technological masterpiece in petroleum refining today. The vast challenges need to be reorganized and restructured as science trudges a visionary as well as a weary path in this century. Catalytic cracking and energy technology in the similar manner are in a juxtaposition of scientific vision and introspection as human civilization moves forward.

14.10 ENGINEERING VISION AND SCIENTIFIC MOTIVATION IN PETROLEUM REFINING

Engineering vision and technological challenges are changing the vast scientific landscape today. Petroleum refining in the similar manner is today replete with scientific vision, scientific candor, and profundity. The scientific motivation in petroleum engineering science needs to be reenvisioned as human civilization treads into a new century. At this crucial juncture of scientific history and visionary timeframe, motivation and intellectual prowess in research pursuit are of immense importance. Petroleum refining and petroleum engineering science are veritably changing the global energy scenario. Scientific history and scientific vision in the engineering endeavor of the global energy research scenario today, needs to be restructured and reorganized. In this treatise, the author rigorously points toward the vast scientific potential and scientific sagacity in the research pursuit of petroleum refining and for the true realization of engineering science today. Today, petroleum refining needs to be redefined. Each unit needs to be revamped with respect to design and operation. Advanced mathematical tools such as multi-objective optimization, genetic algorithm and differential evolution are paving the path toward the greater emancipation of science and engineering of petroleum refining.[10–12] Science and engineering of applied mathematics and chemical process engineering are veritably challenging the scientific domain. Today evolutionary computation and genetic algorithms are reframing the scientific scenario.[8–11] Engineering vision today is in a state of immense regeneration and scientific rejuvenation.

14.11 CHEMICAL REACTION ENGINEERING AND THE DOMAIN OF CATALYTIC CRACKING

Chemical reaction engineering is a vast research frontier of chemical process engineering today. Human scientific endeavor, the vast scientific discernment, and the scientific girth are the forerunners toward a newer engineering eon in present-day human civilization. Chemical process engineering and petroleum engineering are the scientific imperatives toward the holistic development of a developed as well as a developing country today. Futuristic vision of chemical reaction engineering and petroleum refining are vastly changing the scientific landscape. In this treatise, the author pointedly focuses on the immense scientific frontiers, the scientific revelation, and the wide scientific girth in the pursuit of chemical engineering science and petroleum refining in particular. Validation of science and technology are the veritable challenges of scientific research pursuit in the present day human civilization. Deep scientific understanding and the futuristic vision of catalytic cracking are the pillars of this scientific treatise. The startling avenue of chemical process engineering today is chemical reaction engineering. Scientific prudence is in a state of mess with the evergrowing concerns of fossil fuel depletion. Breach of environmental safety and chemical process safety are the other avenues of global concern today. Environmental protection and environmental engineering science in a similar manner are vastly challenged today. The subtleties and the discernment of science need to be reenvisioned and enshrined with the passage of human scientific history and time. Chemical reaction engineering and the domain of catalytic cracking are veritably changing the face of scientific endeavor and the scientific panorama. In this treatise, the author reiterates the scientific success, the scientific revelation, and the deep scientific understanding behind chemical process engineering and petroleum engineering science.

14.12 FUTURE RESEARCH TRENDS, FUTURE FLOW OF THOUGHTS AND ENERGY SUSTAINABILITY

Future of chemical process engineering and petroleum engineering science are wide and bright. The challenges of technology and engineering science are today in the path of newer scientific regeneration. Science today is a huge colossus with a vast vision of its own. Technology has few answers to the intricacies of petroleum engineering science today. Future research trends and the future flow of thoughts today need to be totally reenvisioned and restructured with the passage of scientific history. Scientific innovation and scientific adjudication are today in a state of immense revamping in petroleum engineering science and petroleum refining. Alternate energy resources and energy sustainability are the next generation answers toward the global energy and fossil fuel scenario. The success of today's human civilization lies in the hands of scientists and the validation of science. Future research trends should be directed toward the implementation of sustainable development goals in major social and economic policies of the nations. Energy sustainability, holistic sustainable development, and the success of petroleum industry are the forerunners toward the greater emancipation and true realization of petroleum refining and renewable energy technology today. Solar power, wind power and biomass energy are the scientific imperatives and the holistic scientific vision of today. In this treatise, the author repeatedly urges toward the deep scientific success and the scientific potential of catalytic cracking and other petroleum refining units.[13,14]

14.13 FUTURE SCIENTIFIC FRONTIERS

Technological validation, scientific motivation, and scientific vision are the pillars of scientific research pursuit today. Energy and environmental crisis are the most important concerns of human civilization today. Future frontiers should be targeted toward the vast scientific vision and the scientific applications of petroleum refining toward global energy sustainability and the emancipation of global energy security. Scientific success, scientific fortitude, and scientific sagacity are the torchbearers toward a greater understanding of petroleum refining and petroleum engineering in present day human civilization. Today there are few answers to the petroleum engineering crisis as mankind crosses one visionary boundary over another. Design of petroleum refining units such as FCC unit is a veritable scientific frontier of today. The challenge and the vision of scientific candor and scientific fervor are changing the scientific mindset of the research pursuit in petroleum refining. The vision of Dr. Gro Harlem Brundtland, former Prime Minister of Norway on the science and engineering of sustainability needs to be reenvisioned and restructured at the present global energy situation.[13,14] Energy and environmental sustainability are the veritable scientific imperatives of today's research pursuit. Future research frontiers and future scientific perspectives need to be reenvisaged with the passage of scientific history and time. Research focus and validation of science in present day human civilization should be veritably directed toward the provision of basic human needs such as food, water, and energy. Here comes the vast and versatile importance of energy and environmental sustainability. The vast scientific intellect of human civilization, the immense intellectual prowess, and the immense challenges of scientific research pursuit will all lead a long and visionary way in the true realization of catalytic cracking, other petroleum refining units and the wider world of petroleum refining. The success of science and technology in present day human civilization needs to be redefined and answered again as human civilization faces one environmental and energy catastrophe over another. This century will certainly usher in a visionary eon in the field of sustainability and petroleum engineering science.[13,14]

14.14 SUMMARY, CONCLUSION, AND SCIENTIFIC PERSPECTIVES

The world of chemical process engineering and petroleum engineering science is moving from one visionary paradigm over another. The challenge of catalytic cracking and petroleum engineering needs to be rebuilt with the march of human civilization today. The scientific success, the deep scientific profundity, and scientific revelation are the utmost need of the hour in this crucial crisis of fossil fuel depletion. In this treatise, the author rigorously points out toward the need for scientific innovation in design and operation of petroleum refining units such as catalytic cracking. FCC stands in the midst of scientific genesis and deep scientific vision. The futuristic challenges and the technological motivation in petroleum refining are the true backbones of a new era in the field of catalytic cracking, chemical process engineering, and petroleum engineering science. The wide intricacies of catalytic cracking, the vast intellectual prowess behind petroleum refining, and the futuristic endeavor of engineering science will veritably lead a long and visionary way in the true realization of petroleum engineering. As the human civilization is witnessing tremendous challenges today, scientific perspectives in petroleum refining need to be reenvisioned and overhauled. In such a crucial juxtaposition of science and engineering, the greatness and scientific fervor of petroleum refining in research endeavor needs to be restructured with the course of scientific history and time.

ACKNOWLEDGMENT

The author wishes to acknowledge the vast contribution of his late father Shri Subimal Palit, an eminent textile engineer from Kolkata, India who taught the author rudiments of chemical engineering.

KEYWORDS

- **catalyst**
- **cracking**
- **fluid**
- **modeling**
- **simulation**
- **control**

REFERENCES

1. Elnashaie, S. S. E. H.; Elshishini, S. S. *Modeling, Simulation and Optimization of Industrial Fixed Bed Reactors,* Gordon and Breach Science Publishers S.A.: United Kingdom, 1993.
2. Pinheiro, C. I. C.; Fernandes, J. L.; Domingues, L.; Chambel, A. J. S.; Graca, I.; Oliveira, N. M. C.; Cerqueira, H. S.; Ribeiro, F. R. Fluid Catalytic Cracking(FCC) Process Modeling, Simulation, and Control. *Ind. Eng. Chem. Res.* 2012, *51*, 1–29.

3. Sadeghbeigi, R. *Fluid Catalytic Cracking Handbook,* 2nd ed.; Gulf Professional Publishing: Houston, USA, 2000.
4. Deb, K. Multi-objective Optimization Using Evolutionary Algorithms: An introduction, Kanpur Genetic Algorithm Report No.-2011003, 2011.
5. Ramteke, M.; Gupta, S. K. Kinetic Modeling and Reactor Simulation and Optimization of Industrial Important Polymerization Processes: A Perspective. *Int. J. Chem. React. Eng.* **2011,** *9*, R1.
6. Occelli, M. L. *Advances in Fluid Catalytic Cracking—Testing, Characterization and Environmental Regulations*; CRC Press, (Taylor and Francis Group): New York, USA, 2010.
7. Khandalekar, P. D. Control and optimization of fluidized catalytic cracking process, Master of Science in Chemical Engineering Thesis, 1993.
8. Sankararao, B., Gupta, S. K. Multi-objective Optimization of an Industrial Fluidized-Bed Catalytic Cracking Unit (FCCU) Using Two Jumping Gene Adaptations of Simulated Annealing. *Comput. Chem. Eng.* **2007,** *31*, 1496–1515.
9. Kasat, R. B.; Kunzru, D.; Saraf, D. N.; Gupta, S. K. Multiobjective Optimization of Industrial FCC Units Using Elitist Nondominated Sorting Genetic Algorithm. *Ind. Eng. Chem. Res.* **2002,** *41*, 4765–4776.
10. Kasat, R. B.; Gupta, S. K. Multiobjective Optimization of an Industrial Fluidized-Bed Catalytic Cracking Unit (FCCU) Using Genetic Algorithm (GA) with the Jumping Genes Operator. *Comput. Chem. Eng.* **2003,** *27*, 1785–1800.
11. Bandyopadhyay, S.; Saha, S.; Maulik, U.; Deb, K. A Simulated Annealing-Based Multiobjective Optimization Algorithm: AMOSA. *IEEE Trans. Evol. Comput.* **2008,** *12*(3), 269–283.
12. Chaudhuri, U. R. *Fundamentals of Petroleum and Petrochemical Engineering;* CRC Press (Taylor and Francis Group): New York, USA, 2011.
13. www.google.com (accessed Sep 1, 2017).
14. www.wikipedia.com (accessed Sep 1, 2017).

CHAPTER 15

DIELECTRIC PROPERTIES OF NEMATIC LIQUID CRYSTALS

M. KHADEM SADIGH and M. S. ZAKERHAMIDI*

Research Institute for Applied Physics and Astronomy, University of Tabriz, Tabriz, Iran

**Corresponding author. E-mail: Zakerhamidi@tabrizu.ac.ir, zakerhamidi@yahoo.com*

CONTENTS

ABSTRACT

Todays, liquid crystals, as a distinct phase of matter, are widely used in various field of science. The growing importance of these groups of materials is related to their specific ordering characteristics. There are different types of liquid crystals and their classification is based on their orientational and positional ordering characteristics. The most common and widely studied ones are nematic liquid crystals and their dielectric properties are investigated in this chapter. According to our studies, nematics with orientational order characteristics show different behaviors in parallel and perpendicular directions to externally applied fields. Moreover, these behaviors are highly dependent on molecular constituents, their chemical structures, and the properties of applied external fields. Therefore, the investigation of dielectric properties of nematic liquid crystals will be useful for increasing their applications in various science and technology.

15.1 INTRODUCTION

Liquid crystals have a history of more than a century. The first observation of liquid crystalline behavior was made by Reinitzer (1888) and Lehmann (1889).[1–2] Since then, liquid crystals have been considered as a major branch of science. Their applications in displays, nonlinear optics, optical communication, and data/signal/image processing have drawn considerable attention during the last years. Moreover, they occupy a key position in modern life. They are used in clocks, computer displays, television screens, calculators, temperature sensors, and so forth. Extensive efforts, empirical studies in this area continue to provide effective solutions to many different problems in science and technology.

The aim of this chapter is to describe dielectric properties of nematic liquid crystals. This physical parameter reveals the responses of liquid crystals to electric, magnetic, and optical fields and reflects their electronic and orientational properties. Therefore, the knowledge of dielectric properties of liquid crystal systems will give fruitful information about their applications in various fields of science.

15.2 NEMATIC PHASE

Liquid crystals are an intermediate state of matter in between the ordinary liquids and crystals. Materials that exhibit such unusual phases are often called mesogens, and could exist in various phases. The various phases that they could exist are called mesophases. The most common and widely studied ones are nematic liquid crystals. These mesophases can be distinguished from other condensed phases by the existence of their positional and orientational order characteristics.

Nematic phase is described by organic molecules that have no positional order but tend to align along a preferred direction (Fig. 15.1). Thus, similar to isotropic liquids, molecules are free to flow but still maintain their long-range directional order.

FIGURE 15.1 Orientationally ordered nematic mesophase.

Most of the nematics are uniaxial with one preferred axis and two equivalent axes (can be approximated as cylinders or rods). In this case, the short axes of the molecules have no preferred orientation. Therefore, these molecules indicate anisotropic behaviors and their physical properties will be modified by the average alignment with the preferred axis.

15.2.1 ANISOTROPIC BEHAVIORS OF NEMATIC LIQUID CRYSTAL

A dielectric material has the anisotropic behaviors if it indicates direction-dependent characteristics. Of course, these features depend on the shape, orientation, and position of individual molecules. Therefore, depending on positional and orientational orders in materials, their behaviors can be classified as follows:[3]

- If molecules are arranged in regular periodic patterns and oriented in the same directions, as crystals, the material is anisotropic.
- If molecules are located in random positions and are themselves isotropic or are oriented totally in random directions, the material is isotropic.
- Polycrystalline materials consist of disjoined crystalline domains that are randomly oriented to each other. The domains are themselves anisotropic but their average macroscopic behaviors are isotropic.
- If the molecules are inherently anisotropic and their orientations are not random, the material is anisotropic, even if positions are completely random. These properties are related to nematic liquid crystals that have orientational order but lack positional order.

Liquid crystals owe their remarkable features to anisotropy in shape of their molecules that capable them to form rod-like structures. Furthermore, the tendency of liquid crystal molecules to align parallel to each other leads to long-range orientational order, which causes the anisotropy of physical properties with two principal values, measured parallel and perpendicular to director.

Therefore, uniaxial nematic liquid crystals are anisotropic and their physical properties are different for two directions. Once the long axis of the anisotropic molecules orients with respect to the director, the refractive indices along and perpendicular to this direction will be different. Thus, liquid crystal molecules seem birefringent when they are observed through crossed polarizers.[4–6] The crystal part of their names is also because of their similar behavior for crystalline solids.

15.2.2 ALIGNMENT AND ORDERING OF NEMATIC LIQUID CRYSTAL

The physical properties of liquid crystals are usually characterized in terms of so-called order parameter. For specifying the degree of orientational ordering properties of liquid crystals, suppose rigid and rod-like nematic molecules that express a liquid crystalline system with cylindrical or rotational symmetry. In this case, the average direction of the long molecular axis is called the liquid crystal director and is denoted by $\vec{n}$. In addition, the direction of $\vec{n}$ and $\vec{n}$ is indistinguishable for defined uniaxial liquid crystal.

In order to investigate the orientation of rod-like molecules, we indicate the long axis of the molecule by a unit vector $\hat{a}$. As can be seen in Figure 15.2, the orientation of nematic molecules is determined by the polar angle θ and the azimuthal angle Φ. Thus, the orientation of molecules can be described by orientational distribution function $f(\underline{\theta}, \Phi)$.

$f(\theta, \Phi)\, d\Omega$ ($d\Omega = \sin\theta \cdot d\theta \cdot d\Phi$) indicates the probability of rod-like molecules to align along the preferred axis within the solid angle $d\Omega$. In uniaxial liquid crystals that have no preferred orientation around the short axis of molecules, $f(\theta)$ depends only on polar angle θ. In other words, the cylindrical symmetry of liquid crystal molecules allows the order only in the angle θ.

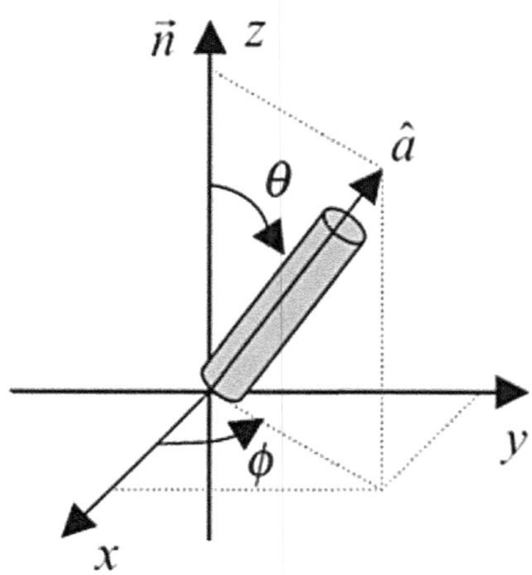

FIGURE 15.2 Coordinate system that defines the microscopic order parameter of a molecule. x, y, and z are the molecular axes, whereas director $\vec{n}$ is considered aligned to z-axis in the laboratory coordinate.[15]

By assuming that order parameter is zero in isotropic phase and nonzero in the low-temperature phase, we can consider the average value of the projection of $\hat{a}$ along the director $\vec{n}$ by Equation (15.1).

$$\langle \cos\theta \rangle = \frac{\int_0^{\pi} \cos\theta f(\theta) \sin\theta d\theta}{\int_0^{\pi} f(\theta) \sin\theta d\theta} \tag{15.1}$$

The average <·> is taken over the whole ensemble and cos θ is the first-order Legendre polynomial. When molecules are randomly oriented and in isotropic phase, $\langle \cos\theta \rangle = 0$. In nematic phase, the probability of orientation of a molecule at the angles θ and $\pi-\theta$ is the same Fig. 15.3; therefore, $\langle \cos\theta \rangle = 0$.

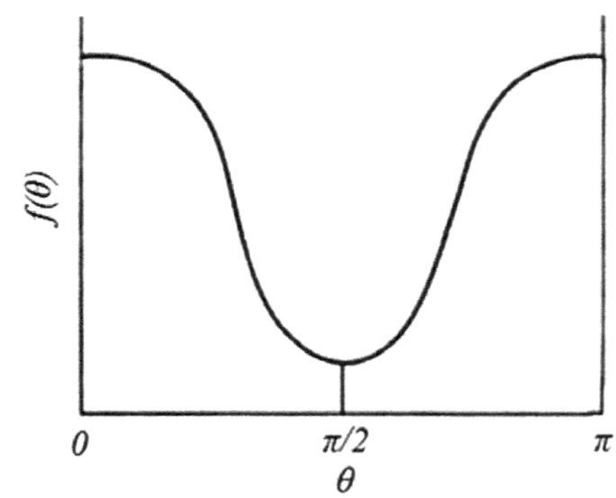

FIGURE 15.3 Molecular orientational distribution function as a function of the angle between long molecular axis and director (θ). (Reproduced with permission from reference 7. © 2010 Springer Nature.)

In real, above definition for order parameter cannot provide precise information about the molecular orientational order. According to our assumptions for axially symmetric systems, $f(\theta)$ can be expanded in series of the Legendre polynomials $P_i(\cos\theta)$,

$$f(\theta) = \frac{1}{2}(1 + a_1 P_1(\cos\theta) + a_2 P_2(\cos\theta) + a_3 P_3(\cos\theta) + ...) \tag{15.2}$$

where Legendre polynomials are

$$P_0(\cos\theta) = 1,\ P_1(\cos\theta) = \cos\theta, \tag{15.3}$$

$$P_2(\cos\theta) = \frac{1}{2}(3\cos^2\theta - 1),\ P_3(\cos\theta) = \frac{1}{2}(5\cos^3\theta - 3\cos\theta). \tag{15.4}$$

In order to find numerical coefficients a_L, both sides of Equation 15.2 are multiplied by P_L (cosθ), and then, using the orthogonality of Legendre polynomials, they are obtained according to Equation 15.5.

$$a_L = (2L+1)\frac{\int_{\pi}^{0} P_L(\cos\theta) f(\theta)\sin\theta d\theta}{\int_{\pi}^{0} f(\theta) d(\cos\theta)} \tag{15.5}$$

Finally, using the normalization condition, molecular orientational function $f(\theta)$ can be written in the form of Equation 15.6.

$$\begin{aligned} f(\theta) = \frac{1}{2}[1 &+ 3\langle P_1(\cos\theta)\rangle P_1(\cos\theta) + 5\langle P_2(\cos\theta)\rangle P_2(\cos\theta) \\ &+ 7\langle P_3(\cos\theta)\rangle P_3(\cos\theta) + 9\langle P_4(\cos\theta)\rangle P_4(\cos\theta) + ...] \end{aligned} \tag{15.6}$$

For uniaxial molecules with head-to-tail symmetry ($\vec{n} = -\vec{n}$), the odd terms can be removed:

$$f(\theta) = \frac{1}{2}[1 + 5\langle P_2(\cos\theta)\rangle P_2(\cos\theta) + 9\langle P_4(\cos\theta)\rangle P_4(\cos\theta) + ...] \tag{15.7}$$

Therefore, microscopic orientational order is expressed according to Equation 15.8.

$$S_2 = \langle P_2(\cos\theta)\rangle = \left\langle \frac{1}{2}(3\cos^2\theta - 1)\right\rangle = \frac{\int_0^{\pi} \frac{1}{2}(3\cos^2\theta - 1) f(\theta)\sin\theta d\theta}{\int_0^{\pi} f(\theta)\sin\theta d\theta} \tag{15.8}$$

In isotropic phase, molecules have an equal probability of orientation in any directions. In this case, orientation function has a constant value and there is no orientational order in system (S=0) Fig. 15.4).

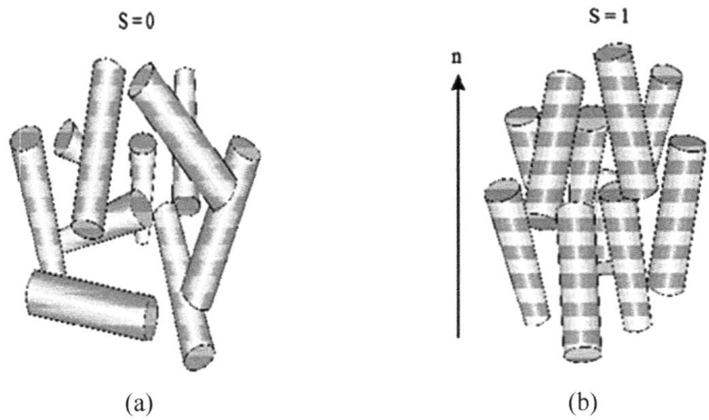

FIGURE 15.4 The value of orientational parameter in the (a) random state and (b) perfectly ordered system.

For perfectly ordered system, we can assume:

$$f(\theta) = \delta(\theta)$$

$$\theta = 0 \sin\theta\delta(\theta) = \infty \tag{15.9}$$

Therefore, the order parameter is equal to one and all the long axes of the molecules are parallel to each other and to director (Fig. 15.4) Also, for an intermediate phase, the order parameter can change discontinuously from a nonzero value to zero.

The defined scalar order parameter is usually termed microscopic long-range order and is sufficient to describe liquid crystalline systems that possess cylindrical or rotational symmetry. For molecules lacking such symmetry, a more general tensor order parameter is needed.[4,7]

15.3 DIELECTRIC PROPERTIES OF ANISOTROPIC FLUIDS

Dielectric constant is a physical parameter that characterizes the electronic response of liquid crystals to externally applied fields. These property and effective parameters on it are investigated based on fundamental ideas about isotropic liquids. Hence, we start our studies on isotropic liquids.

15.3.1 BASIC CONCEPTS

Dielectrics are insulating materials that are polarized under influence of external fields, thereby modifying the dielectric function of the vacuum. Based on a classical experiment, when a dielectric material is placed between the plates of a capacitor, the capacitance is increased from a value C to a value εC due to the polarization of material. In this case, ε is relative permittivity of material.

At the atomic level, the materials consist of positively and negatively charged particles that are in balance with each other in the absence of external fields. When an electric field is applied, the balances between the charges are perturbed by the various polarization mechanisms. The polarization mechanisms depend on material characteristics. In a material with nonpolar molecules, electric polarization and ionic polarization have a contribution in response to applied external fields. In polar materials, an additional polarization is also observed due to the tendency of permanent dipole moments to orient parallel to external fields. In isotropic liquids and gases, the dielectric constant is isotropic. In solid materials, although they can indicate anisotropic behaviors, the contribution of orientation polarization is less important because of the rather fixed orientation of molecules. This situation is more complicated in liquid crystals; while in this case, dipolar polarization can have a dominant contribution in their responses to the applied external fields.

Therefore, in the presence of various polarization mechanisms, the polarization in sample will be given by

$$\vec{P} = \varepsilon_0 \chi_e \vec{E} \tag{15.10}$$

$$P = Np \tag{15.11}$$

$$\vec{p} = \alpha \vec{E}_{local} \tag{15.12}$$

where N, α, and ε_0 symbols are the number density of molecules, the polarizability of the molecule, and permittivity in free space, respectively.

In general, one has to distinguish between the macroscopic electric field and the microscopic electric field (local field) that each atom experiences in sample. The distinction between these two fields is important except for the case of materials such as gases that internal fields can be negligible.

Therefore, if we find E_{loc} (internal field), we could calculate P, and then the value of dielectric constant also. This field can be calculated by using the method described by Lorentz.[8] Consider a dielectric medium that is under influence of a uniform electric field E. In this case, imagine a cavity created by removing the molecule under study that can be regarded as a sphere as shown in Fig. 15.5. The internal field action on this molecule will have four components as follows:

$$E_{local} = E_0 + E_1 + E_2 + E_3 \tag{15.13}$$

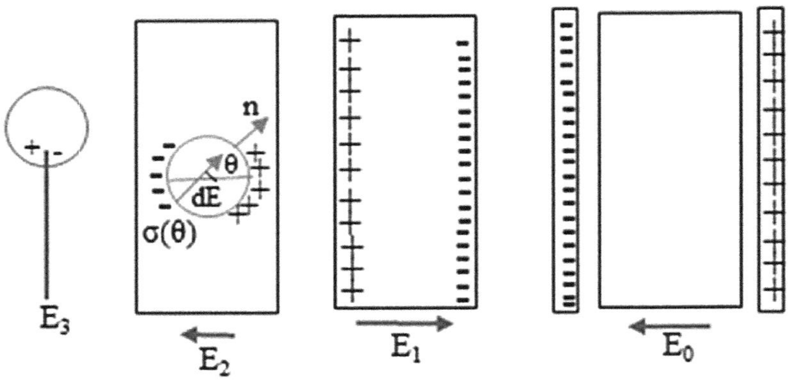

FIGURE 15.5 Calculation of the Lorentz local field.

Here, E_0 is the external field, E_1 with opposite direction to E_0 is a depolarizing field created by the charges on the external surfaces and its value depends on the shape of the sample. E_2 is produced by the dipole moment inside the sphere that can be replaced by the charges on the surface of the sphere. The density of external field that induces surface charges is $\sigma(\theta) = -P \cdot n$, and n is a unit length vector perpendicular to the surface of the cavity at the point θ. Also, the positive direction of the normal n is outward. Therefore, this field is easily calculable from Equation (15.14).[9]

$$E_2 = -\frac{P}{4\pi\varepsilon_0}\int_0^{\pi} \cos^2(\theta)\sin(\theta)d\theta \int_0^{2\pi} d\varphi = \frac{P}{3\varepsilon_0} \tag{15.14}$$

E_3 is the field from all molecules inside the cavity except the molecule under the consideration. Due to high symmetry, E_3 can be ignored and that it is true for the isotropic liquids and cubic crystals.

Therefore, E_{local} is

$$E_{\mathrm{local}} = E_0 + E_1 + \frac{P}{3\varepsilon_0} = E + \frac{P}{3\varepsilon_0} \tag{15.15}$$

and electric susceptibility is given by

$$P = N\alpha E_{local} = N\alpha(E + \frac{P}{3\varepsilon_0}) \tag{15.16}$$

$$\chi_e = \varepsilon / \varepsilon_0 - 1 = \frac{N\alpha / \varepsilon_0}{1 - N\alpha / 3\varepsilon_0} \tag{15.17}$$

Finally, from last three equations, we arrive to Clausius–Mossotti relation for dielectric constant as expressed in Equation (15.18).

$$\frac{\varepsilon - 1}{\varepsilon + 2} = \frac{N\alpha}{3\varepsilon_0} \tag{15.18}$$

By this equation, macroscopic parameter of material (dielectric constant) is related to molecular polarizability as a microscopic parameter.

As discussed previously, molecular polarizability can be included in electronic, ionic, and orientational polarizations, but Clausius–Mossotti equation cannot be used for dipolar materials. Dipolar materials indicate that orientational polarization can exist even in the absence of the external fields. This kind of polarization is acquired of the orientation of permanent dipole moment of molecules, which is heavier than atoms or electrons. Hence, the orientational polarization occurs at lower frequencies than atomic and electronic polarizations.

Although polar molecules tend to align parallel to the external field, complete alignment does not occur except in the presence of extremely strong fields. In this case, the orientation of molecules is in competition with the thermal effects. If the energy put into orientation is dissipated due to the randomizing effects, the dielectric constant is likely to reduce by increasing temperature.

In order to investigate the contribution of temperature-dependent orientational polarization on dielectric constant, Debye used classical Boltzmann statistics and the Langevin function from the theory of paramagnetism, to estimate the thermal effects on polarization. Assuming that there is no interaction between polar molecules, Debye derived the following equation for the orientational polarizability:[9–10]

$$\alpha_{\text{dipolar}} = \frac{\mu^2}{3KT} \tag{15.19}$$

where μ, K, and T are permanent dipole moment, Boltzmann constant, and temperature, respectively.

Therefore, this polarization contribution is added to Clausius–Mossotti equation.

$$\frac{\varepsilon - 1}{\varepsilon + 2} = \frac{N}{3\varepsilon_0}(\alpha_e + \frac{\mu^2}{3KT}) \tag{15.20}$$

Although this result has been used successfully in many polar materials such as polar liquids, it is not applicable to the condensed state of matter. The reason for this breakdown is related to assumption, which is considered in deriving the Clausius–Mossotti equation ($E_3 = 0$). Moreover, in the condensed phase, the tendency of permanent dipoles to interact with their surroundings should be taken into account. To avoid this problem, Onsager modified the Debye theory by introducing a cavity. In this method, the electric field was considered as the combination of a cavity field and a reaction field.

In general, Onsager's model is assumed as a sphere with a permanent dipole moment and an isotropic polarizability. In this case, directing field as a sum of fields in spherical cavity is different from internal field. The dipole polarizes its surrounding and leads to the formation of reaction fields. This reaction field is always parallel to the dipole and can have a contribution in the internal field. Internal field is now expressed according to Equation 15.21.[11]

$$E_i = E_d + \bar{R} \tag{15.21}$$

where E_d and $\bar{R}$ are directing field and the average of reaction field, respectively. Reaction field of a dipole moment with polarizability α is given by Equation 15.22, while directing field can be calculated as a sum of the fields in spherical cavity with radius a and reaction field of induced dipole moment. Therefore, it is expressed according to Equation 15.23.

$$\bar{R} = fF\mu$$
$$f = \frac{(\varepsilon - 1)}{(2\pi\varepsilon_0 a^3(2\varepsilon + 1))}, F = \frac{1}{(1 - f\alpha)} \tag{15.22}$$

$$E_d = hFE$$
$$h = \frac{3\varepsilon}{(2\varepsilon + 1)} \tag{15.23}$$

By substituting the obtained results in Equation 15.24, and replacing molecular polarizability with its average, Onsager's equation is acquired.[7,11]

$$P = N(\langle \alpha.E_i \rangle + \langle \bar{\mu} \rangle) \tag{15.24}$$

$$\varepsilon = 1 + \frac{NFh}{\varepsilon_0}(\bar{\alpha} + \frac{F\mu^2}{3KT}) \tag{15.25}$$

The deviation of Onsager's model can occur in two cases: (1) when molecules cannot be considered in the shape of sphere and (2) when specific interactions play an important role in understudy system.

Therefore, from the obtained results about dielectric properties of isotropic liquids, in the next section, we will discuss the dielectric characteristics of nematic liquid crystals.

15.3.2 DIELECTRIC ANISOTROPIES

The most important properties of liquid crystals that distinguish them from the other liquids are related to their orientational order. In contrast to ordinary liquids, they indicate various behaviors in different directions. This direction-dependent characteristic manifests themselves in electric, magnetic, and optical measurements of liquid crystals. Dielectric constant, magnetic susceptibility, and refractive index of liquid crystals indicate different behaviors depending on the direction in which the measurements are performed.

It is convenient to investigate the anisotropic properties of liquid crystals based on their responses to the externally applied fields. Suppose an electric field is applied to system. In this case, dipole moments are induced in nematic liquid crystal. The dipole moment per unit volume is called polarization P and is along the average molecular dipole moment. For small electric fields, polarization is proportional to the applied electric field and the constant of proportionality is called electric susceptibility of material χ_e.

$$\vec{P} = \varepsilon_0 \chi_e \vec{E} \tag{15.26}$$

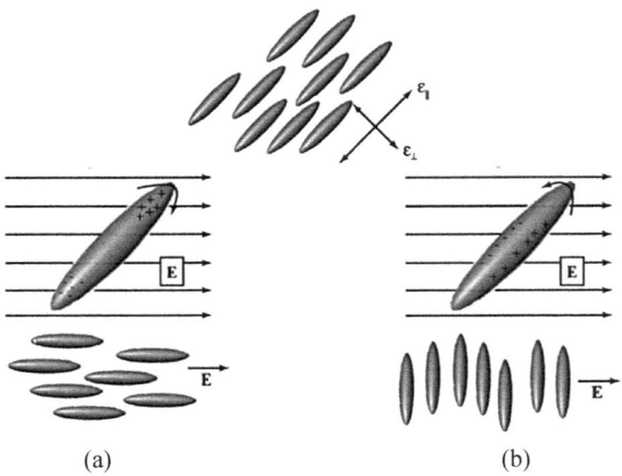

FIGURE 15.6 Schematic diagram of (a) positive and (b) negative dielectric anisotropy.[12]

As shown in Figure 15.6, in anisotropic nematic liquid crystals, the responses to external fields depend on their directions. Therefore, in a specific direction of director, the polarization of liquid crystal molecules depends on whether the applied electric field is parallel or perpendicular to the director. Thus, electric susceptibility is a tensor quantity with different values in parallel and perpendicular directions.

We restrict our study to rod-like nematic liquid crystals and use a macroscopic coordinate system x, y, z, with the z-axis along the director. Hence, it can be written as:

$$\begin{pmatrix} P_x \\ P_y \\ P_z \end{pmatrix} = \varepsilon_0 \begin{pmatrix} \chi_\perp & 0 & 0 \\ 0 & \chi_\perp & 0 \\ 0 & 0 & \chi_\parallel \end{pmatrix} \begin{pmatrix} E_x \\ E_y \\ E_z \end{pmatrix} \tag{15.27}$$

The electric susceptibility is usually expressed as an average part and an anisotropic part as can be seen in Equation 15.28.

$$\vec{\chi} = \bar{\chi}\begin{pmatrix} 1 & 0 & 0 \\ 0 & 1 & 0 \\ 0 & 0 & 1 \end{pmatrix} + \frac{1}{3}\Delta\chi \begin{pmatrix} -1 & 0 & 0 \\ 0 & -1 & 0 \\ 0 & 0 & 2 \end{pmatrix} \tag{15.28}$$

where $\bar{\chi} = \frac{(\chi_{\parallel} + 2\chi_{\perp})}{3}$ and $\Delta\chi = \chi_{\parallel} - \chi_{\perp}$

Although above discussion was provided about polarization anisotropy, the same discussion can be considered for dielectric anisotropy.

By using Maxwell equations and linear relation between electric displacement $\vec{D}$ and polarization $\vec{P}$ vectors, the permittivity of materials is defined according to Equation 15.31.

$$\vec{D} = \vec{\varepsilon}.\vec{E} \tag{15.29}$$

$$\vec{D} = \varepsilon_0\vec{E} + \vec{P} \tag{15.30}$$

$$\vec{\varepsilon} = \varepsilon_0(\vec{I} + \vec{\chi}_e) \tag{15.31}$$

$\vec{I}$ is unit tensor, that is,

$$\vec{I} = \begin{pmatrix} 1 & 0 & 0 \\ 0 & 1 & 0 \\ 0 & 0 & 1 \end{pmatrix} \tag{15.32}$$

$\varepsilon/\varepsilon_0$ is also relative permittivity or dielectric constant of material that is considered as a tensor quantity in anisotropic liquid crystals.

From the obtained results, it is observed that rod-like nematic liquid crystals have different dielectric constants in parallel and perpendicular directions to director. When the external field is parallel to director, the permittivity is $\varepsilon_{\parallel}$ and in the perpendicular state, the permittivity is $\varepsilon_{\perp}$. The induced anisotropy is given by

$$\Delta\vec{\varepsilon} = \varepsilon_{\parallel} - \varepsilon_{\perp} \tag{15.33}$$

Depending on the values of $\varepsilon_{\parallel}$ and $\varepsilon_{\perp}$, anisotropy in nematic liquid crystals will be either positive or negative. In order to measure the parallel and perpendicular components of permittivity, the electric field must be applied parallel and perpendicular to director. In practice, this is acquired by homeotropic and planar alignment in the liquid crystal cells. In this case, the calculated capacitance of the sample is used for measuring $\varepsilon_{\parallel}$ and $\varepsilon_{\perp}$ according to Equation 15.34.[4]

$$\varepsilon_{\parallel} = \frac{c_{\perp}}{c_0} \cdot \varepsilon_{\perp} = \frac{c_{\parallel}}{c_0} \tag{15.34}$$

where $c_{\parallel}$ and $c_{\perp}$ are the capacitances of the alignment liquid crystal in parallel and perpendicular directions of the cell surface, respectively. C_0 is also related to the capacitance of the empty cell.

One way to understand how liquid crystals respond to electric field is based on energy considerations. The electric energy of the liquid crystal per unit volume is approximately given by

$$U_{\text{electric}} = -\frac{1}{2}\vec{P}.\vec{E} \tag{15.35}$$

When the applied field is neither parallel nor perpendicular to director, the applied electric field can be decomposed into parallel and perpendicular components to director (Fig. 15.7). In this case, induced polarization is expressed according to following relation.

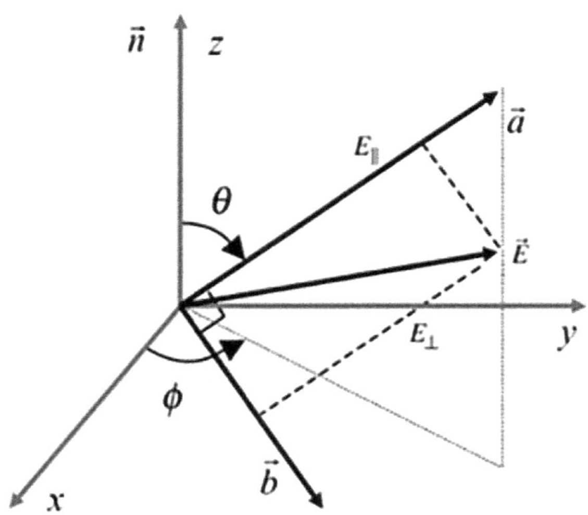

FIGURE 15.7 The electric field decomposition into parallel and perpendicular components to the long molecular axis: a, unit vector parallel to the long molecular axis, b: unit vector perpendicular to the long molecular axis.

$$\vec{P} = \varepsilon_0 \chi_{\parallel} (\vec{E}.\vec{n})\vec{n} + \varepsilon_0 \chi_{\perp} [\vec{E} - (\vec{E}.\vec{n})\vec{n}]$$
$$\vec{P} = \varepsilon_0 (\chi_{\perp} \vec{E} + \Delta\vec{\chi}(\vec{E}.\vec{n})\vec{n}) \quad (15.36)$$

Therefore, electric energy is equal to

$$U_{electric} = -\frac{1}{2}(\varepsilon_0(\chi_{\perp}\vec{E} + \Delta\vec{\chi}(\vec{E}.\vec{n})\vec{n}).\vec{E} = -\frac{1}{2}\varepsilon_0\chi_{\perp}E^2 - \frac{1}{2}\varepsilon_0\Delta\vec{\varepsilon}(\vec{E}.\vec{n})^2 \quad (15.37)$$

In contrast to first term of Equation 15.37, second term depends on molecular reorientation with respect to applied electric field. If liquid crystal indicates positive anisotropic behaviors, electric energy is minimized and liquid crystal molecules tend to align parallel to applied field. Conversely, if dielectric anisotropy is negative, tendency of liquid crystal molecules to align perpendicularly is increased.

In general, when an external field is applied to a liquid crystal, internal fields (local fields) must be considered. In this case, when an external electric field is applied to the uniaxial liquid crystal, polarization is given by

$$\vec{P} = N\varepsilon_0\alpha_{\parallel}(\vec{a}.\vec{E}_{\text{local}})\vec{a} + N\varepsilon_0\alpha_{\perp}(\vec{E}_{\text{local}} - (\vec{a}.\vec{E}_{\text{local}})\vec{a})$$
$$= N\varepsilon_0\alpha_{\perp}\vec{E}_{\text{local}} + N\varepsilon_0\Delta\alpha(\vec{a}\vec{a}).\vec{E}_{\text{local}} \quad (15.38)$$

where N is the number density of molecules, α is molecular polarization, and E_{local} is the field resulting from all external sources and the electric field produced by the dipole moments of other molecules.

$$\Delta\alpha = \alpha_{\parallel} - \alpha_{\perp} \quad (15.39)$$

$\alpha_{\parallel}$ and $\alpha_{\perp}$ are the molecular polarizabilities parallel and perpendicular to the director, respectively. Therefore, our task is to investigate the dielectric characteristics of liquid crystals in the presence of internal fields. As discussed previously, macroscopic electric field is related to the microscopic local field by internal field constant.

$$\vec{E}_{local} = f\vec{E} \quad (15.40)$$

It should be noted that in previous section, E_{local} was calculated for spherical molecules. In general, local field in an anisotropic environment must be considered as a tensor quantity.[13–15] Therefore, induced polarization and dielectric constant in the liquid crystal system is expressed by

$$\vec{P} = N\varepsilon_0\alpha_\perp \langle \vec{f} \rangle.\vec{E} + N\varepsilon_0\Delta\alpha \langle (\vec{f}.\vec{a}\vec{a}) \rangle.\vec{E} \tag{15.41}$$

$$\vec{\varepsilon} = \vec{I} + N\alpha_\perp \langle \vec{f} \rangle + N\Delta\alpha \langle (\vec{f}.\vec{a}\vec{a}) \rangle \tag{15.42}$$

$$\vec{a} = \sin\theta\cos\phi i + \sin\theta\sin\phi j + \cos\theta k \tag{15.43}$$

For calculation of second term of Equation 15.42, internal field tensor should be written in laboratory coordinate with the z-axis parallel to the liquid crystal director.

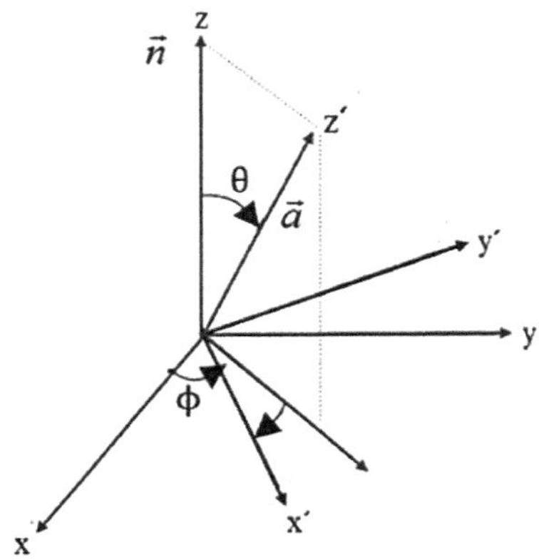

FIGURE 15.8 (See color insert.) The transformation between the molecular coordinate (x′y′z′) and the laboratory coordinate (xyz).

The molecular coordinate (x′y′z′) is acquired from the first rotating of the laboratory coordinate xyz around the z-axis and then rotating of it around the y′-axis through the angles ϕ and θ, respectively Fig. 15.8.[15] Therefore, internal field tensor is

$$\vec{f} = R.\begin{pmatrix} f_\perp & 0 & 0 \\ 0 & f_\perp & 0 \\ 0 & 0 & f_\parallel \end{pmatrix}.R^{-1} = \begin{pmatrix} f_\perp + \Delta f \sin^2\theta\cos\phi^2 & \Delta f \sin^2\theta\sin\phi\cos\phi & \Delta f \sin\theta\cos\theta\cos\phi \\ \Delta f \sin^2\theta\sin\phi\cos\phi & f_\perp + \Delta f \sin^2\theta\sin^2\phi & \Delta f \sin\theta\cos\theta\sin\phi \\ \Delta f \sin\theta\cos\theta\cos\phi & \Delta f \sin\theta\cos\theta\sin\phi & f_\perp + \Delta f \cos^2\theta \end{pmatrix} \tag{15.44}$$

Here, Δf and R are

$$R = \begin{pmatrix} \cos\theta\cos\phi & -\sin\phi & \sin\theta\cos\phi \\ \cos\theta\sin\phi & \cos\phi & \sin\theta\sin\phi \\ -\sin\theta & 0 & \cos\theta \end{pmatrix} \tag{15.45}$$

$$\Delta f = f_{\parallel} - f_{\perp} \tag{15.46}$$

$$(\vec{f}.\vec{a}\vec{a}) = \begin{pmatrix} f_{\parallel}\sin^2\theta\cos^2\phi & f_{\parallel}\sin^2\theta\sin\phi\cos\phi & f_{\parallel}\sin\theta\cos\theta\cos\phi \\ f_{\parallel}\sin^2\theta\sin\phi\cos\phi & f_{\parallel}\sin^2\theta\sin\phi & f_{\parallel}\sin\theta\cos\theta\sin\phi \\ f_{\parallel}\sin\theta\cos\theta\cos\phi & f_{\parallel}\sin\theta\cos\theta\sin\phi & f_{\parallel}\cos^2\theta \end{pmatrix} \tag{15.47}$$

By assuming that[7]

$$\begin{aligned} &\langle\cos^2\theta\rangle = (2S+1)/3, \langle\sin^2\theta\rangle = (2-2S)/3, \langle\sin^2\phi\rangle = \langle\cos^2\phi\rangle = 1/2, \\ &\langle\sin\theta\rangle = \langle\cos\theta\rangle = \langle\sin\phi\rangle = \langle\cos\phi\rangle = \langle\sin\phi.\cos\phi\rangle = 0 \end{aligned} \tag{15.48}$$

Dielectric constant of nematic liquid crystal is

$$\varepsilon = \begin{pmatrix} 1+\frac{N}{3}[\alpha_{\perp}f_{\perp}(2+S)+\alpha_{\parallel}f_{\parallel}(1-S)] & 0 & 0 \\ 0 & 1+\frac{N}{3}[\alpha_{\perp}f_{\perp}(2+S)+\alpha_{\parallel}f_{\parallel}(1-S)] & 0 \\ 0 & 0 & 1+\frac{N}{3}[\alpha_{\perp}f_{\perp}(2-2S)+\alpha_{\parallel}f_{\parallel}(1+2S)] \end{pmatrix} \tag{15.49}$$

Finally, dielectric anisotropy is expressed by

$$\Delta\varepsilon = \varepsilon_{\parallel} - \varepsilon_{\perp} = NS(\alpha_{\parallel}f_{\parallel} - \alpha_{\perp}f_{\perp}) \tag{15.50}$$

which is proportional to the order parameter of liquid crystal, internal field constant, and molecular polarizability. This indicates how the macroscopically observed anisotropy is acquired from the anisotropic behaviors of molecular quantities.

For liquid crystals with polar molecules, the contribution of orientational polarization must be considered as well. Hence, the induced polarization is

$$\vec{P} = N(\langle\vec{\alpha}.\vec{E}_d\rangle + \langle\vec{\mu}\rangle) \tag{15.51}$$

that in this case, directing field (E_d) is different from internal field. Following Onsager's theory of isotropic liquids, Maier and Meier[16] considered the contribution of orientational polarization in nematic liquid crystals. They considered a molecule with anisotropic polarizability that its permanent dipole moment made an angle β with long molecular axis Fig. 15.9.

- In calculation of anisotropic behaviors of nematic liquid crystals, they pursue the following assumptions:
- Nematic liquid crystal molecules have a center of symmetry and it is characterized by orientational order parameter *S*.
- In the calculation of cavity factor *h* and reaction field factor *f*, the anisotropy of permittivity was not considered. In this case, the mean dielectric constant <ε> was taken instead of ε. This approximation is justified as long as $\Delta\varepsilon \ll \bar{\varepsilon}$
- The tensor nature of electronic polarizability was neglected in the calculation of *F* factor. In this case, <α> is used instead of α.

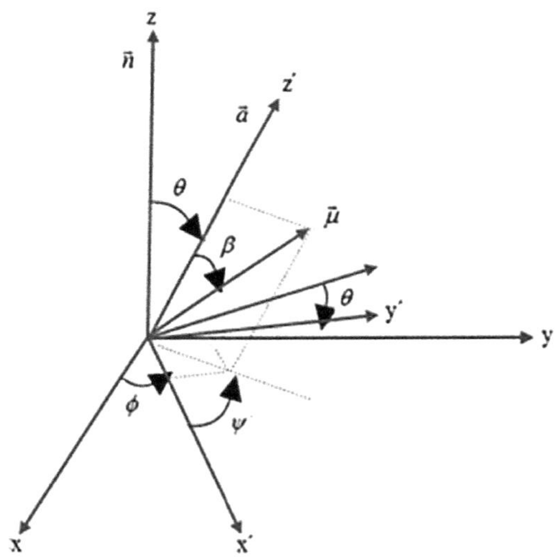

FIGURE 15.9 Orientation of the permanent dipole moment in the molecular coordinate (x′y′z′) and the laboratory coordinate (xyz).

By using the assumptions listed, the orientational polarization will be calculated. As shown in Figure 15.9, in molecular coordinate the components of $\vec{\mu}$ that made an angle β with long molecular axis are

$$\vec{\mu} = \mu \sin\beta \cos\psi i' + \mu \sin\beta \sin\psi j' + \mu \cos\beta k' \tag{15.52}$$

Hence, they should be acquired from laboratory frame by using rotation matrix similar to Equation 15.44. Applying an external electric first parallel to the director and then perpendicular to it leads to appearance of anisotropic behaviors in molecular responses as can be seen in Equation 15.53.

$$\begin{aligned}\mu_{\parallel} &= -\mu(\sin\beta\sin\psi\sin\theta + \cos\beta\cos\theta) \\ \mu_{\perp} &= \mu(\sin\beta\cos\psi\cos\phi - \sin\beta\sin\psi\cos\theta\sin\phi - \cos\beta\sin\theta\sin\phi)\end{aligned} \tag{15.53}$$

Also, the energy of a dipole in the directing electric field E_d is also,

$$U_{\gamma} = -\mu_{\gamma}.E_d(\gamma = \parallel, \perp) \tag{15.54}$$

Therefore, the average value of μ_{γ} in the presence of external field is calculated by Equation 15.55 and leads to Equation 15.57.

$$\langle \mu_{\gamma} \rangle = \frac{\int \mu_{\gamma}^2 e^{-(U+V(\theta))/KT} \sin\theta d\theta d\phi d\psi}{\int e^{-(U+V(\theta))/KT} \sin\theta d\theta d\phi d\psi} \tag{15.55}$$

$$-U \ll KT \rightarrow e^{-(U+V(\theta))/KT} \approx (1 - \frac{U}{KT})e^{-V(\theta)/KT} \tag{15.56}$$

$$\begin{aligned}\langle \mu_{\parallel} \rangle &= \frac{E_{d\parallel}\mu^2}{KT}(1 - (1 - 3\cos^2\beta)S) \\ \langle \mu_{\perp} \rangle &= \frac{E_{d\perp}\mu^2}{KT}(1 + \frac{1}{2}(1 - 3\cos^2\beta)S)\end{aligned} \tag{15.57}$$

After similar analysis, we found

$$\langle \alpha_{\parallel} \rangle = \frac{1}{3}[(\alpha_{\perp}(2 - 2S) + \alpha_{\parallel}(1 + 2S)] \tag{15.58}$$

$$\langle \alpha_{\perp} \rangle = \frac{1}{3}[(\alpha_{\perp}(2 + S) + \alpha_{\parallel}(1 - S)] \tag{15.59}$$

Then, by inserting the contribution of cavity and reaction fields in Equation 15.51, it can be written as:

$$\varepsilon_{\parallel} = 1 + \frac{NFh}{3\varepsilon_0}[(\alpha_{\perp}(2 - 2S) + \alpha_{\parallel}(1 + 2S) + \frac{F\mu^2}{KT}(1 - (1 - 3\cos^2\beta)S)] \tag{15.60}$$

$$\varepsilon_{\parallel} = 1 + \frac{NFh}{3\varepsilon_0}[(\alpha_{\perp}(2 - 2S) + \alpha_{\parallel}(1 + 2S) + \frac{F\mu^2}{KT}(1 - (1 - 3\cos^2\beta)S)] \tag{15.61}$$

Finally, the dielectric anisotropy is expressed according to Equation 15.62.

$$\varepsilon_{\parallel} = 1 + \frac{NFh}{3\varepsilon_0}[(\alpha_{\perp}(2-2S) + \alpha_{\parallel}(1+2S) + \frac{F\mu^2}{KT}(1-(1-3\cos^2\beta)S)] \quad (15.62)$$

This equation gives some important information about anisotropic behaviors of nematic liquid crystals. For dipolar nematic liquid crystals in a special position, where $(1 - 3cos^2 \beta) = 0$ (β =54.7°), the contribution of orientational polarization on dielectric anisotropy is completely vanished. In this case, rod-like molecules with high polarizability along the large molecular axis indicate positive anisotropy. This is in agreement with experimental results where they express the sign of dielectric anisotropy change by changing of the angle β and long molecular axis. This has been observed in certain fluorinated cyclohexyl ethanyl biphenyls[17] that the sign of dielectric anisotropy changed from negative to positive by increasing temperature.

For β<54.7°, the contribution of the dipolar term to $\Delta\varepsilon$ is positive, and for β>54.7°, this contribution is negative. Hence, $\Delta\varepsilon$ can be positive or negative depending on the contribution of induced polarization (electronic and ionic polarization) and orientational polarization in liquid crystal system. Moreover, these contributions have different temperature dependences. The contribution of induced polarization to anisotropy varies with temperature like *S*, while temperature dependence in orientational polarization is proportional to *S*/*T*. Typical temperature dependencies of dielectric constants are seen in Figure 15.10.

Therefore, the anisotropic behavior of liquid crystal systems can be controlled by the magnitude of permanent dipole moment, angle between molecular dipole moment with respect to long molecular axis, order parameter, and thermal effects.

Although this model can explain some essential properties of dipolar nematic liquid crystals, there is a basic inconsistency with the model. It cannot predict a discontinuity of average dielectric permittivity by the transition of the nematic phase to the disorder isotropic phase.

Obviously, these behaviors are related to all approximations considered in description of cavity and reaction fields. Despite considerable efforts on the description of internal fields and local dipole correlations in isotropic liquids, the formulation of these effects for a macroscopically anisotropic dipolar fluid is a formidable problem and various approximate models cannot completely describe their behaviors.

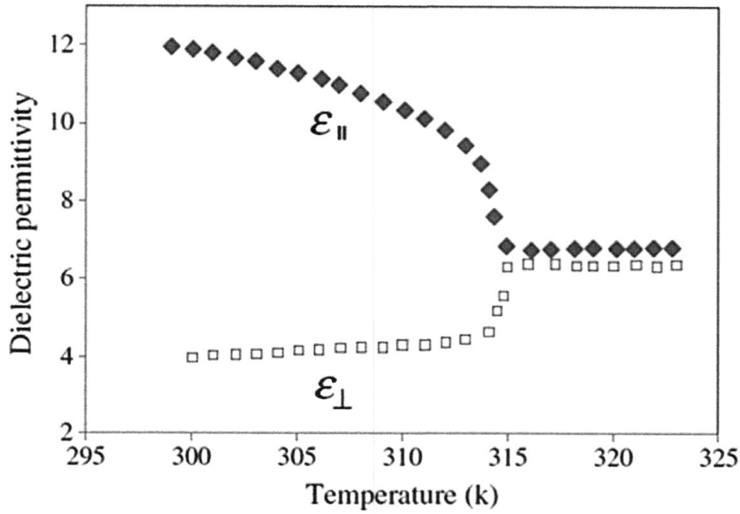

FIGURE 15.10 Temperature dependence of dielectric permittivities for nematic and isotropic phases.[6]

15.3.3 FREQUENCY DEPENDENCE OF THE DIELECTRIC PROPERTIES

As shown in Figure 15.11, the response of materials to the external fields depends on the applied frequencies. In this case, the electronic polarization, atomic polarization, and dipolar polarization will have various contributions on molecular responses, while dipolar polarization occurs only in polar materials. These contributions come from the deformation of the electron clouds of atoms, relative displacement of the atoms, and reorientation of the dipole moment, respectively.[18]

The rotation of the molecule is a slow process, and therefore, the orientational polarization can occur up to a frequency of megahertz. The displacement of the atoms in molecules is faster and this kind of polarization can occur up to the frequency of infrared light. The motion of electrons is the fastest and the electronic polarization can occur up to the frequency of ultraviolet light. Therefore, if electric field is removed, decay of orientational polarization takes place slower than others. By removing the electric field, reorientation of molecules occur in the finite time τ defined as the time when dipolar polarization reduces to $1/e$ its original value.

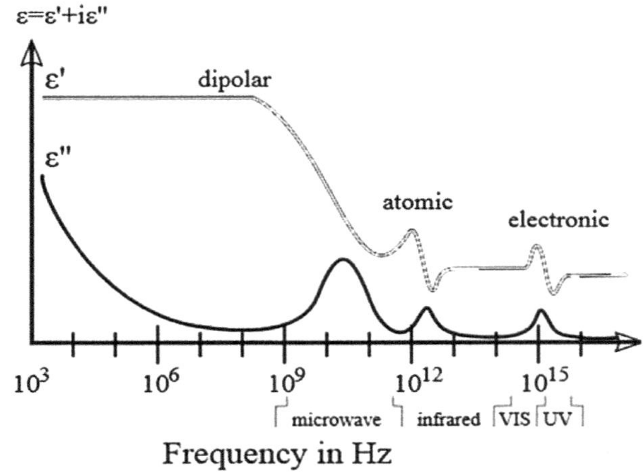

FIGURE 15.11 Frequency spectrum of dielectric permittivity.

When the frequency of applied field is not too high, in the limit of $\omega<\tau^{-1}$, dipoles can follow the electric field without any retardation phase, while measured total polarization (induced polarization and orientational polarization) can be related to static dielectric permittivity $\varepsilon(0)$. At the frequencies of the order of relaxation time ($\omega \approx \tau^{-1}$), relaxation process leads to some lag between average polarization of dipoles and applied field. At the higher frequencies ($\omega \approx \tau^{-1}$), orientational polarization cannot follow electric field, therefore, induced polarization has effective contribution in dielectric permittivity $\varepsilon(\infty)$.

In the general theory of linear response of dielectric materials to electric field, phase retardation between displacement and electric field ($D = \varepsilon^* \cdot E$) in dispersion region can be expressed by a complex dielectric constant (Eq. 15.63). The imaginary part of permittivity describes the energy losses due to molecular friction.

$$\varepsilon^* = \varepsilon' + i\varepsilon'' \tag{15.63}$$

According to Debye dispersion law,[19] the frequency dependence of ε^* can be written as:

$$\varepsilon^* - \varepsilon(\infty) = \frac{\varepsilon(0) - \varepsilon(\infty)}{1 - i\omega\tau_D} \tag{15.64}$$

where really and imaginary parts of complex permittivity are given by

$$\varepsilon' = \varepsilon(\infty) + \frac{\varepsilon(0) - \varepsilon(\infty)}{1 + \omega^2 \tau_D^2} \quad \varepsilon'' = \frac{[\varepsilon(0) - \varepsilon(\infty)]\omega\tau_D}{1 + \omega^2 \tau_D^2} \tag{15.65}$$

Frequency dependence of a real and imaginary parts of electric permittivity is shown in Figure 15.12, while ratio of ε'' to $\varepsilon' - \varepsilon(\infty)$ determines the phase angle of dielectric losses.

$$\tan \varphi = \omega\tau \tag{15.66}$$

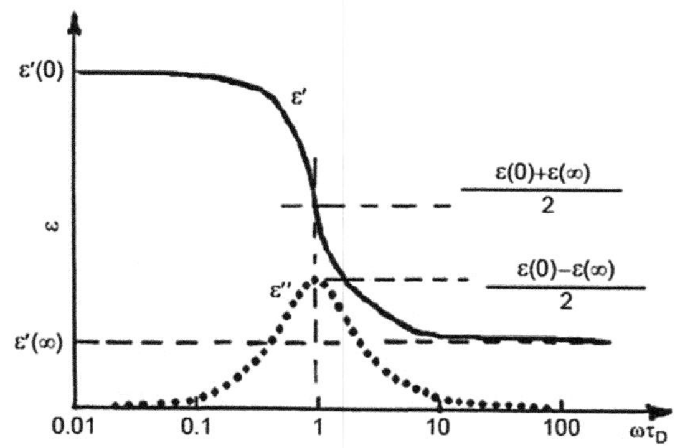

FIGURE 15.12 Frequency dependence of real (solid line) and imaginary parts (dash curve) of complex dielectric permittivity of an isotropic liquid.[7]

According to Debye's theory, relaxation time of dipoles depends on viscosity of liquids and temperature that can be expressed according to Equation 15.67.[20]

$$\tau_D = \frac{4\pi\eta a^3}{KT} \tag{15.67}$$

Here, η is viscosity and a is the radius of molecules.

Although Debye equation has successfully described the dielectric relaxation of a large number of systems, some deviations may be observed for other systems such as mixtures. An alternative method for calculation of relaxation time is based on Cole–Cole diagram.

Based on Cole–Cole model, frequency dependence of complex permittivity can be described by Equation 15.68 and decreased to Debye equation for $h = 0$.

$$\varepsilon^*(\omega) - \varepsilon(\infty) = \frac{\varepsilon(0) - \varepsilon(\infty)}{1 + (i\omega\tau_D)^{1-h}} \tag{15.68}$$

In this method, $\varepsilon'(\omega)$ is found versus $\varepsilon''(\omega)$ and in the case of simple dipole relaxation, the functional relation can be shown in the form of semicircle as shown in Figure 15.13.

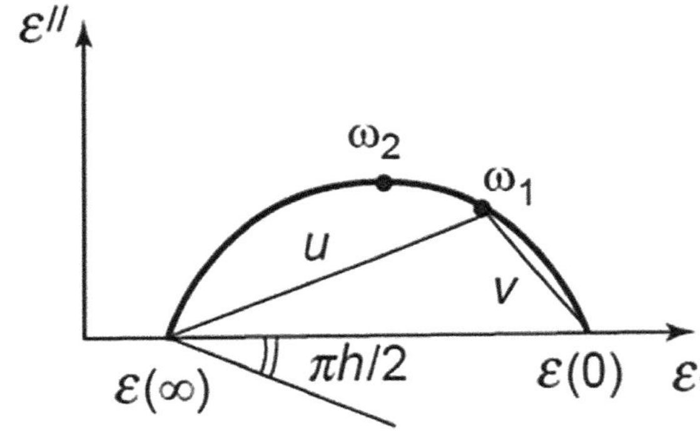

FIGURE 15.13 The typical Cole–Cole diagram.[7]

Here, the angle $\pi h/2$ defines the position of the center and the exponent parameter $\underline{h}$ that takes value between 0 and 1 describe different spectral shapes.

In this case, the relaxation time[7] can be also found from $\omega\tau_D = (\frac{v}{u})^{1-h}$.

By Cole–Cole diagrams, the calculations of relaxation times in complex systems can be facilitated. In a complex system, mixture of different dipolar molecules with different relaxation times, several semicircles can be plotted, and based on the obtained diagrams, relaxation times are calculated. Furthermore, this model can be expanded to liquid crystals for the estimation of relaxation times in liquid crystals.[21–22]

As discussed previously, nematic liquid crystals are described based on their long-rang order characteristics. For a rod-like molecule with anisotropic characteristics, there are three rotational modes: the rotation of molecules about short molecular axes (ω_1), the precession of long molecular axes about the director (ω_2), and the rotation of molecules about long molecular axes (ω_3) as shown in Figure 15.14.

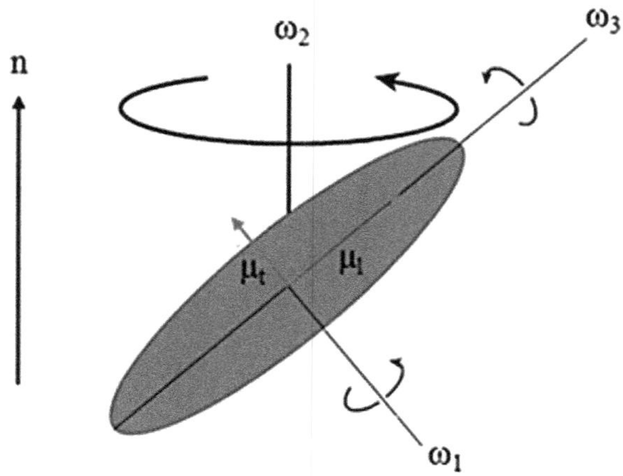

FIGURE 15.14 Three rotational modes in nematic liquid crystals with different contribution in dielectric relaxation process.

In a uniaxial nematic liquid crystal, it is expected that the rotation of molecules about long molecular axis occurs in the higher frequencies than other rotational modes. Hence, the magnitude of the characteristic frequencies for three described modes will be in the order $\omega_1 \ll \omega_2 < \omega_3$.[22] Therefore, the anisotropic characteristics of liquid crystals are reflected in the frequency dependence of parallel and perpendicular components of the permittivity. The variation of parallel and perpendicular components of dielectric constant with frequency is shown in Figure 15.15.[7] This indicates that relaxation process occurs at low and high frequencies, while perpendicular components of permittivity show relaxations at higher frequencies. As shown in Figure 15.16, this is in agreement with experimental results. As a typical result, the components of complex permittivity of 4-pentylphebyl4-propylbenzoate in different frequencies are shown in Figure 15.16.

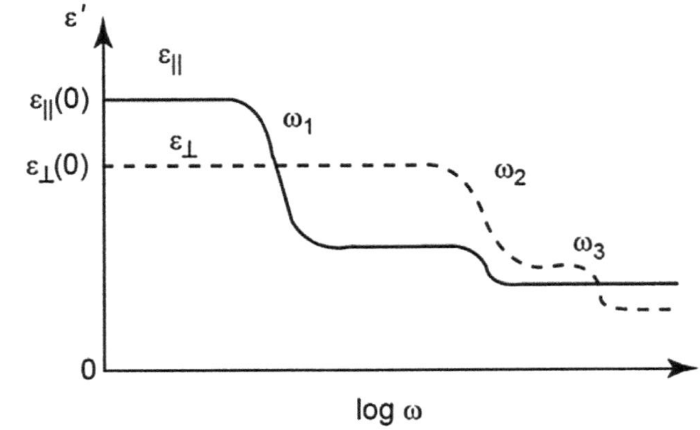

FIGURE 15.15 Frequency dependence of dielectric permittivities.[7]

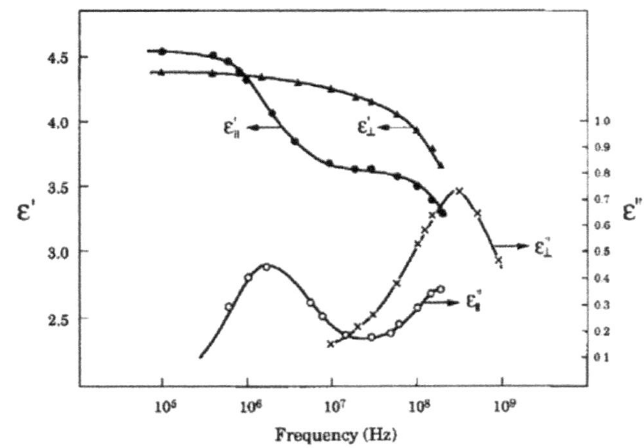

FIGURE 15.16 The components of complex permittivity for 4-pentylphebyl4-propy lbenzoate.[22]

A general method for the description of these behaviors is based on Martin, Meier, and Saupe[22] theory, which is an extension of Debye's model to nematic liquid crystals. In this model, they obtained pseudopotential for a nematic liquid crystal according to Equation 15.69.

$$u = -bS\cos^2\theta \tag{15.69}$$

Here, θ is the angle between the long molecular axis and the director and bS is identified as the height of the potential barrier.

In real, it is assumed that each molecule moves in a potential well and its rotations depend on the well depth and angle between the long molecular axis and director.

In this case, rotation of molecules along their long molecular axis is not affected by nematic potential. Therefore, frequency ω_3 corresponds to a fast molecular rotation. By increasing the angle between the long molecular axis and director from 0 to $\pi/2$, the molecule has to overcome a high barrier, and because of this, molecular frequency decreases to ω_1. In other words, the orientational order of nematic phase hinders the rotation of molecules about the short axes, and because of this, relaxation time is increased.

When nematic molecules precess about the director at θ angles, they are more or less free. Therefore, frequency ω_2 is considered as an intermediate molecular rotation.

Three dispersion regions can be investigated by dielectric spectroscopic techniques as shown in Figure 15.15.

Molecular relaxation times are usually expressed in terms of their relaxation times in the absence of a nematic potential by

$$\tau_i = j_i \tau_0 (i = \|, \perp) \tag{15.70}$$

The quantity j is defined as a retardation factor and depends on potential barrier characteristics. Approximately, it can be considered in the form of Equation 15.71.

$$j_{\|} = \frac{KT}{bS}\left[\exp\left(\frac{bS}{KT}\right) - 1\right] \tag{15.71}$$

These studies indicate that temperature, and on the other hand, the order of nematic liquid crystals have a significant effect on relaxation of $\varepsilon_{\|}$ and $\varepsilon_{\perp}$. For instance, in para-azoxyanisole, the order parameter decreases the relaxation frequency for rotation about short molecular axis (ω_1) and increases it for rotation about the long molecular axis (ω_3).[23]

Moreover, frequency-dependent characteristics of nematic liquid crystals lead to the appearance of interesting phenomena. For nematic liquid crystals with positive anisotropy ($\Delta\varepsilon > 0$), at a certain frequency (f_{inv}), low-frequency dispersion leads to a reversal sign of anisotropy ($\Delta\varepsilon < 0$). This interesting characteristic of nematic liquid crystals is known as dual frequency addressing and is notable in various electro-optical applications.

Therefore, an investigation of a frequency-dependent characteristic of liquid crystals gives important information about their dynamic behaviors that is valuable in designing of various electro-optical devices.

15.3.4 MOLECULAR STRUCTURE AND ALIGNMENT EFFECTS ON DIELECTRIC PROPERTIES OF NEMATIC LIQUID CRYSTAL

The physical and optical properties of liquid crystals depend strongly on the features of core system, side-chain, linking, and terminal groups. Anisotropy properties, absorption spectra, optical nonlinearity, and existence of mesophases are the consequences of how they are chemically synthesized together. Since these molecules consist of various groups with different characteristic, it seems complex the investigation of all physical properties by the possible variations in the molecular architecture. During the last years, chemists have tried to understand the relation between molecular structures and observed physical properties. Despite successful results for some properties such as transition temperatures, there is still much work to be performed. There are some general observations about the molecular constituents and structural effects on the physical properties of liquid crystal molecules, which we will discuss in this section.

In general, liquid crystals are described based on molecular constituents and their chemical structures. The basic structure of the most commonly used liquid crystals has been shown in Figure 15.17.

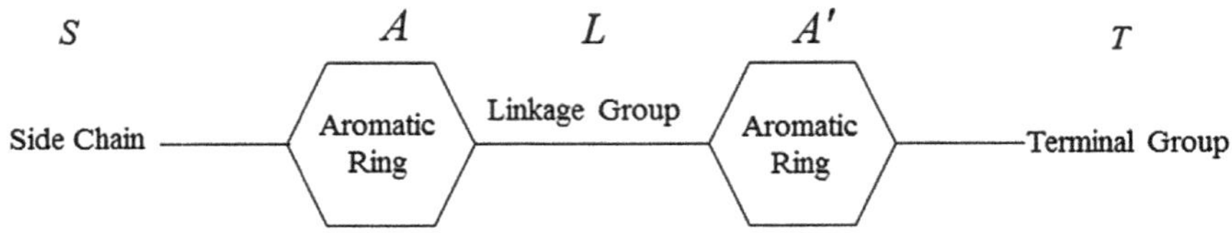

FIGURE 15.17 Molecular structure of a typical nematic liquid crystal.

In this figure, T indicates terminal groups; A, A' are the aromatic rings; L indicates linking groups; and S is called a side-chain group. These groups are selected in such a way so as to provide rigidity, linearity, and necessary fluidity to the molecule. Some of the most commonly encountered core systems, linking groups, side-chain groups, and terminal groups are shown below.

15.3.4.1 AROMATIC RINGS

The liquid crystals usually present two or more of aromatic rings in their chemical structures. These units comprise saturated cyclohexane, unsaturated phenyl, biphenyl, and terphenyl or their various combinations. The saturated groups consist of purely σ-electrons, whereas combination of σ- and π-electrons is found in unsaturated forms. Hence, their effects on various properties of liquid crystals are different. In Figure 15.18, the effect of saturated and unsaturated ring systems on the magnitude of dielectric anisotropy is shown. These results indicate that saturated groups decrease the value of dielectric anisotropy.

In addition to this, by substituting of hydrogen atoms by polar groups such as cyano and fluoro groups in the different position of phenyl rings, one can modify the dielectric properties of liquid crystals significantly. In this case, the sign of dielectric anisotropy depends highly on the selected substituents in a certain position of phenyl rings. For instance, the substitution of 2- and 3- lateral hydrogen atoms by fluoro groups can lead to the appearance of negative dielectric anisotropic behaviors, whereas positive dielectric anisotropy can be observed by replacing 4- (terminal) and 3- (lateral) hydrogen atoms by fluoro groups.[24]

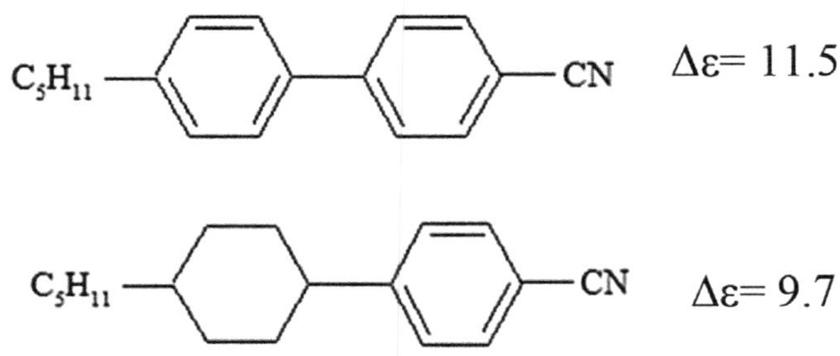

FIGURE 15.18 A typical liquid crystal with different ring systems.[25]

15.3.4.2 TERMINAL GROUPS

A terminal group is an important group which can have important contribution to dielectric anisotropy of liquid crystals. The examples of terminal groups are alkyl (C_nH_{2n+1}), alkoxy ($C_nH_{2n+1}O$), sulfide, and others such as halides (F, Cl, CF_3, and OCF_3), cyano (CN), and isocyanate (NCS). Among them, alkyl, alkoxy, and sulfide are low polar terminal groups; therefore, their effects on dielectric anisotropy are weak.

As discussed in Section 15.3.2, the magnitude of dielectric anisotropy depends on molecular polarizability, molecular dipole moment, the angle between molecular dipole moment and long molecular axis, and order parameter. Therefore, terminal groups with a large dipole moment can increase $\Delta\varepsilon$. In Table 15.1, the effect of terminal groups (Fig 15.19) on the value of dielectric anisotropy is indicated. It should be noted that $\Delta\varepsilon$ is measured at 20°C.[25]

The importance of terminal groups is not restricted to dielectric anisotropy. Moreover, the polar groups such as cyano with π-electrons can also modify the absorption characteristic of liquid crystals.

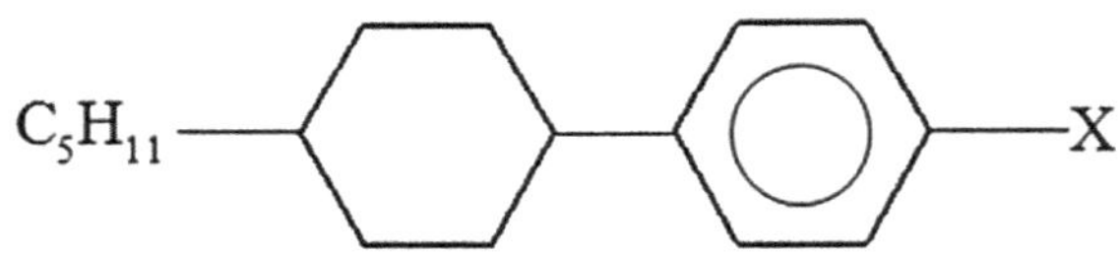

FIGURE 15.19 A typical liquid crystal with different terminal groups.

TABLE 15.1 The Effect of Terminal Groups on Dielectric Anisotropy.

X	$\Delta\varepsilon$
CN	+13
CF_3	+11
NCS	+11
OCH_3	-0.5
OCF_3	+7
F	+3
CH_3	+0.3

15.3.4.3 SIDE-CHAIN GROUPS

The length and flexibility of these groups determine the type of mesogenic phase. From the most commonly used side-chain groups, we can point to alkyl (C_nH_{2n+1}) and alkoxy ($C_nH_{2n+1}O$) groups.

The studies on cyano-biphenyl homologs (n-CB) have shown that for $n = 1$ and 2, the components cannot indicate mesogenic phase, while for $n = 3$ and 4, monotropic phase, and then for $n = 5$, 6, and 7, the nematic phase was observed.[24]

15.3.4.4 LINKING GROUPS

Linking groups that connect aromatic rings to each other can affect polarizability and flexibility of molecules. Moreover, these groups should preserve the linearity of molecules. Some of these groups are C_2H_4 (ethylene), –COO– (ester), –CH=CH (stilbene), –C≡C– (acetylene), –CH_2CH_2– (diphenylethane), –CH_2O– (formaldehyde), –CH=N– (Schiff base), and –N=N– (azoy).

If ring systems are linked by unsaturated groups, they can contribute their π-electrons to the phenyl rings on the side. This effect leads to increment of the molecular dipole moment by a push–pull system.[22] Hence, liquid crystals with high positive dielectric anisotropies can be produced by these methods as shown in Figure 15.20.

In other words, unsaturated linking groups increase the conjugation; therefore, electronic transitions can occur at longer wavelengths. Therefore, absorption characteristic of liquid crystals such as their dichroic ratio may be modified as well.

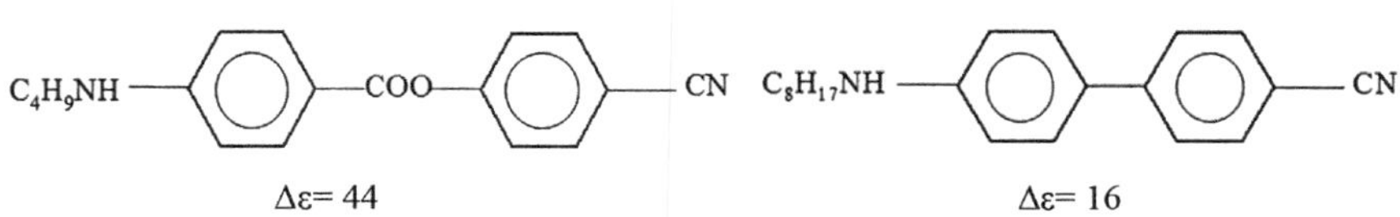

FIGURE 15.20 A typical liquid crystal with different linking groups.[26]

Therefore, dependence on molecular structures and the presence of various active groups in their structures and nematic molecules may indicate positive or negative dielectric anisotropy with different magnitudes.

It is obvious that the combination of various active groups and substitutions leads to more complicated chemical structures, and in this case, the prediction of different physical properties will be a hard work.

15.4 DIELECTRIC PROPERTIES AND APPLICATION OF NEMATIC LIQUID CRYSTAL

The orientational order of nematic liquid crystals and the fact that bulk orientation can be controlled by various methods increase their applications in various fields of science. In this section, we will highlight some of these applications due to unique dielectric characteristics of nematics that depend strongly on the molecular dipole moment, angle between molecular dipole moment with respect to long molecular axis, order parameter, and thermal effects.

Nematic liquid crystals with anisotropy dielectric characteristics are considered as an excellent solvent media for studying various physico-chemical properties of guest molecules. In this case, anisotropic guest–host interactions lead to an appropriable orientation of guest molecules along the preferred direction of host nematic molecules. The average orientation of guest molecules is related to the anisotropic characteristics of liquid crystal molecules and intermolecular interactions. Depending on anisotropy properties of liquid crystals, liquid crystal solutions can be aligned parallel ($\Delta\varepsilon>0$) or perpendicular ($\Delta\varepsilon<0$) to applied external fields. Hence, guest molecules with highly oriented characteristics can be prepared. This indicates that anisotropic solvents are good environments for controlling of the orientation of solute molecules.

Moreover, liquid crystals as anisotropic solvents may change the reactivity of incorporated solute molecules as compared with those in isotropic solvents.[22] Liquid crystals with particular alignment characteristics can be considered as an ideal medium for operation of biological systems. Therefore, the prediction of specific effects of a liquid crystal on given reaction can be useful, especially in biological systems. In 1916, Svedberg showed that the rate of decomposition of picric acid, trinitroresorcinol, and pyrogallol increases slightly by increasing temperature when the reaction is performed in the nematic phase of p-azoxyphenetole.[27] After that, the anisotropic media effects on orientational, conformational, diffusional, and aggregative characteristics of a wide variety of solute molecules were investigated.[28–33] Depending on the structure of the solute molecules, these effects originate due to nature and degree of various solute–solvent interactions. The intensity of general solute–solvent interactions depends highly on the dielectric constants of liquid crystals as anisotropic solvents.[34–35] These interactions are also employed for the effective separation of positional and geometric isomers. Chromatography is of the great importance in physicochemical investigations of materials. The use of liquid crystals as stationary phase in gas chromatograph was proposed by Kelker[36] for the first time. Since then, liquid crystals have been used successfully as anisotropic stationary phases for separation of composite mixtures and isomers. Among the various liquid crystal phases, nematics and polymeric liquid crystals are the most applicable ones, and polymeric liquid crystals are usually used for higher temperature experiments. The anisotropy properties of nematic liquid crystals and their strong interactions with solute molecules increase their abilities for analyzing the complex mixtures that are important in biochemical, environmental, and industrial applications.

By exploiting the fact that reorientation of liquid crystals director has a notable effect on their physical properties, the induced changes in their alignment characteristics can be used in molecular sensing and detection process as well. In this case, a nematic liquid crystal cell with special alignment is exposed to the sensing targets and the induced changes in director axis orientation profile can be determined by measuring their physical properties such as dielectric constants. This method can be also used for detecting ligand–receptor and organoamines–carboxylic acids bindings.[4] Hence, the sensitivity of physical properties of liquid crystals to their alignment characteristics can increase their abilities as sensors.

Moreover, liquid crystals with unique properties characteristics are widely used in display devices. Although liquid crystals occupy a small portion of these devices, they have a major contribution in their performance. For instance, the operating voltage of displays depends on liquid crystals alignment and dielectric characteristics. The dielectric constants specify the threshold voltage, whereas by increasing the dielectric anisotropy ($\Delta\varepsilon$), the threshold voltage decreases.[15]

Therefore, the investigation of dielectric properties of nematic liquid crystals will give valuable information about their applications in various field of science.

15.5 CONCLUSION

In this chapter, by exploiting the different theories, the dielectric properties of nematic liquid crystals were investigated. The studies indicate that unique properties of these materials are related to their orientational order characteristics. Due to this type of order, nematics show different behaviors in parallel and perpendicular directions to externally applied fields. Various factors such molecular constituents, their chemical structures and the properties of applied external fields can affect the dielectric anisotropy characteristics. Hence, the investigation of dielectric properties of nematic liquid crystals and determination of effective factors on their dielectric constants will improve their operations and applications in various science and technology.

KEYWORDS

- **liquid crystal**
- **orientational order**
- **nematic**
- **dielectric properties**

REFERENCES

1. Chandrasekhar, S. *Liquid Crystals,* 2nd ed.; Cambridge University Press: Cambridge, 1992.
2. Reinitzer, F. Beiträge zur Kenntniss des Cholesterins. *Monatschr. Chem.* **1898,** *9,* 421–441.
3. Saleh, B. E. A.; Carl Teich, M. *Fundamentals of Photonics;* John Wiley & Sons: New York, 1991.
4. Khoo, I. C. *Liquid Crystals,* 2nd ed.; John Wiley & Sons: New York, 2007.
5. Collings, P. J.; Hird, M. *Introduction to Liquid Crystals, Chemistry and Physics;* Taylor & Francis: London, 1997.
6. Zakerhamidi, M. S.; Majles Ara, M. H.; Maleki, A. Dielectric Anisotropy, Refractive Indices and Order Parameter of W-1680 Nematic Liquid Crystal. *J. Mol. Liq.* **2013,** *181,* 77–81.
7. Blinov, L. M. *Structure and Properties of Liquid Crystals;* Springer: Dordrecht, 2011.
8. Kittel, C. *Introduction to Solid State Physics,* 4th ed.; Wiley: New York, 1971.
9. Raju, G. G. *Dielectrics in Electric Fields;* Marcel Dekker. Inc.: New York, 2003.
10. Williams, D. *Methods of Experimental Physics (volume III), Molecular Physics;* Academic Press: New York, 1962.
11. de Jeu, W. H. *Physical Properties of Liquid Crystalline Materials;* Gordon and Breach: New York, 1980.

12. Woltman, J.; Crawford, G. P.; Jay, G. D. *Liquid Crystals: Frontiers in Biomedical Applications;* World Scientific: Singapore, 2007.
13. de Jeu, W. H. *Physical Properties of Liquid Crystal Materials, Liquid Crystal Monographs;* Gordon and Breach: London, 1980; vol 1.
14. de Jeu, W. H.; Bordewijk, P. Physical Studies of Nematic Azoxybenzenes. II. Refractive Indices and the Internal Field. *J. Chem. Phys.* **1978,** *68*, 109.
15. Yang, D. K.; Wu, S. T. *Fundamentals of Liquid Crystal Devices;* John Wiley & Sons: New York, 2006.
16. Maier, W.; Meier, G. Eine einfache Theorie der dielektrischen Eigenschaften homogeny orientierter kristallinfluussiger Phasen des nematischen Typs. *Z. Naturforschg.* **1961,** *16a*, 262–267; Hauptdielektrizitatskonstanten der homogen geordneten kristallinflussigen Phase des p-Azoxyanizols, ibid, 16a, 470–477.
17. Dunmur, D. A.; Hitchen, D. A.; Hong, X.-J. The Physical and Molecular Properties of Some Nematic Fluorobiphenylalkanes. *Mol. Cryst. Liq. Cryst.* **1986,** *140*, 303–318.
18. Gao, W.; Sammes, N. M. *An Introduction to Electronic and Ionic Materials;* World Scientific: Singapore, 1999.
19. Frolich, H. *Theory of Dielectrics,* 2nd ed.; Clarendon Press: London, 1978.
20. Bartnikas, R. *Engineering Dielectrics: Electrical Insulating Liquids;* ASTM: USA, 1994; vol. 3.
21. Aliev, F. M.; Breganov, M. N. Electric Polarization and Dynamics of Molecular Motion of Polar Liquid Crystals in Micropores and Macropores. *Sov. Phys. JETP.* **1989,** *68*, 70–79.
22. Demus, D.; Goodby, J.; Gray, G. W.; Spiess, H. W.; Vill, V. *Handbook of Liquid Crystals, Fundamentals;* WILE-YVCH: New York, 1998, vol. 1.
23. Nordio, P. L.; Rigatti, G. Segre, U. Dielectric Relaxation Theory in Nematic Liquids. *Mol. Phys.* **1973,** *25*, 129–136.
24. Khoo, I. C.; Wu, S. T. *Optics and Nonlinear Optics of Liquid Crystals;* World Scientific: Singapore, 1993.
25. Bahadur, B. *Liquid Crystals: Applications and Uses;* World Scientific: Singapore, 1990; vol. 1.
26. Demus, D.; Goodby, J.; Gray, G. W; Spiess, H. W; Vill, V. *Handbook of Liquid Crystals, Low Molecular Weight Liquid Crystals I;* WILE-YVCH: New York, 1998; vol. 2A.
27. Paleos, C. M. *Polymerization in Organized Media;* Gordon and Breach: Philadelphia, 1992.
28. Fahie, B. J.; Mitchell, D. S.; Workentin, M. S.; Leigh, W. S. Organic Reactions in Liquid Crystalline Solvents. 9. Investigation of the Solubilization of Guest Molecules in a Smectic (Crystal B) Liquid Crystal by Deuterium NMR, Calorimetry, Optical Microscopy, and Photoreactivity Methods. *J. Am. Chem. Soc.* **1989,** *111*, 2916–2929.
29. Workentin, M. S.; Leigh, W. J.; Jeffrey, K. R. Organic Reactions in Liquid Crystalline Solvents. 10. Studies of the Ordering and Mobilities of Simple Alkanophenones in CCH-n Liquid Crystals by Deuterium NMR Spectroscopy and Norrish II Photoreactivity. *J. Am. Chem. Soc.* **1990,** *112*, 7329–7336.
30. Hrovat, D. A.; Liu, J. H.; Turro, N. J.; Weiss, R. G. Photolyses of Dibenzyl Ketones in Liquid-Crystalline Media. The Fate of Benzyl Radical Pairs in Various Anisotropic Environments. *J. Am. Chem. Soc.* **1984,** *106*, 7033–7037.

31. Deval, P.; Singh, A. K. Photoisomerization of All-Trans-Retinal in Organic Solvents and Organized Media. *J. Photochem. Photobiol. A: Chem.* **1988,** *42*, 329–336.
32. Kreysig, D.; Stumpe, J. *Selected Topics in Liquid Crystal Research;* Akademie-Verlag: Berlin, 1990.
33. Kunieda, T.; Takahashi, T.; Hirobe, M. Highly Stereoselective Photodimerization of 1, 3-dimethylthymine in Liquid Crystalline Media. *Tet. Lett.* **1983,** *24*, 5107–5108.
34. Tajalli, H.; Ghanadzadeh Gilani, A.; Zakerhamidi, M. S.; Tajalli, P. The Photophysical Properties of Nile Red and Nile Blue in Ordered Anisotropic Media. *Dyes Pigm* **2008,** *78,* 15–24.
35. Ghanadzadeh Gilani, A.; Moghadam, M.; Zakerhamidi, M. S. Estimation of Ground- and Excited-State Dipole Moments of Oxazine 1 in Liquid and Liquid Crystalline Media. *Spectrochim. Acta Part A* **2011,** *79*, 148–155.
36. Bahadur, B. Liquid Crystals—Applications and Uses; World Scientific, Singapore, 1991; Vol. 2.

INDEX